Self-Assembly of Polymers

Self-Assembly of Polymers

Special Issue Editors

Dmitry Volodkin
Anna Vikulina

MDPI • Basel • Beijing • Wuhan • Barcelona • Belgrade • Manchester • Tokyo • Cluj • Tianjin

Special Issue Editors
Dmitry Volodkin
Nottingham Trent University
UK

Anna Vikulina
Nottingham Trent University
UK

Editorial Office
MDPI
St. Alban-Anlage 66
4052 Basel, Switzerland

This is a reprint of articles from the Special Issue published online in the open access journal *Micromachines* (ISSN 2072-666X) (available at: https://www.mdpi.com/journal/micromachines/special_issues/Self_Assembly_Polymers#Research).

For citation purposes, cite each article independently as indicated on the article page online and as indicated below:

LastName, A.A.; LastName, B.B.; LastName, C.C. Article Title. *Journal Name* **Year**, *Article Number*, Page Range.

ISBN 978-3-03928-506-8 (Pbk)
ISBN 978-3-03928-507-5 (PDF)

Contents

About the Special Issue Editors

Dmitry Volodkin is Associate Professor at Nottingham Trent University and heads the group Active-Bio-Coatings. He has studied Chemistry at the Lomonosov Moscow State University in Russia and further research stays brought him to France (University of Strasbourg) and Germany (Max-Planck Institute of Colloids and Interfaces, TU Berlin, Fraunhofer Institute for Cell Therapy and Immunology). His research activities are focused on design of advanced stimuli-responsive biomaterials for applications in tissue engineering, diagnostics, toxicology, drug delivery. His group engineers self-assembled polymer-based 2D and 3D structures with tailor-made properties: multilayer films, microcapsules and beads, liposome-polymer composites, polymeric scaffolds, etc. Dmitry Volodkin has published ca 80 peer-reviewed articles (citations without self-citations ¿3000) and received prestigious scientific awards such as Sofja Kovalevskaja Award, Richard-Zsigmondy Price, Alexander von Humboldt and Marie Curie Fellowships. He is organizing committee member of a number of conferences and delivered more than 10 plenary/keynote lectures at prestigious conferences and symposia.

Anna Vikulina has completed her PhD in the field of Biophysics in 2015. Her research interests are in the development of drug delivery carriers, deciphering the pathways of biological action and transport of macromolecules and therapeutic drugs of various nature, and design of functional materials for tissue engineering and biomedical applications. She has been awarded Alexander von Humboldt and Marie Curie Individual Fellowships. Currently she is an independent research fellow in Fraunhofer Institute for Cell Therapy and Immunology, Germany.

micromachines

Editorial

Editorial for the Special Issue on Self-Assembly of Polymers

Anna S. Vikulina [1,*] and Dmitry Volodkin [2,*]

1 Fraunhofer Institute for Cell Therapy and Immunology, Branch Bioanalytics and Bioprocesses, Am Mühlenberg 13, 14476 Potsdam-Golm, Germany
2 School of Science and Technology, Nottingham Trent University, Clifton Lane, Nottingham NG11 8NS, UK
* Correspondence: Anna.Vikulina@izi-bb.fraunhofer.de (A.S.V.); Dmitry.Volodkin@ntu.ac.uk (D.V.)

Received: 2 August 2019; Accepted: 4 August 2019; Published: 5 August 2019

The self-assembly of polymers is a powerful tool for producing various functional materials with a high precision from nano- to macroscale. Nowadays, the polymer self-assembly has become extremely attractive for both biological (drug delivery, tissue engineering, advanced cell culture) and non-biological (packaging, semiconductors) applications. Besides this, a number of key biological processes in nature are driven by self-assembly of macromolecules.

This special issue gives insights into diverse spectrum of peculiar tailor-made self-assembled polymer structures and hybrid materials that have been designed through modern powerful techniques and approaches such as the micellization [1], hydrogel formation [2,3], self-assembly at the air-liquid interface [4], the layer-by-layer deposition and the hard templating [5,6], solvent evaporation [7,8], photo-induced [9] and enzyme-mediated cross-linking [10].

In particular, significant attention has been attracted to the layer-by-layer assembly of biopolymers. Thus, Balabushevich et al. have presented the study of mucin-based layer-by-layer assemblies that are mainly composed from the natural component of mammalian mucus and therefore show high potential for mucosal drug delivery [6]. In the recent years, collagen is gaining interest as a component of modern biomaterials [2]; some novel sources of the biopolymers to be used in the self-assemblies, e.g. biopolymers of marine origin, have also been described [2]. Yang et al. reviewed the current state of biocompatible and biodegradable polypeptide-based nanomaterials [9].

Another issue in the field of polymer self-assembled structures that has been reflected in this special issue is the design of stimuli-responsive materials. This includes both, chemical stimuli such as pH and physical stimuli including light. In this special issue, Xu et al. presented an elegant way for fabrication of pH-responsive amphiphilic block copolymer-based assembles [1]; Yang et al. paid special attention on photo-triggered assembly of polymer-based nanostructures [9].

Bridging the gap between the science of the polymer self-assembly and the industrial application of this technology, Nelson et al. have designed and constructed a new solvent vapour annealing chamber that allowed the potential scaling-up of the production of block copolymer thin films [7].

The next equally interesting aspect that has been deciphered in this issue is organisation of self-assembled polymer-based and hybrid structures in 3D. Herein, microfluidics-assisted assembly of porous calcium carbonate templates and their covering with polymer coatings opened new avenues for the fabrication of tailor-made biocompatible scaffolds for advanced tissue engineering and 3D cell culture as reviewed in [3]. Surface patterning with the polymers [8] and crystalline colloidal monolayers [4] also have been demonstrated.

That is rather interesting to note that while the papers listed above mostly aimed at the invention and investigation of the new types of materials, the paper presented by Jeannot et al. took under the consideration a well-known poly(styrene-sulfonate)/poly(diallyldimethylammonium chloride) polyelectrolyte pair and the authors have shown principally new insights into the classical understanding of the internal structure of multilayer capsules [5]. Their findings potentially have high importance for the field of controlled drug release.

Finally, it is worth to emphasize the review of Savoca et al. that summarises the up to date knowledge on the biocatalysis by transglutaminases, enzymes that crosslink polymer-bound glutamine and lysine residues and plays a special role in the formation of extracellular polymer matrix [10]. The review highlights how bioinspired polymer cross-linking with transglutaminase allows fabrication of biocompatible scaffolds and hydrogels for biotechnological and bioengineering applications.

We would like to take the opportunity to thank all the authors for submitting their papers to this special issue. We also want to thank all the reviewers for dedicating their time and helping to improve the quality of the submitted papers.

Conflicts of Interest: The authors declare no conflict of interest.

References

1. Xu, Y.; He, K.; Wang, H.; Li, M.; Shen, T.; Liu, X.; Yuan, C.; Dai, L. Self-Assembly Behavior and pH-Stimuli-Responsive Property of POSS-Based Amphiphilic Block Copolymers in Solution. *Micromachines (Basel)* **2018**, *9*, 258. [CrossRef] [PubMed]
2. Norris, K.; Mishukova, O.I.; Zykwinska, A.; Colliec-Jouault, S.; Sinquin, C.; Koptioug, A.; Cuenot, S.; Kerns, J.G.; Surmeneva, M.A.; Surmenev, R.A.; et al. Marine Polysaccharide-Collagen Coatings on Ti6Al4V Alloy Formed by Self-Assembly. *Micromachines (Basel)* **2019**, *10*, 68. [CrossRef] [PubMed]
3. Sergeeva, A.; Vikulina, A.S.; Volodkin, D. Porous Alginate Scaffolds Assembled Using Vaterite CaCO3 Crystals. *Micromachines (Basel)* **2019**, *10*, 357. [CrossRef] [PubMed]
4. Feng, D.; Weng, D.; Wang, J. A Facile Interfacial Self-Assembly of Crystalline Colloidal Monolayers by Tension Gradient. *Micromachines (Basel)* **2018**, *9*, 297. [CrossRef] [PubMed]
5. Jeannot, L.; Bell, M.; Ashwell, R.; Volodkin, D.; Vikulina, A.S. Internal Structure of Matrix-Type Multilayer Capsules Templated on Porous Vaterite $CaCO_3$ Crystals as Probed by Staining with a Fluorescence Dye. *Micromachines (Basel)* **2018**, *9*, 547. [CrossRef] [PubMed]
6. Balabushevich, N.G.; Sholina, E.A.; Mikhalchik, E.V.; Filatova, L.Y.; Vikulina, A.S.; Volodkin, D. Self-Assembled Mucin-Containing Microcarriers via Hard Templating on $CaCO_3$ Crystals. *Micromachines (Basel)* **2018**, *9*, 307. [CrossRef] [PubMed]
7. Nelson, G.; Drapes, C.S.; Grant, M.A.; Gnabasik, R.; Wong, J.; Baruth, A. High-Precision Solvent Vapor Annealing for Block Copolymer Thin Films. *Micromachines (Basel)* **2018**, *9*, 271. [CrossRef] [PubMed]
8. Li, X.; Zhu, X.; Wei, H. Microstructure Formation of Functional Polymers by Evaporative Self-Assembly under Flexible Geometric Confinement. *Micromachines (Basel)* **2018**, *9*, 124. [CrossRef] [PubMed]
9. Yang, L.; Tang, H.; Sun, H. Progress in Photo-Responsive Polypeptide Derived Nano-Assemblies. *Micromachines (Basel)* **2018**, *9*, 296. [CrossRef] [PubMed]
10. Savoca, M.P.; Tonoli, E.; Atobatele, A.G.; Verderio, E.A.M. Biocatalysis by Transglutaminases: A Review of Biotechnological Applications. *Micromachines (Basel)* **2018**, *9*, 562. [CrossRef] [PubMed]

 *micromachines*

Review

Porous Alginate Scaffolds Assembled Using Vaterite $CaCO_3$ Crystals

Alena Sergeeva [1], Anna S. Vikulina [1,2,*] and Dmitry Volodkin [2]

1 Fraunhofer Institute for Cell Therapy and Immunology, Branch Bioanalytics and Bioprocesses, Am Mühlenberg 13, 14476 Potsdam-Golm, Germany; alenasergeeva@mail.ru
2 School of Science and Technology, Nottingham Trent University, Clifton Lane, Nottingham NG11 8NS, UK; dmitry.volodkin@ntu.ac.uk
* Correspondence: anna.vikulina@izi-bb.fraunhofer.de; Tel.: +49-331-58187-122

Received: 26 April 2019; Accepted: 23 May 2019; Published: 29 May 2019

Abstract: Formulation of multifunctional biopolymer-based scaffolds is one of the major focuses in modern tissue engineering and regenerative medicine. Besides proper mechanical/chemical properties, an ideal scaffold should: (i) possess a well-tuned porous internal structure for cell seeding/growth and (ii) host bioactive molecules to be protected against biodegradation and presented to cells when required. Alginate hydrogels were extensively developed to serve as scaffolds, and recent advances in the hydrogel formulation demonstrate their applicability as "ideal" soft scaffolds. This review focuses on advanced porous alginate scaffolds (PAS) fabricated using hard templating on vaterite $CaCO_3$ crystals. These novel tailor-made soft structures can be prepared at physiologically relevant conditions offering a high level of control over their internal structure and high performance for loading/release of bioactive macromolecules. The novel approach to assemble PAS is compared with traditional methods used for fabrication of porous alginate hydrogels. Finally, future perspectives and applications of PAS for advanced cell culture, tissue engineering, and drug testing are discussed.

Keywords: calcium alginate; porous hydrogel; polymer scaffold; calcium carbonate; encapsulation; drug delivery; cell culture

---

1. Introduction

At present, in the field of biomedical technologies, researchers have been attracted to the development of novel multifunctional structures with structure and properties well-controlled on both the micro- and nano-scales. One of the major focuses in tissue engineering and regenerative medicine is the formation of polymer scaffolds, temporal or permanent constructions providing both mechanical support for cell seeding and growth, as well as encapsulation/protection and controlled delivery of bioactive molecules (for instance, growth factors and enzymes) in order to guide tissue organization. Besides the tissue engineering for regenerative medicine, such scaffolds can serve as a platform for animal-free drug testing. Up to date, fabrication of multifunctional scaffolds that have a well-defined structure remains challengeable due to a high level of complexity in the composition of such scaffolds and the need to employ sophisticated methods for the scaffold assembly. In addition, conditions of scaffold preparation often require intolerably high costs of exclusive techniques and often result in a loss of bioactivity of bioactives loaded into scaffolds. Thus arises the need to develop simple strategies for the manufacture of *intelligent* polymer-based scaffolds possessing a well-controlled internal structure, efficient encapsulation, protection, and controlled release of desired bioactives (recent reviews [1–4]).

Polymeric 3D scaffolds serve as the supports to guide the growth of biological cells and the development of a microtissue; often these scaffolds are biodegradable. Polymeric scaffolds are usually designed as porous structures with highly developed internal surfaces to ensure cell infiltration/growth and to avoid diffusion limitations for transport of nutrition and metabolites. This also mimics the architecture of natural tissues well (Figure 1).

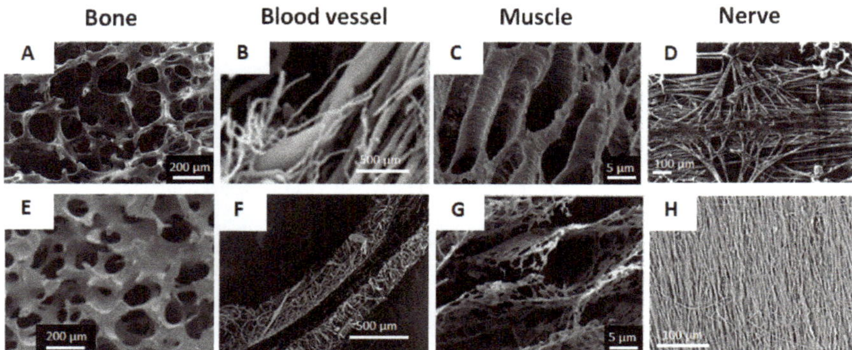

Figure 1. SEM/cryo-SEM images of mammalian tissues (**a**–**d**) and porous polymer-based scaffolds that mimic corresponding tissues (**e**–**h**): (**a**,**e**) human trabecular bone. Reproduced with permission from [5], published by Springer Nature, 2011 and [6], published by Elsevier B.V., 2012; (**b**,**f**) human uterine peripheral vessels. Reproduced with permission from [7], published by Via Medica, 2004 and [8], published by Elsevier B.V., 2005; (**c**,**g**) porcine muscular tissue. Reproduced with permission from [9], published by Springer Nature, 2018; (**d**,**h**) human transverse cervical nerves. Reproduced with permission from [10], published by Springer-Verlag, 1977 and [11], published by Springer Nature, 2019

Among the others, alginate hydrogels are one of the pivotal materials used for the fabrication of polymeric scaffolds due to alginate biocompatibility, opportunity to shape alginate hydrogels into a variety of sophisticated geometries and topologies. This is possible in both 2D (e.g., thin films patterned with microwells [12] and gel grids [13]) and in 3D (sponge-like structures [14] and gels possessing tube-like [15] or spherical pores [16,17]).

In recent years, a novel bench-top method has been proposed for the fabrication of porous alginate scaffolds (PAS) [18,19]. These scaffolds are produced by formulation of alginate mixture with vaterite calcium carbonate microcrystals (cores), followed by elimination of the $CaCO_3$ cores under mild conditions including physiological pH. This is accompanied by the release of Ca^{2+} ions inducing the cross-linking of the alginate gel and formation of hollow pores as inverse replica of the cores. Schematic representation of the process of PAS formation using calcium carbonate is given and discussed in details in Section 4.1 of this review. PAS have highly developed internal structure and offer unique opportunities to host bioactive molecules of a different nature via proper localization of them in the scaffold and to release these molecules in a controlled manner [18,19].

This review summarizes different aspects of the PAS formation discussing current achievements and challenges in this field. Critical comparison of PAS with other approaches proposed to tackle the problems associated with the design of multifunctional scaffolds allows for manifesting high potential of the novel developed technology for tissue engineering, regenerative medicine, drug testing and other applications where multifunctional polymer-based scaffolds are currently employed and strongly required.

2. Ca^{2+}-Alginate Gel Based Scaffolds

2.1. Chemistry of Alginate

Alginic acid is a naturally-derived block copolymer forming polyanions which macromolecule is composed of α-L-glucoronate and β-D-mannuronate units (Figure 2a) [20–25]. Sequence, and hence the properties of alginic acid, depend on the natural source. Commercially available alginic acid is produced from brown algae and usually available as in the form of the salt called alginate; molecular weights of these alginates typically vary from 32 to 400 kDa [22]. Since dissociation constants (pK_a) of carboxylic groups of alginate are 3.65 and 3.38 for α-L-glucoronate and β-D-mannuronate residues, respectively, [21,26] in order to dissolve alginate, it is essential to achieve pH above a certain critical value, higher than the pK_a. Besides this, viscosity of alginate depends on the ionic strength and, remarkably, the addition of some ions causes alginate gelation. The latest provides great advantages to alginate compared with the other polysaccharides (gelatine, agar) because alginate is able to form a gel in the range 0–100 °C. Moreover, alginate gels are highly hydrated, having water content > 95%, and can be heated without melting (phase transition) [24].

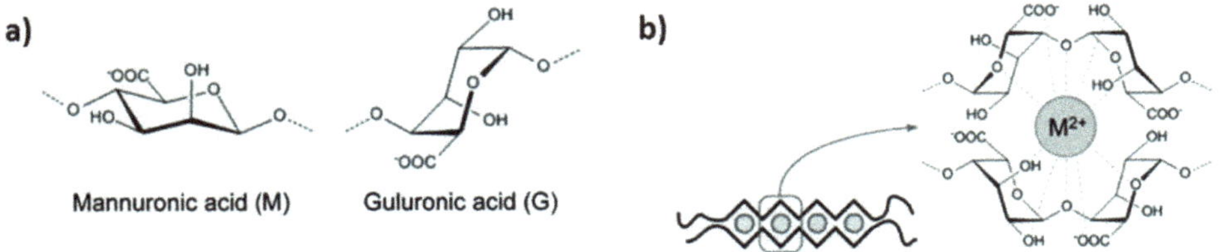

Figure 2. (**a**) Chemical structure of α-L-guluronic and β-D-mannuronic acids (residues of alginate macromolecules). (**b**) Schematics of the binding between G-blocks of alginate molecules by means of divalent metal ions (M^{2+}).

Chemical modification of alginate is widely employed to provide the polymer with novel desired properties (solubility, hydrophobicity, affinity to specific molecules, etc.) [20,22]. Thus, the phosphorylation of alginate results in the enhancement of hydroxyapatite nucleation and growth. Alginate sulfonation has been applied to provoke anticoagulant activity of the alginate. The attempts to transform hydrophilic alginate to a molecule with hydrophobic or amphiphilic properties were also demonstrated. The other way to provide alginate with new properties is based on the graft copolymerization. Alginates can also be functionalized with specific cell-targeting ligands in order to strengthen the affinity of alginate gels to biological cells. In the next section, the gelation and properties of alginate hydrogels will be discussed.

2.2. Alginate Hydrogels: Formation and Structure

Alginate hydrogels are highly hydrated 3D cross-linked polymer networks [20,22,24]. In general, alginate molecules chelate with multivalent cations. This process leads to the gelation occurring via the precipitation of alginate-cation complex and hence the formation of ionically cross-linked gels, also widely called physical gels. The chemistry behind this process is based on the cooperative binding between glucoronates (G-blocks), between mannuronates (M-blocks) and between glucoronates with mannuronates (MG-blocks). Of note, the binding between G-blocks is the most pronounced, although all types of the binding strongly depend on the type of the gel-forming cation. For instance, divalent ions Ca^{2+}, Ba^{2+}, and Sr^{2+} bound mainly to GG–dimers, while trivalent lanthanide ions such as La^{3+}, Pr^{3+}, and Nd^{3+} prone to bind to both GG– and MM–segments. This molecular organisation results in the formation of a diamond-shaped hole consisting of a hydrophilic cavity with the multivalent cation that coordinates oxygen atoms from the carboxyl groups of alginate (Figure 2b) [20–22,25,27,28]. The size of the cooperative unit is estimated to consist of more than 20 monomers [25]. It has been demonstrated that alginate affinity to cations increases in the order of Mn < Zn, Ni, Co < Fe < Ca < Sr < Ba < Cd < Cu < Pb [24,29]. This is directly related to the ionic radius and coordination number of cross-linking cations [23] and can be used to tune the properties of the hydrogel. Among others, Ca^{2+} cross-linked alginate gels have an advantage of a high biocompatibility, while the use of other cations may be limited due to the toxicity issue. Ca^{2+} cross-linked alginates gels are predominantly formed via the binding of glucuronic segments. Because of this, the strength of alginate gel is significantly influenced by the content of G-blocks. In general, the higher the gel strength, the lower its elasticity [21]. Hydrogels fabricated from alginate enriched with G-blocks form stiff and brittle gels, while high M content results in the formation of rather soft elastic gels [24].

External and internal gelation, as well as gelation upon cooling, represent three main techniques used to formulate ionic alginate gels [20,22,24]. External gelation which is also called a "diffusion method" is based on direct exposure of alginate into a solution containing cross-linking ions (e.g., $CaCl_2$). Ca^{2+} ions diffuse from the continuous phase into alginate droplets cross-linking them and forming gel particles [21]. The main disadvantage of this method is the formation of non-uniform alginate hydrogels due to the establishment of a gradient of Ca^{2+} concentration towards the boundary of the hydrogel where it is in contact with the solution of Ca^{2+} and an extremely high rate of the cross-linking reaction [30–32]. To some extent, the problem of non-uniform distribution of the cations in external gelation method can be eliminated using alginates of higher molecular weights or carrying out gelation in the buffer solutions containing phosphate ions that also bind calcium ions and, in this way, compete with alginate.

On the other hand, the internal gelation (so-called "*in situ* gelation") can be applied to avoid the gel inhomogeneity. For this approach, the source of Ca^{2+} ions (usually particles of low-soluble $CaCO_3$ or other salts of Ca^{2+}) is distributed within the precursor solution of alginate. Slow dissolution of these particles is generally induced by changing pH (e.g., by addition of self-hydrolysing polymer as D-glucono-δ-lactone, GDL), providing constant flow of crosslinking ions to surrounding alginate molecules. This method results in a uniform ion concentration throughout the gel. As an alternative method, gelation upon cooling [23] is based on consequent dissolution of alginate solution and calcium salt in a hot medium of 90 °C followed by cooling. At the temperature of 90 °C, high thermal energy of alginate chains prevents alignment of the polymers required for gelation (Figure 2b) and irreversibly obstructs cooperative binding of the monomers. Further cooling facilitates the formation of an ordered inter-polymer structure that results in the formation of a homogeneous gel matrix [29]. However, the elevated temperatures used in this approach are unsuitable when using labile and fragile bio-macromolecules (e.g., growth factors).

Hydrogel properties are easily tunable via adjusting alginate composition and gel fabrication approach [33–38], as well as the type of crosslinking counterions, the ionic strength, the molecular weight of alginate (reflecting its viscosity), pH and the temperature [39–44].

Besides ionically cross-linked (physical) hydrogels, alginate can also form covalently cross-linked (chemical) hydrogels. In general, covalently cross-linked alginate gels have higher mechanical and chemical resistance compared to those of physical hydrogels [20]. Chemical hydrogels possess a high stability in a wide range of pH (between 1 and 13), temperature (from 0 up to 100 °C), and various polar solvents and high ionic strength as well [20]. On the other hand, physical hydrogels are reversible because they are formed due to conformational changes, whereas chemical hydrogels form a permanent structure that is irreversible because of configurational changes occurring during hydrogel formation. Largely, this reversibility of physical hydrogels makes them favourable candidates for a variety of biomedical applications. Other advantages of physical and chemical hydrogels are described elsewhere [45]. Further in this review, physical hydrogels will be considered if not mentioned otherwise.

2.3. Alginate Gels as Drug Carriers: Encapsulation and Release

The tailor-made structure and widely tuneable properties of alginate hydrogels, as well as their biocompatibility, make them favourable candidates for versatile biological and medical applications. Thus, alginate hydrogels have been extensively developed as nano- or micro-formulations (in a form of gel particles, beads, or capsules) for controlled drug delivery, as well as materials for wound care and engineering of microtissues [20–24]. The opportunity to encapsulate bioactives into alginate hydrogels under mild conditions, as well as release them in a controlled manner, plays a crucial role for all of these applications and will further be considered more in detail.

Alginate hydrogels are versatile matrices allowing to encapsulate living cells, macromolecules (proteins, growth factors, enzymes, etc.), therapeutic molecules or functional nanomaterials into the gel network preserving their bioactivity and functions [22,23,25,46–49]. First achieved almost four decades ago, the encapsulation of islet cells into alginate hydrogel gave rise to their use for cell culture and opened new avenues for tissue engineering [29]. Nearly at the same time, alginate particles have been proposed as containers for encapsulation of molecules [24]. To form alginate beads, at first the alginate solution is usually mixed with the solution of the molecules of interest or the suspension of cells. Then, two scenarios for the preparation of alginate beads are possible. The straightforward way is based on the further exposure of this mixture to the solution of cross-linking ions (employing one of the methods described above) that leads to the formation of the large piece of a hydrogel. Further, smaller alginate beads can be obtained via mechanical breakdown of a bulk gel into the particles of a desired size. However, this approach concedes the opposite way that is usually employed. Namely, the alginate solution mixed with an encapsulated component is immersed into the cross-linking solution drop by drop [21,24]. Depending on the droplet fabrication approach, formed alginate particles range from macro dimensions (> 1 mm) down to nano-beads (< 0.2 μm). Notably, encapsulated components can be either homogeneously distributed over the whole bead volume or concentrated into the centre of a gel bead (e.g., one cell per a single bead) [21,23,24,48–50]. Thus, alginate macro-beads (1–2 mm) can be fabricated using the extrusion method when alginate is dripped into the $CaCl_2$ bath using a syringe. Modification of this method via employment of an electric field, mechanical vibration, or by using a rotating device results in the formation of microbeads (0.2–1000 μm). A variety of other methods for the fabrication of alginate micro-beads have also been reported and are described elsewhere [21,23,24,48–50]. Alginate nano-beads (200 nm and less) are typically produced employing nano-vesicles and emulsion droplets as sacrificial templates. This templating strategy allows for designing not only matrix-type but also hollow structures (nano-capsules) that are formed after elimination of the template. Herein, among others we would like to highlight the use of insoluble vaterite $CaCO_3$ crystals as they will be a key for the formulation of PAS. The originality of the use of these crystals arises from its ability to simultaneously serve as sacrificial templates and a source of Ca^{2+}.

A high water content and porous nature of alginate gels (pore sizes in the range 5–200 nm) result in a relatively fast diffusion of biomolecules and drugs within the gel [21,22]. Indeed, the release kinetics directly depend on the gel porosity which can be well-tuned by varying the number of cross-linking cations and its type, composition (source and chemical modification, if applicable) of alginate and the size of alginate beads [21]. As a general rule, smaller pores of 12–16 nm are typical for alginate gels prepared using the diffusion method of gelation, while hydrogels prepared via in situ gelation have larger pores.

Strong electrostatic interaction of alginate matrix with the charged encapsulates also affects the release kinetics. For instance, simultaneous encapsulation of multiple drugs (methotrexate, doxorubicin, and mitoxantrone) has been demonstrated in [51]. It was found that methotrexate that does not interact with alginate rapidly liberates from the hydrogel while covalently bound doxorubicin releases with lower rates via chemical hydrolysis of the cross-linker, and mitoxantrone that is ionically bound to alginate releases only after dissociation of the hydrogel. Mild conditions used during the encapsulation and the gelation minimize protein denaturation and degradation, making alginate an excellent candidate for loading of protein-based bioactives. This stimulated a number of studies aimed at loading/release of a wide range of proteins and nucleic acids [22,25].

It is important to note that mammalian cells have no enzymes to cleave alginate chains which make alginate hydrogels non-degradable in mammals. Therefore, molecular diffusion and erosion of the polymer network are the only two factors that determine the release kinetics of bioactives. While the first scenario of erosion-mediated release is typically observed for prolonged release, the second one (diffusion-mediated release) is usually rather fast and accompanied by a low loading efficiency. The latest results in spontaneous leakage of bioactives from alginate beads. Deceleration of diffusion-mediated release has been reported to be achieved via additional protection of the hydrogel beads using the layer-by-layer (LbL) assembled polymer multilayer shell [21,22]. Due to their chemical structure, alginate gels shrink at a low pH and swell at a neutral pH. At a very high pH or in the presence of EDTA or citric acid as cation chelators the cross-linking ions are released that leads to the dissolution of the gel inducing erosion-mediated release. This phenomenon has been widely used for pH-induced release from alginate hydrogels [21].

2.4. Alginate Gels for the Design of Porous Scaffolds

Traditionally, polymer scaffolds for tissue engineering are fabricated using naturally derived biomaterials. Among them, alginate hydrogels have been extensively developed due to their similarity to extracellular matrix of mammalian tissues in terms of mechanical properties and widely tunable kinetics of hydrogel degradation, as well as controlled release of molecules at various pH values including neutral pH [33,52,53]. A wide range of bio-applications of alginate hydrogels includes but not limits to cell transplantation, wound healing, encapsulation and controlled or programmed delivery of drugs and biomacromolecules, and the use as anti-adhesive and repair materials [22,23,32,46,54–57]. Recent progress in the development of alginate hydrogels for the fabrication of scaffolds showed the employment of a number of advanced techniques including gas foaming [58], 3D printing [59,60], electrospinning [60–62], emulsion freeze drying [63], microfluidics [58,64], etc. Alginate gels serve as platforms for cell culture and growth of microtissues [65], cardiovascular muscles [66], bones [67], liver [68,69], etc.

Design of hydrogels on the macroscopic level assumes control over the size and porous structure of the gels [70]. Hydrogel matrices can be either non-porous (having only small pores that are typically in the range of tens of nm for the alginate gel network [71]) or contain macroscopic pores that are typically in the range of 10–500 μm [72] (Figure 3). Dual nano- and macro-porosity is essential for controlled growth of a tissue and drug delivery [70].

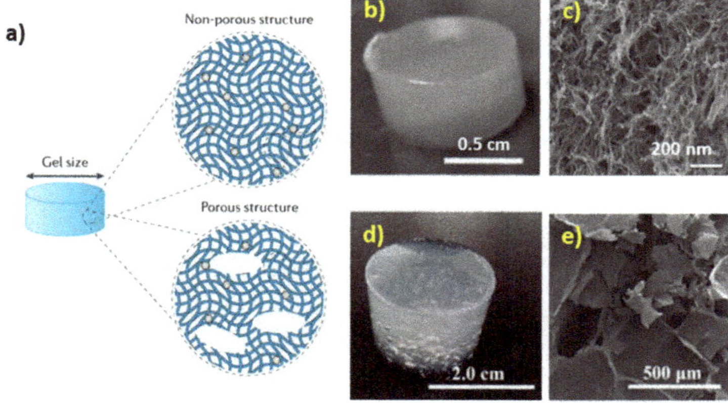

Figure 3. (**a**) Scheme of the architecture of porous and non-porous hydrogels. Reproduced with permission from [70], published by Springer Nature, 2016; (**b**,**d**) Photographs of wet alginate scaffolds and (**c**,**e**) SEM images of dry alginate scaffolds: (**b**,**c**) — non-porous. Reproduced with permission from [71], published by John Wiley & Sons, 2016; (d,e) — porous alginate scaffold. Reproduced with permission from [72], published by Springer-Verlag Berlin Heidelberg, 2018.

High stability of ionically cross-linked alginate gels makes it possible to fabricate a gel with defined dimensions and geometries using different patterning techniques [12,55,73] including light-triggered pattering and employment of microfluidics [54,56,74], electrochemical methods [75], etc. However, utilization of lithography and 3D printing technologies are usually required for design of any hydrogel. The use of harsh conditions during scaffold fabrication (e.g. high or low temperatures, exposure to gas-liquid or solid-liquid interface, the use of organic solvents and aggressive media, surfactants) still remains an obstacle. This often does not allow us to encapsulate bioactives during the hydrogel synthesis that may be crucial for utilization of scaffolds and limits the control over the scaffold internal structure [47].

The latest issue is typically accompanied by scarce pore interconnectivity that is essential for cell colonization in the entire volume of the scaffold. This problem has partially been solved by A. Barbetta et al. [76]. Therein, two methodologies for the formulation of PAS with highly interconnected pores in different size ranges have been proposed. Emulsion templating of the hydrogel allowed producing PAS with the pores of about 10–20 μm that are interconnected via the channels of 2–5 μm [76]. An approach based on the foam templating results in the formation of alginate gels with large 100–300 μm pores and interconnections in the range of 30–80 μm [76]. Both methods allow one to produce PAS with a highly developed internal macro-sized structure that is crucial for cell growth and proliferation due to a need for a free space for cell colonization and requirements of a non-restricted transport of cell metabolites and essential nutrients. However, both approaches [76] lack precise control over the pore distribution that appears random and does not give any options for loading of therapeutics and growth factors, important for cell attachment, growth and proliferation.

This seems to be one of the major challenges for further progress in this field. To the best of our knowledge, only a few works reported the formation of alginate hydrogels with an opportunity to host macromolecules, and no alginate gels possessing both a well-defined internal structure and loading of bioactives at desired doses have been reported. In this sense, one of the most promising methodologies is the templating of alginate hydrogels on mesoporous vaterite $CaCO_3$ crystals (the strategy that has been briefly mentioned above). Explored by Wang et al. [16] and by Roberts and co-workers [17], this idea has been implemented employing model molecules (ibuprofen [16] and bovine serum albumin (BSA) [17]) that have been pre-loaded to $CaCO_3$ vaterite crystals. A suspension of these crystals has been mixed with the solution of alginate followed by addition of glucono-δ-lactone that slowly doped H^+ due to its hydrolysis. The acidification caused a mild dissolution of $CaCO_3$ and release of Ca^{2+} that cross-link the alginate forming a gel. Macro-sized pores have been formed as a result of calcium carbonate dissolution in the places where $CaCO_3$ crystals have been initially located. Consequently, ibuprofen [16] and BSA [17] have been rapidly liberated from the PAS. Additional coating of $CaCO_3$ with a polyelectrolyte multilayer shell resulted in slowing down the release rate by ca 50 times as compared with alginate gels formed using bare $CaCO_3$ crystals [17]. These works manifested that the utilization of $CaCO_3$ crystals as soluble cores for templating alginate hydrogels is a powerful approach promising for the development of scaffolds towards cell-based applications. However, the approach above is rather sophisticated and involves multiple steps including a rather long procedure of multilayer coating bringing additional costs. The way proposed to avoid these issues and novel achievements in $CaCO_3$-assistant formation of PAS will be addressed in the Section 4 of this review. However, prior to this, the following Section 3 will describe the structure, principles of the fabrication, and featured properties of vaterite $CaCO_3$ crystals.

3. $CaCO_3$ Vaterite Crystals: Loading and Release Opportunities

3.1. *Morphology of Vaterite Caco$_3$ Crystals*

Calcium carbonate mainly exists in the form of one of three polymorphs: calcite, vaterite and aragonite (Figure 4a–c). All the polymorphs have different shapes and morphologies that can be distinguished from each other employing various methods, for instance X-ray diffraction (Figure 4d) or Raman spectroscopy (Figure 4e). Among all of the polymorphs, vaterite is the most attractive for biomedical applications because it has a highly developed internal structure ideal for microencapsulation/release of bio-macromolecules and drugs. Vaterite $CaCO_3$ crystals can easily be formed upon mixing of aqueous solutions of precursor salts of Ca^{2+} and carbonate ions. The mechanism of crystal growth is expletively described elsewhere [77,78]. Briefly, spherical vaterite crystals comprise of small nanocrystallines interconnected to each other forming mesoporous structure of the crystal. The use of organic additives [79], some protein/polymer matrices [80] or nanoparticles [81] can direct the growth of vaterite crystals of specific shape and morphology. The porosity of the crystals can also be controlled, e.g. via the variation of crystal preparation temperature [82]. The typical sizes of crystals range from 3 to 20 μm, although a number of recent studies proposed novel ways for the fabrication of nano-crystals [79,80,83] or large vaterite of sizes in sub-millimeter range [84].

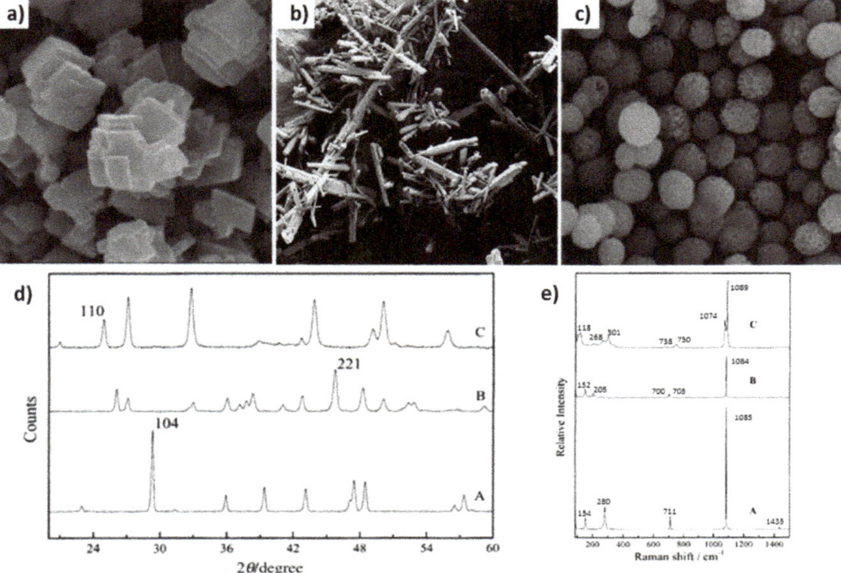

Figure 4. Scanning electron microscopy (SEM) images of $CaCO_3$ polymorphs: (**a**) rhombohedral calcilte, (**b**) needle aragonite and (**c**) spherical vaterite. Reproduced with permission from [85], published by Elsevier, 2016. (**d**) XRD and (**e**) Raman spectra of the synthetically prepared calcite (A), aragonite (B) and vaterite (C). Reproduced with permission from [86], published by Royal Society of Chemistry (Great Britain), 2000.

3.2. Vaterite $CaCO_3$ as Decomposable Templates for Microencapsulation

Nowadays, inorganic crystals of the vaterite polymorph of $CaCO_3$ are classified as advanced biodegradable and biocompatible materials to be employed for a wide range of bio-applications such as biomedical engineering, biosensors and controlled drug delivery. The growing interest in vaterite $CaCO_3$ has emerged based on crystal highly porous nature, easy adjustment of dimensions and porosity during the crystal synthesis, cost-effective formulation and marginal toxicity [87–91]. Indeed, the internal structure of vaterite crystals is mesoporous with the typical pore size of tens of nanometers that is highly favorable for the loading of bio-macromolecules and drugs as well as functional materials such as inorganic nanoparticles (e.g., magnetite [92–95], silver [95,96]), carbon nanotubes and halloysites [97]. $CaCO_3$ crystals can be loaded with the low-molecular-weight molecules, e.g., small drugs (doxorubicin [98]) and photosensitizer [99], as well as with high-molecular-weight macromolecules, e.g., dextrans [88,90], polymers (alginate [90], mucin [100,101]), and proteins (catalase, BSA, insulin [88–90,102,103]). The functionalization of $CaCO_3$ vaterite crystals with inorganic nanoparticles brings new properties desired for the use of crystals in surface enhanced Raman microscopy [96,104–106], making crystals sensitive to external stimuli (e.g., electrical/magnetic fields, light irradiation [107–110]). The fabrication of pure protein [102,111,112] or polymer [113–115] particles can be achieved via hard templating on the vaterite cores. The templating is based on filling the crystal pores with material of interest followed by the crystal elimination that results in the formation of the inverted crystal replica (in case of a full filling of the pores). This opens new avenues for the utilization of vaterite $CaCO_3$ crystals and hybrid structures assembled on them as sacrificial templates [116], for controlled release, targeted drug delivery [83], surface patterning [117], and reconstitution of artificial cellular compartments [118].

Impregnation of the encapsulates into the interior of the vaterite $CaCO_3$ crystals can be performed at mild conditions in one of two ways: i) during the crystal growth (so-called co-precipitation or co-synthesis) or via the post-loading of the pores of pre-formed crystals (by means of adsorption or via solvent evaporation) [81,89,90,102,119,120]. All methods for the encapsulation have their advantages and disadvantages; the choice of appropriate approach mainly depends on the nature of the encapsulate. Thus, post-loading by means of adsorption represents the mildest method suitable for the encapsulation of fragile macromolecules that are highly sensitive to their microenvironment and can easily lose their bio-activity [120,121]. In its turn, the co-precipitation approach is based on the inclusion of encapsulates into one of the precursor salt solutions to make the crystals, further mixing of the salts and entrapment of encapsulates inside the growing crystals. This leads to higher encapsulation efficiencies if compared to the adsorption method, but may result in a partial loss of the bio-activity of the encapsulated molecules caused by the influence of crystal growth conditions [121]. On the other hand, the co-precipitation provides a homogeneous distribution of molecules within an interconnected internal volume of vaterite crystals. Finally, solvent evaporation grants the highest encapsulation efficiencies, yet it is however the harshest method among three approaches described above due to conditions of solvent removal and it is thus less suitable for labile [121] and sensitive molecules [83].

Besides the integration of molecules of interest inside the crystals, deposition of additional coatings onto the external surface of the crystals can also be achieved, e.g. via the LbL assembly of the polyelectrolytes [88,98,109,122–124]. Importantly, the multilayer shells assembled on the crystals are fully permeable for ions and small molecules that allows for a complete decomposition of $CaCO_3$ cores when lowering pH or using chelating agents (e.g., EDTA). This results in the formation of completely hollow polyelectrolyte capsules or capsules of a matrix type that contain a polymer matrix inside [125]. The most attractive feature of multilayer capsules assembled on $CaCO_3$ cores is a selective permeability of a multilayer shell that can also be functionalized with some stimuli-sensitive materials (e.g., those responsive to pH [126], infra-red light [127,128]), so the encapsulated molecules can be released from the capsule in a controlled manner [129].

3.3. Release from Vaterite Caco$_3$ Crystal: Dissolution and Recrystallization

If not considering the case of molecular release from functionalized vaterite $CaCO_3$ crystals that is mediated by external stimuli, one can distinguish two main mechanisms of the release from bare $CaCO_3$ crystals: dissolution- and recrystallization-mediated release. $CaCO_3$ can easily be dissolved at a slightly acidic pH [116] or upon the addition of chelating agents, e.g., EDTA or citric acid (corresponding constants of the binding to Ca^{2+} in the $CaCl_2$ solution are $K_a \sim 2 \times 10^8$ M^{-1}, and $K_a \sim 10^{3.5}$ M^{-1} at pH 7 for EDTA and citric acid, respectively). While acidic pH is not desirable for sensitive compounds such as proteins or growth factors, dissolution of $CaCO_3$ crystals at neutral pH has a crucial importance for bio-applications providing a complete release of the loaded molecules.

On the other hand, the immersion of mesoporous vaterite crystals into aqueous media results in a phase transition and spontaneous recrystallization of vaterite to thermodynamically more stable but non-porous calcite. If the crystals have been laden with some molecules of interest, the transformation of the vaterite to calcite provokes the liberation of these molecules from the porous interior of vaterite crystals to external medium [130,131].

It is known that vaterite → calcite recrystallization is to a large extent a surface-mediated process [78,132–134] and recrystallization kinetics usually exhibits an exponential-like behavior [81]. The recrystallization kinetics can be controlled via the use of additives. For instance, $CaCO_3$-Fe_3O_4

vaterite microparticles recrystallize significantly faster if compared with pure vaterite crystals [81]. The LbL assembled polyelectrolyte coating of pre-loaded $CaCO_3$ crystals can also effectively regulate molecular and ion transport on the crystal-liquid interface allowing us to program the release kinetics [81].

As a short summary of the described above, unique properties of vaterite crystals are in their i) biocompatibility, ii) ability to trap and retain huge amounts of small and large molecules and nanoparticles of various nature, (iii) opportunity to encapsulate bioactives at mild conditions and neutral pH; and iv) wide range of options for programmed and controlled slow/fast release that is either regulated via crystal dissolution and recrystallization (for the bare crystals) or by external stimuli (for functionalized crystals). In recent years, these features stimulated the idea to utilize vaterite calcium carbonate as sacrificial templates for the fabrication of polymer-based alginate scaffolds. In principle, this strategy can provide simultaneous i) cross-linking and adjustment of hydrogel nanoporosity; ii) control over the macroporosity of porous scaffolds and iii) encapsulation and preservation of fragile bioactives in the entire volume of the scaffold. This makes the use of mesoporous $CaCO_3$ crystals for the fabrication of PAS a beneficial and superior approach. Latest achievements in this area are discussed below in Section 4.

4. Vaterite $CaCO_3$-Assistant Porous Alginate Scaffolds (PAS)

4.1. *Fabrication Strategy*

In a majority of works focused on composite $CaCO_3$-alginate gel materials, $CaCO_3$ crystals are used as a source for mineralization [135] of the scaffolds and/or as a hardening component for the scaffolds utilized in hard tissue engineering (e.g., [136,137]). Therefore, there was no need to eliminate $CaCO_3$ crystals in order to form the pores, and even vice versa, the crystals have been kept in the final scaffold architecture. A straightforward approach for the fabrication of composite $CaCO_3$-alginate gel materials was first employed nearly one decade ago and has been based on the simultaneous growth of $CaCO_3$ crystals and gelation of alginate hydrogel [138–142]. In this design, calcium carbonate crystals grow in the presence of the gel and get entrapped inside this polymer matrix. This strategy showed its promise for the controlled crystallization of $CaCO_3$ crystals. However, it has serious limitations. The major one is a lack of control over the internal structure of the scaffold. Although the structure of the growing crystals can be manipulated via the variation of environmental conditions (polymer concentration, gel composition, etc.), the final distribution of crystals and the macrostructure of the scaffold cannot be controlled since the crystallization of $CaCO_3$ is a spontaneous and highly sensitive process. In addition, the presence of calcite and sometimes amorphous calcium carbonate has been detected [139]. For some cases, a significant decrease in the size of $CaCO_3$ crystals resulted in the formation of nano-$CaCO_3$ that found its application in drug delivery but was not suitable for the fabrication of macro-porous scaffolds [141].

In contrast, the utilization of vaterite $CaCO_3$ crystals as sacrificial cores for the formation of alginate scaffolds in microfluidics set-up allows one to design stable PAS that have a well-defined and highly developed porous structure. The high potential of this fabrication strategy was recently manifested for the precise control over the scaffold porosity [18,19] and a high performance of the encapsulation/controlled release of biomolecules [19]. The work [18] introduced the method offering the fabrication of 2D $CaCO_3$-assistant alginate scaffolds at acidic conditions (Figure 5a). Therein, $CaCO_3$ crystals suspended in the alginate solution have been spread over a glass substrate (Figure 5ai) followed by the addition of HCl that resulted in the dissolution of calcium carbonate and the release of Ca^{2+} ions that induces physical cross-linking of the hydrogel (Figure 5aii–iii). Control over the concentration of Ca^{2+} and as a result over the cross-linking degree has been achieved via variation of the ionic strength (Figure 5aiii).

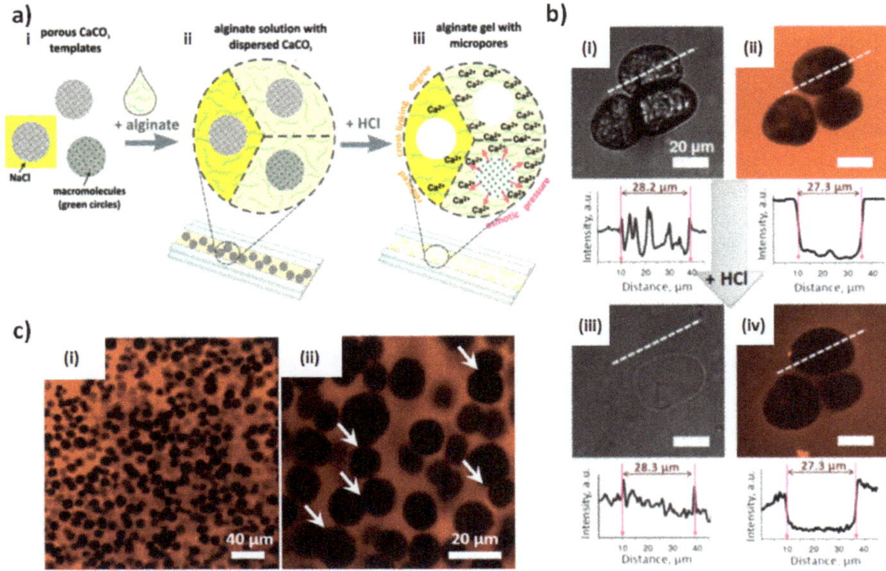

Figure 5. (**a**) Schematics of formation of porous alginate hydrogels. (i,ii) Dispersion of CaCO3 crystals in alginate solution followed by deposition of the suspension onto a glass substrate. (ii,iii) Formation of porous hydrogel by addition of HCl, which induces CaCO3 dissolution. The dissolution process is accompanied by alginate cross-linking and formation of hollow pores. (**b**) Optical transmission and fluorescent images of 33 µm vaterite $CaCO_3$ crystals dispersed in alginate before (i,ii) and after (iii,iv) addition of HCl. Fluorescence profiles are given to each image and taken along the white interrupted lines. (**c**) CSLM images of porous alginate gel formed at compact packing of 11 µm $CaCO_3$ templates. White arrows in (c,ii) indicate interconnected pores. The gel (b,c) has been stained with rhodamine 6G. Reproduced with permission from [18], published by John Wiley & Sons, 2015.

4.2. PAS Porosity and Mechanical Properties

Gel cross-linking and osmotic pressure generated by the released calcium ions have been shown to play a pivotal role in the formation of the micro-sized pores in PAS [18]. Notably, the pores of the formed PAS are hollow (Figure 5b). The pH used for dissolution of the crystal core is the key to manipulate the stability and size of the pores during $CaCO_3$ elimination. Thus, a less acidic pH (that can be achieved by addition of relatively low HCl concentration) results in the slow dissolution of $CaCO_3$ that is accompanied by the collapse and closing of the pores. The use of high HCl concentration provokes fast dissolution of vaterite cores that results in the uncontrolled spontaneous formation of CO_2 bubbles and the enlargement of the formed micro-sized pores. Under optimal acidic conditions, pores keep the size equal to that of the $CaCO_3$ crystals used (Figure 5b). This allows a rather easy control over the pore size distribution via the utilization of vaterite crystals of desired dimensions [18].

The structure of the $CaCO_3$-assistant PAS assembled on $CaCO_3$ crystals of about 11 µm shown in Figure 5c clearly reveals the presence of both closed and interconnected pores. These scaffolds are soft, having the Young modulus of tens of kPa. A highly developed and tunable internal structure and soft nature of $CaCO_3$-assistant PAS make them promising for the use in biomedical applications, e.g., for bio-engineering of soft tissues and organs (Figure 6). Another advantage is the opportunity to load and release bioactive molecules into the PAS. Therefore, the last section of this review will highlight recent achievements in the loading/release of non-charged dextrans, charged bio-macromolecules (proteins) and small molecules (dyes) into/out of PAS assembled using vaterite $CaCO_3$ crystals.

4.3. PAS as Reservoirs for Encapsulation and Controlled Release

Fluorescein isothiocyanate-labeled dextrans (FITC-dextran) are known as model macromolecules widely employed for investigation of the release performance and kinetics of various carriers. Dextrans have slightly negative zeta-potentials closed to zero at neutral pH and can have molecular weight variable in a wide range [143]. The kinetics of the release of dextrans from vaterite $CaCO_3$ crystals has been extensively studied in recent years. The study [19] investigated the release of dextrans of different molecular weights from PASs assembled on ca 8 µm-sized $CaCO_3$ cores (Figure 7a). The release rate of FITC-dextran has been demonstrated to be directly related to its molecular weight that is rather typical for homogeneous matrices and indicates a significant role of spontaneous molecular diffusion. Interestingly, alginate concentration had no influence on FITC-dextran release [19]. Assuming the absence of strong electrostatic interaction between dextrans and PAS, it has been concluded that there is a cut off for the molecules of 7–16 nm, so larger macromolecules are retarded by the alginate network and small molecules can freely diffuse the gel outward.

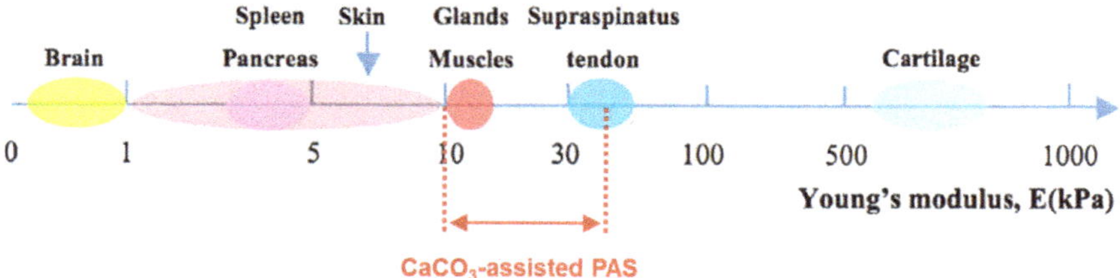

Figure 6. Comparison of Young's modulus of natural soft tissues and organs and Young's modulus of $CaCO_3$-assisted PASs. Adopted with the permission from [144]. Reproduced with permission from [144], published by MDPI, Basel, 2015.

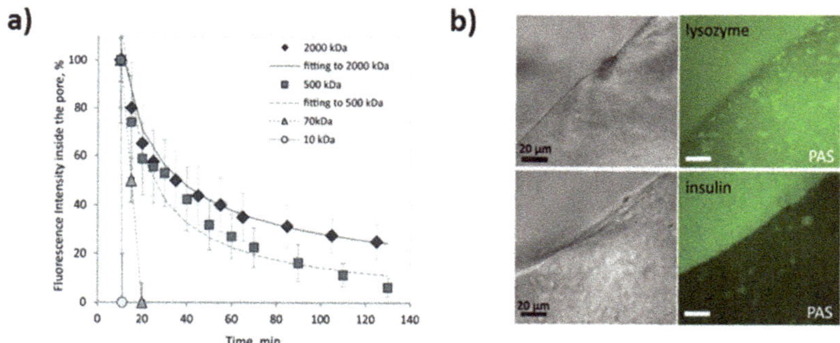

Figure 7. (**a**) Release kinetics of FITC-dextran (different MW) release from the macro-pores of PAS. (**b**) Mass transport of proteins from the solution into/through the PAS gel network. Reproduced with permission from [19], published by American Chemical Society, 2015.

Strong interaction between charged macromolecules (proteins) and alginate gel has also been reported. As opposed to the study of dextran release kinetics, the protein-PAS interaction has been examined assessing the loading of proteins into prepared PAS. Alginate gel itself possesses a negative charge due to a low pK_a of the alginic acid. Small protein lysozyme (oppositely charged compared to alginate gel) accumulates inside PAS (Figure 7b) while negatively charged insulin reaches much lower internal concentration inside the PAS although its diffusion is also rather fast (a scale of minutes) [19]. This clearly indicates a high potential for the encapsulation of macromolecules that possess a positive net charge into the negatively charged PAS. At the same time, the retention of macromolecules processing a negative net charge can be awkward. The scenario described above can, however, be different in some special cases as it has been shown, for instance, for small anionic dye [145].

Therein, the dissolution of phthalocyanine-loaded $CaCO_3$ crystals covered by an alginate matrix and re-distribution/release of CuPcTs dye molecules [146] has been monitored by Raman spectroscopy [145]. Notably, small CuPcTs molecules pre-encapsulated into sacrificial $CaCO_3$ cores can be retained inside the macro-sized pores formed at the places of eliminated $CaCO_3$ crystals that are probably due to the repulsion between negatively charged dye molecules and similarly charged ALG gel surrounding the pores. From the other side, the reason could be the aggregation of small dye molecules inside the pores of $CaCO_3$ during the co-synthesis, so such large molecular aggregates cannot escape from the pores due to sterical limitations. In any case, these results are promising for the loading and retention of small and/or for anionic molecules inside the pores of the PAS.

5. Summary and Perspectives

Fabrication of porous biopolymer-based scaffolds is rapidly developing field of biomedical engineering. In this field, porous alginate scaffolds built up employing mesoporous vaterite $CaCO_3$ microcrystals are extremely promising due to i) highly porous PAS structure that can be well-tuned and ii) the ability to load the scaffolds with bioactive molecules of a diverse nature and release them on demand. Microfluidics-based design of $CaCO_3$-assisted PASs utilizing pre-formed $CaCO_3$ crystals offers a high degree of control over the internal PAS structure. As opposed to that, the simultaneous crystal growth and alginate gelation lacks the control over the PAS structure and does not provide an opportunity to pre-load $CaCO_3$ cores with the desired active compounds.

Despite the high potential of $CaCO_3$-assisted PASs, their fabrication and use have not been investigated well yet, and it is better to say that nowadays this approach is only just emerging. Therefore, the design of $CaCO_3$-assisted PASs requires a further deep development in terms of fundamental issues raised and applications. Thus, the fabrication of the scaffolds under mild conditions (media with the pH near the physiological one, i.e., pH 7.4) is urgently required as currently employed HCl-mediated leaching of $CaCO_3$ cores may result in the reduction bioactives' activity and cell viability. Thus, the investigation of the interaction of $CaCO_3$-assisted PASs with cells will be a crucial step for further development of PASs. There is an intuitive perception that one of the best options would be the substitution of HCl as a relatively aggressive dissolution agent to weaker acids (e.g., citric acid) or chelating Ca^{2+}-binding agents (e.g., EDTA). In principle, the latest can be achieved at neutral pH [114].

On the other hand, the strategies used to encapsulate bioactive compounds into $CaCO_3$ cores and to protect the scaffolds from undesired spontaneous leakage of these bioactives should further be addressed, verified and improved. Herein, the entrapment of small and/or anionic compounds can turn up the challenge due to the relatively large nano-pores of alginate matrix of the PAS (7–16 nm) and the negative charge of the alginate gel due to carboxylic groups on the alginate backbone. Here, one of the strategies might be the co-loading of these small drugs with large oppositely charged biopolymers. The formation of drug-biopolymer complexes inside the pores of $CaCO_3$ promotes the entrapment of the drugs and allows one to substantially increase encapsulation efficiency (e.g., [147,148]). Formation of LbL capsules on $CaCO_3$ cores could be alternative strategy.

Pioneering studies on the design of $CaCO_3$-assisted PASs indicate that all obstacles mentioned above can potentially be overcome. We believe that the described PASs can become a new generation of biopolymer scaffolds with tailor-made architecture and controlled porosity, high pore interconnection and an opportunity to load and release biomolecules of interest. This allows one to use the terms *intelligent* or *smart* for the PASs, and opens a new avenue for further successful PAS employment towards tissue engineering and regenerative medicine.

Author Contributions: Conceptualization, all authors; investigation, A.S. and A.S.V.; writing—original draft preparation, A.S. and A.S.V.; writing—review and editing, all authors; supervision, project administration and funding acquisition, D.V.

Funding: This work was partially supported by the Alexander von Humboldt Foundation (Sofja Kovalevskaja Program) and by the DAAD (German Academic Exchange Service, Mikhail Lomonosov program). Dmitry Volodkin acknowledges Quality Research (QR) Fund from Nottingham Trent University.

Conflicts of Interest: The authors declare no conflict of interest.

References

1. Chaudhari, A.A.; Vig, K.; Baganizi, D.R.; Sahu, R.; Dixit, S.; Dennis, V.; Singh, S.R.; Pillai, S.R. Future Prospects for Scaffolding Methods and Biomaterials in Skin Tissue Engineering: A Review. *Int. J. Mol. Sci.* **2016**, *17*, 1974. [CrossRef] [PubMed]
2. Dong, C.; Lv, Y. Application of Collagen Scaffold in Tissue Engineering: Recent Advances and New Perspectives. *Polymers* **2016**, *8*, 42. [CrossRef] [PubMed]
3. Raeisdasteh Hokmabad, V.; Davaran, S.; Ramazani, A.; Salehi, R. Design and fabrication of porous biodegradable scaffolds: A strategy for tissue engineering. *J. Biomater. Sci. Polym. Ed.* **2017**, *28*, 1797–1825. [CrossRef] [PubMed]
4. Van Bochove, B.; Grijpma, D.W. Photo-crosslinked synthetic biodegradable polymer networks for biomedical applications. *J. Biomater. Sci. Polym. Ed.* **2019**, *30*, 77–106. [CrossRef]
5. Chappard, D.; Baslé, M.F.; Legrand, E.; Audran, M. New laboratory tools in the assessment of bone quality. *Osteoporos. Int.* **2011**, *22*, 2225–2240. [CrossRef] [PubMed]
6. Santamaría, V.A.; Deplaine, H.; Mariggió, D.; Villanueva-Molines, A.R.; García-Aznar, J.M.; Ribelles, J.G.; Doblaré, M.; Ferrer, G.G.; Ochoa, I. Influence of the macro and micro-porous structure on the mechanical behavior of poly(l-lactic acid) scaffolds. *J. Non-Cryst. Solids* **2012**, *358*, 3141–3149. [CrossRef]
7. Walocha, J.A.; Miodoński, A.J.; Szczepański, W.; Skrzat, J.; Stachura, J. Two types of vascularisation of intramural uterine leiomyomata revealed by corrosion casting and immunohistochemical study. *Folia Morphol.* **2004**, *63*, 37–41.
8. Vaz, C.M.; van Tuijl, S.; Bouten, C.V.C.; Baaijens, F.P.T. Design of scaffolds for blood vessel tissue engineering using a multi-layering electrospinning technique. *Acta Biomater.* **2005**, *1*, 575–582. [CrossRef]
9. Ivashchenko, O.; Przysiecka, Ł.; Peplińska, B.; Jarek, M.; Coy, E.; Jurga, S. Gel with silver and ultrasmall iron oxide nanoparticles produced with Amanita muscaria extract: Physicochemical characterization, microstructure analysis and anticancer properties. *Sci. Rep.* **2018**, *8*, 13260. [CrossRef]
10. Wasserthal, L.; Wasserthal, W. Innervation of heart and alary muscles in Sphinx ligustri L. (Lepidoptera). *Cell Tissue Res.* **1977**, *184*. [CrossRef]
11. Baklaushev, V.P.; Bogush, V.G.; Kalsin, V.A.; Sovetnikov, N.N.; Samoilova, E.M.; Revkova, V.A.; Sidoruk, K.V.; Konoplyannikov, M.A.; Timashev, P.S.; Kotova, S.L.; et al. Tissue Engineered Neural Constructs Composed of Neural Precursor Cells, Recombinant Spidroin and PRP for Neural Tissue Regeneration. *Sci. Rep.* **2019**, *9*, 3161. [CrossRef] [PubMed]
12. Ozawa, F.; Ino, K.; Arai, T.; Ramón-Azcón, J.; Takahashi, Y.; Shiku, H.; Matsue, T. Alginate gel microwell arrays using electrodeposition for three-dimensional cell culture. *Lab Chip* **2013**, *13*, 3128–3135. [CrossRef] [PubMed]
13. Zawko, S.A.; Schmidt, C.E. Simple benchtop patterning of hydrogel grids for living cell microarrays. *Lab Chip* **2010**, *10*, 379–383. [CrossRef]
14. Sharma, C.; Dinda, A.K.; Mishra, N.C. Fabrication and characterization of natural origin chitosan-gelatin-alginate composite scaffold by foaming method without using surfactant. *J. Appl. Polym. Sci.* **2013**, *127*, 3228–3241. [CrossRef]
15. Despang, F.; Börner, A.; Dittrich, R.; Tomandl, G.; Pompe, W.; Gelinsky, M. Alginate/calcium phosphate scaffolds with oriented, tube-like pores. *Mater. Und Werkst.* **2005**, *36*, 761–767. [CrossRef]
16. Wang, C.; Liu, H.; Gao, Q.; Liu, X.; Tong, Z. Alginate–calcium carbonate porous microparticle hybrid hydrogels with versatile drug loading capabilities and variable mechanical strengths. *Carbohydr. Polym.* **2008**, *71*, 476–480. [CrossRef]
17. Roberts, J.R.; Ritter, D.W.; McShane, M.J. A Design Full of Holes: Functional Nanofilm-Coated Microdomains in Alginate Hydrogels. *J. Mater. Chem. B* **2013**, *107*, 3195–3201. [CrossRef]
18. Sergeeva, A.; Feoktistova, N.; Prokopovic, V.; Gorin, D.; Volodkin, D. Design of Porous Alginate Hydrogels by Sacrificial $CaCO_3$ Templates: Pore Formation Mechanism. *Adv. Mater. Interfaces* **2015**, *2*, 1500386. [CrossRef]

19. Sergeeva, A.S.; Gorin, D.A.; Volodkin, D.V. In-situ assembly of Ca-alginate gels with controlled pore loading/release capability. *Langmuir* **2015**, *31*, 10813–10821. [CrossRef]
20. Ching, S.H.; Bansal, N.; Bhandari, B. Alginate gel particles-A review of production techniques and physical properties. *Crit. Rev. Food Sci. Nutr.* **2017**, *57*, 1133–1152. [CrossRef]
21. Pawar, S.N.; Edgar, K.J. Alginate derivatization: A review of chemistry, properties and applications. *Biomaterials* **2012**, *33*, 3279–3305. [CrossRef]
22. Lee, K.Y.; Mooney, D.J. Alginate: Properties and biomedical applications. *Prog. Polym. Sci.* **2012**, *37*, 106–126. [CrossRef]
23. Goh, C.H.; Heng, P.W.S.; Chan, L.W. Alginates as a useful natural polymer for microencapsulation and therapeutic applications. *Carbohydr. Polym.* **2012**, *88*, 1–12. [CrossRef]
24. Gombotz, W. Protein release from alginate matrices. *Adv. Drug Deliv. Rev.* **1998**, *31*, 267–285. [CrossRef]
25. Paques, J.P.; van der Linden, E.; van Rijn, C.J.M.; Sagis, L.M.C. Preparation methods of alginate nanoparticles. *Adv. Colloid Interface Sci.* **2014**, *209*, 163–171. [CrossRef] [PubMed]
26. Rehm, B.H.A. *Alginates: Biology and Applications*; Springer: Berlin/Heidelberg, Germany, 2009; ISBN 978-3-540-92678-8.
27. Li, L.; Fang, Y.; Vreeker, R.; Appelqvist, I.; Mendes, E. Reexamining the egg-box model in calcium-alginate gels with X-ray diffraction. *Biomacromolecules* **2007**, *8*, 464–468. [CrossRef]
28. Sikorski, P.; Mo, F.; Skjåk-Braek, G.; Stokke, B.T. Evidence for egg-box-compatible interactions in calcium-alginate gels from fiber X-ray diffraction. *Biomacromolecules* **2007**, *8*, 2098–2103. [CrossRef] [PubMed]
29. Lim, F.; Sun, A. Microencapsulated islets as bioartificial endocrine pancreas. *Science* **1980**, *210*, 908–910. [CrossRef]
30. Higham, A.K.; Bonino, C.A.; Raghavan, S.R.; Khan, S.A. Photo-activated ionic gelation of alginate hydrogel: Real-time rheological monitoring of the two-step crosslinking mechanism. *Soft Matter* **2014**, *10*, 4990–5002. [CrossRef]
31. Larobina, D.; Cipelletti, L. Hierarchical cross-linking in physical alginate gels: A rheological and dynamic light scattering investigation. *Soft Matter* **2013**, *9*, 10005. [CrossRef]
32. Alexander, B.R.; Murphy, K.E.; Gallagher, J.; Farrell, G.F.; Taggart, G. Gelation time, homogeneity, and rupture testing of alginate-calcium carbonate-hydrogen peroxide gels for use as wound dressings. *J. Biomed. Mater. Res. Part B Appl. Biomater.* **2012**, *100*, 425–431. [CrossRef] [PubMed]
33. Jeon, O.; Bouhadir, K.H.; Mansour, J.M.; Alsberg, E. Photocrosslinked alginate hydrogels with tunable biodegradation rates and mechanical properties. *Biomaterials* **2009**, *30*, 2724–2734. [CrossRef] [PubMed]
34. Banerjee, A.; Arha, M.; Choudhary, S.; Ashton, R.S.; Bhatia, S.R.; Schaffer, D.V.; Kane, R.S. The influence of hydrogel modulus on the proliferation and differentiation of encapsulated neural stem cells. *Biomaterials* **2009**, *30*, 4695–4699. [CrossRef] [PubMed]
35. Purcell, E.K.; Singh, A.; Kipke, D.R. Alginate composition effects on a neural stem cell-seeded scaffold. *Tissue Eng. Part C Methods* **2009**, *15*, 541–550. [CrossRef] [PubMed]
36. Østberg, T.; Vesterhus, L.; Graffner, C. Calcium alginate matrices for oral multiple unit administration: II. Effect of process and formulation factors on matrix properties. *Int. J. Pharm.* **1993**, *97*, 183–193. [CrossRef]
37. Østberg, T.; Graffner, C. Calcium alginate matrices for oral multiple unit administration: III. Influence of calcium concentration, amount of drug added and alginate characteristics on drug release. *Int. J. Pharm.* **1994**, *111*, 271–282. [CrossRef]
38. Østberg, T.; Lund, E.M.; Graffner, C. Calcium alginate matrices for oral multiple unit administration: IV. Release characteristics in different media. *Int. J. Pharm.* **1994**, *112*, 241–248. [CrossRef]
39. Dodoo, S.; Steitz, R.; Laschewsky, A.; von Klitzing, R. Effect of ionic strength and type of ions on the structure of water swollen polyelectrolyte multilayers. *Phys. Chem. Chem. Phys.* **2011**, *13*, 10318–10325. [CrossRef]
40. Burmistrova, A.; Richter, M.; Eisele, M.; Üzüm, C.; von Klitzing, R. The Effect of Co-Monomer Content on the Swelling/Shrinking and Mechanical Behaviour of Individually Adsorbed PNIPAM Microgel Particles. *Polymers* **2011**, *3*, 1575–1590. [CrossRef]
41. Karg, M.; Pastoriza-Santos, I.; Rodriguez-González, B.; von Klitzing, R.; Wellert, S.; Hellweg, T. Temperature, pH, and ionic strength induced changes of the swelling behavior of PNIPAM-poly(allylacetic acid) copolymer microgels. *Langmuir* **2008**, *24*, 6300–6306. [CrossRef]

42. Micciulla, S.; Dodoo, S.; Chevigny, C.; Laschewsky, A.; von Klitzing, R. Short versus long chain polyelectrolyte multilayers: A direct comparison of self-assembly and structural properties. *Phys. Chem. Chem. Phys.* **2014**, *16*, 21988–21998. [CrossRef] [PubMed]
43. Volodkin, D.; von Klitzing, R. Competing mechanisms in polyelectrolyte multilayer formation and swelling: Polycation–polyanion pairing vs. polyelectrolyte–ion pairing. *Curr. Opin. Colloid Interface Sci.* **2014**, *19*, 25–31. [CrossRef]
44. Schelero, N.; Hedicke, G.; Linse, P.; Klitzing, R.V. Effects of counterions and co-ions on foam films stabilized by anionic dodecyl sulfate. *J. Phys. Chem. B* **2010**, *114*, 15523–15529. [CrossRef] [PubMed]
45. Ullah, F.; Othman, M.B.H.; Javed, F.; Ahmad, Z.; Md Akil, H. Classification, processing and application of hydrogels: A review. *Mater. Sci. Eng. C Mater. Biol. Appl.* **2015**, *57*, 414–433. [CrossRef] [PubMed]
46. Grulova, I.; Slovinska, L.; Blaško, J.; Devaux, S.; Wisztorski, M.; Salzet, M.; Fournier, I.; Kryukov, O.; Cohen, S.; Cizkova, D. Delivery of Alginate Scaffold Releasing Two Trophic Factors for Spinal Cord Injury Repair. *Sci. Rep.* **2015**, *5*, 13702. [CrossRef]
47. Hashimoto, T.; Suzuki, Y.; Tanihara, M.; Kakimaru, Y.; Suzuki, K. Development of alginate wound dressings linked with hybrid peptides derived from laminin and elastin. *Biomaterials* **2004**, *25*, 1407–1414. [CrossRef]
48. Saha, S.; Chhatbar, M.U.; Mahato, P.; Praveen, L.; Siddhanta, A.K.; Das, A. Rhodamine-alginate conjugate as self indicating gel beads for efficient detection and scavenging of Hg^{2+} and Cr^{3+} in aqueous media. *Chem. Commun.* **2012**, *48*, 1659–1661. [CrossRef]
49. Mierisch, C.M.; Cohen, S.B.; Jordan, L.C.; Robertson, P.G.; Balian, G.; Diduch, D.R. Transforming growth factor-β in calcium alginate beads for the treatment of articular cartilage defects in the rabbit. *Arthrosc. J. Arthrosc. Relat. Surg.* **2002**, *18*, 892–900. [CrossRef]
50. Mazutis, L.; Vasiliauskas, R.; Weitz, D.A. Microfluidic Production of Alginate Hydrogel Particles for Antibody Encapsulation and Release. *Macromol. Biosci.* **2015**, *15*, 1641–1646. [CrossRef]
51. Bouhadir, K.H.; Alsberg, E.; Mooney, D.J. Hydrogels for combination delivery of antineoplastic agents. *Biomaterials* **2001**, *22*, 2625–2633. [CrossRef]
52. Augst, A.D.; Kong, H.J.; Mooney, D.J. Alginate hydrogels as biomaterials. *Macromol. Biosci.* **2006**, *6*, 623–633. [CrossRef] [PubMed]
53. Gao, C.; Liu, M.; Chen, J.; Zhang, X. Preparation and controlled degradation of oxidized sodium alginate hydrogel. *Polym. Degrad. Stab.* **2009**, *94*, 1405–1410. [CrossRef]
54. Chueh, B.-H.; Zheng, Y.; Torisawa, Y.-S.; Hsiao, A.Y.; Ge, C.; Hsiong, S.; Huebsch, N.; Franceschi, R.; Mooney, D.J.; Takayama, S. Patterning alginate hydrogels using light-directed release of caged calcium in a microfluidic device. *Biomed. Microdevices* **2010**, *12*, 145–151. [CrossRef] [PubMed]
55. Esser, E.; Tessmar, J.K.V. Preparation of well-defined calcium cross-linked alginate films for the prevention of surgical adhesions. *J. Biomed. Mater. Res. Part B Appl. Biomater.* **2013**, *101*, 826–839. [CrossRef] [PubMed]
56. Nunamaker, E.A.; Otto, K.J.; Kipke, D.R. Investigation of the material properties of alginate for the development of hydrogel repair of dura mater. *J. Mech. Behav. Biomed. Mater.* **2011**, *4*, 16–33. [CrossRef]
57. Shi, X.-W.; Tsao, C.-Y.; Yang, X.; Liu, Y.; Dykstra, P.; Rubloff, G.W.; Ghodssi, R.; Bentley, W.E.; Payne, G.F. Electroaddressing of Cell Populations by Co-Deposition with Calcium Alginate Hydrogels. *Adv. Funct. Mater.* **2009**, *19*, 2074–2080. [CrossRef]
58. Costantini, M.; Colosi, C.; Mozetic, P.; Jaroszewicz, J.; Tosato, A.; Rainer, A.; Trombetta, M.; Święszkowski, W.; Dentini, M.; Barbetta, A. Correlation between porous texture and cell seeding efficiency of gas foaming and microfluidic foaming scaffolds. *Mater. Sci. Eng. C Mater. Biol. Appl.* **2016**, *62*, 668–677. [CrossRef]
59. You, F.; Wu, X.; Chen, X. 3D printing of porous alginate/gelatin hydrogel scaffolds and their mechanical property characterization. *Int. J. Polym. Mater. Polym. Biomater.* **2016**, *66*, 299–306. [CrossRef]
60. Chen, H.; Xie, S.; Yang, Y.; Zhang, J.; Zhang, Z. Multiscale regeneration scaffold in vitro and in vivo. *J. Biomed. Mater. Res. Part B Appl. Biomater.* **2018**, *106*, 1218–1225. [CrossRef]
61. Kim, M.S.; Kim, G. Three-dimensional electrospun polycaprolactone (PCL)/alginate hybrid composite scaffolds. *Carbohydr. Polym.* **2014**, *114*, 213–221. [CrossRef]
62. Kuznetsov, K.A.; Stepanova, A.O.; Kvon, R.I.; Douglas, T.E.L.; Kuznetsov, N.A.; Chernonosova, V.S.; Zaporozhchenko, I.A.; Kharkova, M.V.; Romanova, I.V.; Karpenko, A.A.; et al. Electrospun Produced 3D Matrices for Covering of Vascular Stents: Paclitaxel Release Depending on Fiber Structure and Composition of the External Environment. *Materials* **2018**, *11*, 2176. [CrossRef] [PubMed]

63. Chen, C.-Y.; Ke, C.-J.; Yen, K.-C.; Hsieh, H.-C.; Sun, J.-S.; Lin, F.-H. 3D porous calcium-alginate scaffolds cell culture system improved human osteoblast cell clusters for cell therapy. *Theranostics* **2015**, *5*, 643–655. [CrossRef] [PubMed]
64. Hu, Y.; Mao, A.S.; Desai, R.M.; Wang, H.; Weitz, D.A.; Mooney, D.J. Controlled self-assembly of alginate microgels by rapidly binding molecule pairs. *Lab Chip* **2017**, *17*, 2481–2490. [CrossRef] [PubMed]
65. Wang, C.-C.; Yang, K.-C.; Lin, K.-H.; Liu, H.-C.; Lin, F.-H. A highly organized three-dimensional alginate scaffold for cartilage tissue engineering prepared by microfluidic technology. *Biomaterials* **2011**, *32*, 7118–7126. [CrossRef] [PubMed]
66. Agarwal, A.; Farouz, Y.; Nesmith, A.P.; Deravi, L.F.; McCain, M.L.; Parker, K.K. Micropatterning Alginate Substrates for in vitro Cardiovascular Muscle on a Chip. *Adv. Funct. Mater.* **2013**, *23*, 3738–3746. [CrossRef] [PubMed]
67. Alsberg, E.; Anderson, K.W.; Albeiruti, A.; Franceschi, R.T.; Mooney, D.J. Cell-interactive alginate hydrogels for bone tissue engineering. *J. Dent. Res.* **2001**, *80*, 2025–2029. [CrossRef] [PubMed]
68. Chung, T.W.; Yang, J.; Akaike, T.; Cho, K.Y.; Nah, J.W.; Kim, S.I.; Cho, C.S. Preparation of alginate/galactosylated chitosan scaffold for hepatocyte attachment. *Biomaterials* **2002**, *23*, 2827–2834. [CrossRef]
69. Dvir-Ginzberg, M.; Gamlieli-Bonshtein, I.; Agbaria, R.; Cohen, S. Liver tissue engineering within alginate scaffolds: Effects of cell-seeding density on hepatocyte viability, morphology, and function. *Tissue Eng.* **2003**, *9*, 757–766. [CrossRef]
70. Li, J.; Mooney, D.J. Designing hydrogels for controlled drug delivery. *Nat. Rev. Mater.* **2016**, 1. [CrossRef]
71. Shahriari, D.; Koffler, J.; Lynam, D.A.; Tuszynski, M.H.; Sakamoto, J.S. Characterizing the degradation of alginate hydrogel for use in multilumen scaffolds for spinal cord repair. *J. Biomed. Mater. Res. A* **2016**, *104*, 611–619. [CrossRef]
72. Yan, H.; Huang, D.; Chen, X.; Liu, H.; Feng, Y.; Zhao, Z.; Dai, Z.; Zhang, X.; Lin, Q. A novel and homogeneous scaffold material: Preparation and evaluation of alginate/bacterial cellulose nanocrystals/collagen composite hydrogel for tissue engineering. *Polym. Bull.* **2018**, *75*, 985–1000. [CrossRef]
73. Lee, S.-H.; Jo, A.R.; Choi, G.P.; Woo, C.H.; Lee, S.J.; Kim, B.-S.; You, H.-K.; Cho, Y.-S. Fabrication of 3D alginate scaffold with interconnected pores using wire-network molding technique. *Tissue Eng. Regen. Med.* **2013**, *10*, 53–59. [CrossRef]
74. Johann, R.M.; Renaud, P. Microfluidic patterning of alginate hydrogels. *Biointerphases* **2007**, *2*, 73–79. [CrossRef] [PubMed]
75. Bruchet, M.; Mendelson, N.; Melman, A. Photochemical Patterning of Ionically Cross-Linked Hydrogels. *Processes* **2013**, *1*, 153–166. [CrossRef]
76. Barbetta, A.; Barigelli, E.; Dentini, M. Porous alginate hydrogels: Synthetic methods for tailoring the porous texture. *Biomacromolecules* **2009**, *10*, 2328–2337. [CrossRef]
77. Bots, P.; Benning, L.G.; Rodriguez-Blanco, J.-D.; Roncal-Herrero, T.; Shaw, S. Mechanistic Insights into the Crystallization of Amorphous Calcium Carbonate (ACC). *Cryst. Growth Des.* **2012**, *12*, 3806–3814. [CrossRef]
78. Gebauer, D.; Völkel, A.; Cölfen, H. Stable prenucleation calcium carbonate clusters. *Science* **2008**, *322*, 1819–1822. [CrossRef]
79. Trushina, D.B.; Bukreeva, T.V.; Antipina, M.N. Size-Controlled Synthesis of Vaterite Calcium Carbonate by the Mixing Method: Aiming for Nanosized Particles. *Cryst. Growth Des.* **2016**, *16*, 1311–1319. [CrossRef]
80. Trushina, D.B.; Sulyanov, S.N.; Bukreeva, T.V.; Kovalchuk, M.V. Size control and structure features of spherical calcium carbonate particles. *Crystallogr. Rep.* **2015**, *60*, 570–577. [CrossRef]
81. Sergeeva, A.; Sergeev, R.; Lengert, E.; Zakharevich, A.; Parakhonskiy, B.; Gorin, D.; Sergeev, S.; Volodkin, D. Composite Magnetite and Protein Containing $CaCO_3$ Crystals. External Manipulation and Vaterite → Calcite Recrystallization-Mediated Release Performance. *ACS Appl. Mater. Interfaces* **2015**, *7*, 21315–21325. [CrossRef]
82. Feoktistova, N.; Rose, J.; Prokopović, V.Z.; Vikulina, A.S.; Skirtach, A.; Volodkin, D. Controlling the Vaterite CaCO3 Crystal Pores. Design of Tailor-Made Polymer Based Microcapsules by Hard Templating. *Langmuir* **2016**, *32*, 4229–4238. [CrossRef] [PubMed]
83. Dong, Z.; Feng, L.; Zhu, W.; Sun, X.; Gao, M.; Zhao, H.; Chao, Y.; Liu, Z. $CaCO_3$ nanoparticles as an ultra-sensitive tumor-pH-responsive nanoplatform enabling real-time drug release monitoring and cancer combination therapy. *Biomaterials* **2016**, *110*, 60–70. [CrossRef] [PubMed]

84. Keowmaneechai, E.; McClements, D.J. Influence of EDTA and Citrate on Physicochemical Properties of Whey Protein-Stabilized Oil-in-Water Emulsions Containing $CaCl_2$. *J. Agric. Food Chem.* **2002**, *50*, 7145–7153. [CrossRef] [PubMed]
85. Al Omari, M.M.H.; Rashid, I.S.; Qinna, N.A.; Jaber, A.M.; Badwan, A.A. Calcium Carbonate. *Profiles Drug Subst. Excip. Relat. Methodol.* **2016**, *41*, 31–132. [CrossRef] [PubMed]
86. Kontoyannis, C.G.; Vagenas, N.V. Calcium carbonate phase analysis using XRD and FT-Raman spectroscopy. *Analyst* **2000**, *125*, 251–255. [CrossRef]
87. Sergeeva, A.S.; Gorin, D.A.; Volodkin, D.V. Polyelectrolyte Microcapsule Arrays: Preparation and Biomedical Applications. *Bionanosci.* **2014**, *4*, 1–14. [CrossRef]
88. Parakhonskiy, B.V.; Foss, C.; Carletti, E.; Fedel, M.; Haase, A.; Motta, A.; Migliaresi, C.; Antolini, R. Tailored intracellular delivery via a crystal phase transition in 400 nm vaterite particles. *Biomater. Sci.* **2013**, *1*, 1273. [CrossRef]
89. Sukhorukov, G.B.; Volodkin, D.V.; Günther, A.M.; Petrov, A.I.; Shenoy, D.B.; Möhwald, H. Porous calcium carbonate microparticles as templates for encapsulation of bioactive compounds. *J. Mater. Chem.* **2004**, *14*, 2073–2081. [CrossRef]
90. Trushina, D.B.; Bukreeva, T.V.; Kovalchuk, M.V.; Antipina, M.N. $CaCO_3$ vaterite microparticles for biomedical and personal care applications. *Mater. Sci. Eng. C Mater. Biol. Appl.* **2014**, *45*, 644–658. [CrossRef]
91. Volodkin, D.V.; Schmidt, S.; Fernandes, P.; Larionova, N.I.; Sukhorukov, G.B.; Duschl, C.; Möhwald, H.; von Klitzing, R. One-Step Formulation of Protein Microparticles with Tailored Properties: Hard Templating at Soft Conditions. *Adv. Funct. Mater.* **2012**, *22*, 1914–1922. [CrossRef]
92. Parakhonskiy, B.V.; Svenskaya, Y.I.; Yashchenok, A.M.; Fattah, H.A.; Inozemtseva, O.A.; Tessarolo, F.; Antolini, R.; Gorin, D.A. Size controlled hydroxyapatite and calcium carbonate particles: Synthesis and their application as templates for SERS platform. *Colloids Surf. B Biointerfaces* **2014**, *118*, 243–248. [CrossRef]
93. Kulak, A.N.; Semsarilar, M.; Kim, Y.-Y.; Ihli, J.; Fielding, L.A.; Cespedes, O.; Armes, S.P.; Meldrum, F.C. One-pot synthesis of an inorganic heterostructure: Uniform occlusion of magnetite nanoparticles within calcite single crystals. *Chem. Sci.* **2014**, *5*, 738–743. [CrossRef]
94. Fakhrullin, R.F.; Bikmullin, A.G.; Nurgaliev, D.K. Magnetically responsive calcium carbonate microcrystals. *ACS Appl. Mater. Interfaces* **2009**, *1*, 1847–1851. [CrossRef] [PubMed]
95. Lengert, E.; Kozlova, A.; Pavlov, A.M.; Atkin, V.; Verkhovskii, R.; Kamyshinsky, R.; Demina, P.; Vasiliev, A.L.; Venig, S.B.; Bukreeva, T.V. Novel type of hollow hydrogel microspheres with magnetite and silver nanoparticles. *Mater. Sci. Eng. C Mater. Biol. Appl.* **2019**, *98*, 1114–1121. [CrossRef] [PubMed]
96. Stetciura, I.Y.; Markin, A.V.; Ponomarev, A.N.; Yakimansky, A.V.; Demina, T.S.; Grandfils, C.; Volodkin, D.V.; Gorin, D.A. New surface-enhanced Raman scattering platforms: Composite calcium carbonate microspheres coated with astralen and silver nanoparticles. *Langmuir* **2013**, *29*, 4140–4147. [CrossRef] [PubMed]
97. Shchukin, D.G.; Sukhorukov, G.B.; Price, R.R.; Lvov, Y.M. Halloysite nanotubes as biomimetic nanoreactors. *Small* **2005**, *1*, 510–513. [CrossRef]
98. Peng, C.; Zhao, Q.; Gao, C. Sustained delivery of doxorubicin by porous CaCO3 and chitosan/alginate multilayers-coated $CaCO_3$ microparticles. *Colloids Surf. A Physicochem. Eng. Asp.* **2010**, *353*, 132–139. [CrossRef]
99. Svenskaya, Y.; Parakhonskiy, B.; Haase, A.; Atkin, V.; Lukyanets, E.; Gorin, D.; Antolini, R. Anticancer drug delivery system based on calcium carbonate particles loaded with a photosensitizer. *Biophys. Chem.* **2013**, *182*, 11–15. [CrossRef]
100. Balabushevich, N.G.; Sholina, E.A.; Mikhalchik, E.V.; Filatova, L.Y.; Vikulina, A.S.; Volodkin, D. Self-Assembled Mucin-Containing Microcarriers via Hard Templating on $CaCO_3$ Crystals. *Micromachines* **2018**, *9*, 307. [CrossRef]
101. Balabushevich, N.G.; Kovalenko, E.A.; Mikhalchik, E.V.; Filatova, L.Y.; Volodkin, D.; Vikulina, A.S. Mucin adsorption on vaterite CaCO3 microcrystals for the prediction of mucoadhesive properties. *J. Colloid Interface Sci.* **2019**, *545*, 330–339. [CrossRef]
102. Volodkin, D.V.; von Klitzing, R.; Möhwald, H. Pure protein microspheres by calcium carbonate templating. *Angew. Chem. Int. Ed Engl.* **2010**, *49*, 9258–9261. [CrossRef]
103. Schmidt, S.; Uhlig, K.; Duschl, C.; Volodkin, D. Stability and cell uptake of calcium carbonate templated insulin microparticles. *Acta Biomater.* **2014**, *10*, 1423–1430. [CrossRef]
104. Volodkin, D.V.; Petrov, A.I.; Prevot, M.; Sukhorukov, G.B. Matrix Polyelectrolyte Microcapsules: New System for Macromolecule Encapsulation. *Langmuir* **2004**, *20*, 3398–3406. [CrossRef]

105. Yashchenok, A.M.; Borisova, D.; Parakhonskiy, B.V.; Masic, A.; Pinchasik, B.; Möhwald, H.; Skirtach, A.G. Nanoplasmonic smooth silica versus porous calcium carbonate bead biosensors for detection of biomarkers. *Ann. Phys.* **2012**, *524*, 723–732. [CrossRef]
106. Kamyshinsky, R.; Marchenko, I.; Parakhonskiy, B.; Yashchenok, A.; Chesnokov, Y.; Mikhutkin, A.; Gorin, D.; Vasiliev, A.; Bukreeva, T. Composite materials based on Ag nanoparticles in situ synthesized on the vaterite porous matrices. *Nanotechnology* **2019**, *30*, 35603. [CrossRef] [PubMed]
107. Bukreeva, T.V.; Orlova, O.A.; Sulyanov, S.N.; Grigoriev, Y.V.; Dorovatovskiy, P.V. A new approach to modification of polyelectrolyte capsule shells by magnetite nanoparticles. *Crystallogr. Rep.* **2011**, *56*, 880–883. [CrossRef]
108. Gorin, D.A.; Portnov, S.A.; Inozemtseva, O.A.; Luklinska, Z.; Yashchenok, A.M.; Pavlov, A.M.; Skirtach, A.G.; Möhwald, H.; Sukhorukov, G.B. Magnetic/gold nanoparticle functionalized biocompatible microcapsules with sensitivity to laser irradiation. *Phys. Chem. Chem. Phys.* **2008**, *10*, 6899–6905. [CrossRef] [PubMed]
109. Luo, R.; Venkatraman, S.S.; Neu, B. Layer-by-layer polyelectrolyte-polyester hybrid microcapsules for encapsulation and delivery of hydrophobic drugs. *Biomacromolecules* **2013**, *14*, 2262–2271. [CrossRef]
110. Skirtach, A.G.; Dejugnat, C.; Braun, D.; Susha, A.S.; Rogach, A.L.; Parak, W.J.; Möhwald, H.; Sukhorukov, G.B. The Role of Metal Nanoparticles in Remote Release of Encapsulated Materials. *Nano Lett.* **2005**, *5*, 1371–1377. [CrossRef]
111. Volodkin, D. Colloids of pure proteins by hard templating. *Colloid Polym Sci.* **2014**, *292*, 1249–1259. [CrossRef]
112. Schmidt, S.; Behra, M.; Uhlig, K.; Madaboosi, N.; Hartmann, L.; Duschl, C.; Volodkin, D. Mesoporous Protein Particles Through Colloidal $CaCO_3$ Templates. *Adv. Funct. Mater.* **2013**, *23*, 116–123. [CrossRef]
113. Feoktistova, N.; Stoychev, G.; Puretskiy, N.; Ionov, L.; Volodkin, D. Porous thermo-responsive pNIPAM microgels. *Eur. Polym. J.* **2015**, *68*, 650–656. [CrossRef]
114. Behra, M.; Schmidt, S.; Hartmann, J.; Volodkin, D.V.; Hartmann, L. Synthesis of porous PEG microgels using $CaCO_3$ microspheres as hard templates. *Macromol. Rapid Commun.* **2012**, *33*, 1049–1054. [CrossRef] [PubMed]
115. Behra, M.; Azzouz, N.; Schmidt, S.; Volodkin, D.V.; Mosca, S.; Chanana, M.; Seeberger, P.H.; Hartmann, L. Magnetic porous sugar-functionalized PEG microgels for efficient isolation and removal of bacteria from solution. *Biomacromolecules* **2013**, *14*, 1927–1935. [CrossRef]
116. Goss, S.L.; Lemons, K.A.; Kerstetter, J.E.; Bogner, R.H. Determination of calcium salt solubility with changes in pH and P_{CO2}, simulating varying gastrointestinal environments. *J. Pharm. Pharmacol.* **2007**, *59*, 1485–1492. [CrossRef] [PubMed]
117. Paulraj, T.; Feoktistova, N.; Velk, N.; Uhlig, K.; Duschl, C.; Volodkin, D. Microporous polymeric 3D scaffolds templated by the layer-by-layer self-assembly. *Macromol. Rapid Commun.* **2014**, *35*, 1408–1413. [CrossRef] [PubMed]
118. Vikulina, A.S.; Skirtach, A.G.; Volodkin, D. Hybrids of Polymer Multilayers, Lipids, and Nanoparticles: Mimicking the Cellular Microenvironment. *Langmuir* **2019**. [CrossRef]
119. Balabushevich, N.G.; Lopez de Guerenu, A.V.; Feoktistova, N.A.; Skirtach, A.G.; Volodkin, D. Protein-Containing Multilayer Capsules by Templating on Mesoporous $CaCO_3$ Particles: POST- and PRE-Loading Approaches. *Macromol. Biosci.* **2016**, *16*, 95–105. [CrossRef] [PubMed]
120. Balabushevich, N.G.; Lopez de Guerenu, A.V.; Feoktistova, N.A.; Volodkin, D. Protein loading into porous $CaCO_3$ microspheres: Adsorption equilibrium and bioactivity retention. *Phys. Chem. Chem. Phys.* **2015**, *17*, 2523–2530. [CrossRef]
121. Vikulina, A.S.; Feoktistova, N.A.; Balabushevich, N.G.; Skirtach, A.G.; Volodkin, D. The mechanism of catalase loading into porous vaterite $CaCO_3$ crystals by co-synthesis. *Phys. Chem. Chem. Phys.* **2018**, *20*, 8822–8831. [CrossRef]
122. Volodkin, D.; von Klitzing, R.; Moehwald, H. Polyelectrolyte Multilayers: Towards Single Cell Studies. *Polymers* **2014**, *6*, 1502–1527. [CrossRef]
123. Saveleva, M.S.; Eftekhari, K.; Abalymov, A.; Douglas, T.E.L.; Volodkin, D.; Parakhonskiy, B.V.; Skirtach, A.G. Hierarchy of Hybrid Materials—The Place of Inorganics-in-Organics in it, Their Composition and Applications. *Front. Chem.* **2019**, *7*, 1129. [CrossRef] [PubMed]
124. Balabushevich, N.G.; Pechenkin, M.A.; Shibanova, E.D.; Volodkin, D.V.; Mikhalchik, E.V. Multifunctional polyelectrolyte microparticles for oral insulin delivery. *Macromol. Biosci.* **2013**, *13*, 1379–1388. [CrossRef] [PubMed]

125. Jeannot, L.; Bell, M.; Ashwell, R.; Volodkin, D.; Vikulina, A.S. Internal Structure of Matrix-Type Multilayer Capsules Templated on Porous Vaterite $CaCO_3$ Crystals as Probed by Staining with a Fluorescence Dye. *Micromachines* **2018**, *9*, 547. [CrossRef] [PubMed]
126. Volodkin, D.V.; Balabushevitch, N.G.; Sukhorukov, G.B.; Larionova, N.I. Model system for controlled protein release: pH-sensitive polyelectrolyte microparticles. *STP Pharma Sci.* **2003**, *13*, 163–170.
127. Volodkin, D.; Skirtach, A.; Madaboosi, N.; Blacklock, J.; von Klitzing, R.; Lankenau, A.; Duschl, C.; Möhwald, H. IR-light triggered drug delivery from micron-sized polymer biocoatings. *J. Control. Release* **2010**, *148*, e70–e71. [CrossRef] [PubMed]
128. Stetciura, I.Y.; Yashchenok, A.; Masic, A.; Lyubin, E.V.; Inozemtseva, O.A.; Drozdova, M.G.; Markvichova, E.A.; Khlebtsov, B.N.; Fedyanin, A.A.; Sukhorukov, G.B.; et al. Composite SERS-based satellites navigated by optical tweezers for single cell analysis. *Analyst* **2015**, *140*, 4981–4986. [CrossRef]
129. Parakhonskiy, B.V.; Yashchenok, A.M.; Möhwald, H.; Volodkin, D.; Skirtach, A.G. Release from Polyelectrolyte Multilayer Capsules in Solution and on Polymeric Surfaces. *Adv. Mater. Interfaces* **2017**, *4*, 1600273. [CrossRef]
130. Hernández-Hernández, A.; Rodríguez-Navarro, A.B.; Gómez-Morales, J.; Jiménez-Lopez, C.; Nys, Y.; García-Ruiz, J.M. Influence of Model Globular Proteins with Different Isoelectric Points on the Precipitation of Calcium Carbonate. *Cryst. Growth Des.* **2008**, *8*, 1495–1502. [CrossRef]
131. Parakhonskiy, B.V.; Yashchenok, A.M.; Donatan, S.; Volodkin, D.V.; Tessarolo, F.; Antolini, R.; Möhwald, H.; Skirtach, A.G. Macromolecule loading into spherical, elliptical, star-like and cubic calcium carbonate carriers. *ChemPhysChem* **2014**, *15*, 2817–2822. [CrossRef]
132. Andreassen, J.-P.; Beck, R.; Nergaard, M. Biomimetic type morphologies of calcium carbonate grown in absence of additives. *Faraday Discuss.* **2012**, *159*, 247. [CrossRef]
133. Guo, X.; Liu, L.; Wang, W.; Zhang, J.; Wang, Y.; Yu, S.-H. Controlled crystallization of hierarchical and porous calcium carbonate crystals using polypeptide type block copolymer as crystal growth modifier in a mixed solution. *CrystEngComm* **2011**, *13*, 2054. [CrossRef]
134. Parakhonskiy, B.V.; Haase, A.; Antolini, R. Sub-Micrometer Vaterite Containers: Synthesis, Substance Loading, and Release. *Angew. Chem. Int. Ed.* **2012**, *51*, 1195–1197. [CrossRef]
135. Lopez-Heredia, M.A.; Łapa, A.; Mendes, A.C.; Balcaen, L.; Samal, S.K.; Chai, F.; van der Voort, P.; Stevens, C.V.; Parakhonskiy, B.V.; Chronakis, I.S.; et al. Bioinspired, biomimetic, double-enzymatic mineralization of hydrogels for bone regeneration with calcium carbonate. *Mater. Lett.* **2017**, *190*, 13–16. [CrossRef]
136. Savelyeva, M.S.; Abalymov, A.A.; Lyubun, G.P.; Vidyasheva, I.V.; Yashchenok, A.M.; Douglas, T.E.L.; Gorin, D.A.; Parakhonskiy, B.V. Vaterite coatings on electrospun polymeric fibers for biomedical applications. *J. Biomed. Mater. Res. A* **2017**, *105*, 94–103. [CrossRef]
137. Ivanova, A.A.; Syromotina, D.S.; Shkarina, S.N.; Shkarin, R.; Cecilia, A.; Weinhardt, V.; Baumbach, T.; Saveleva, M.S.; Gorin, D.A.; Douglas, T.E.L.; et al. Effect of low-temperature plasma treatment of electrospun polycaprolactone fibrous scaffolds on calcium carbonate mineralisation. *RSC Adv.* **2018**, *8*, 39106–39114. [CrossRef]
138. Butler, M.F.; Glaser, N.; Weaver, A.C.; Kirkland, M.; Heppenstall-Butler, M. Calcium Carbonate Crystallization in the Presence of Biopolymers. *Cryst. Growth Des.* **2006**, *6*, 781–794. [CrossRef]
139. Kosanović, C.; Fermani, S.; Falini, G.; Kralj, D. Crystallization of Calcium Carbonate in Alginate and Xanthan Hydrogels. *Crystals* **2017**, *7*, 355. [CrossRef]
140. Olderøy, M.Ø.; Xie, M.; Strand, B.L.; Flaten, E.M.; Sikorski, P.; Andreassen, J.-P. Growth and Nucleation of Calcium Carbonate Vaterite Crystals in Presence of Alginate. *Cryst. Growth Des.* **2009**, *9*, 5176–5183. [CrossRef]
141. Xie, M.; Olderøy, M.Ø.; Andreassen, J.-P.; Selbach, S.M.; Strand, B.L.; Sikorski, P. Alginate-controlled formation of nanoscale calcium carbonate and hydroxyapatite mineral phase within hydrogel networks. *Acta Biomater.* **2010**, *6*, 3665–3675. [CrossRef]
142. Douglas, T.E.L.; Sobczyk, K.; Łapa, A.; Włodarczyk, K.; Brackman, G.; Vidiasheva, I.; Reczyńska, K.; Pietryga, K.; Schaubroeck, D.; Bliznuk, V.; et al. Ca:Mg:Zn:CO_3 and Ca:Mg:CO_3-tri- and bi-elemental carbonate microparticles for novel injectable self-gelling hydrogel-microparticle composites for tissue regeneration. *Biomed. Mater.* **2017**, *12*, 25015. [CrossRef] [PubMed]
143. Zhou, Z.; He, X.; Zhou, M.; Meng, F. Chemically induced alterations in the characteristics of fouling-causing bio-macromolecules—Implications for the chemical cleaning of fouled membranes. *Water Res.* **2017**, *108*, 115–123. [CrossRef] [PubMed]

144. Liu, J.; Zheng, H.; Poh, P.S.P.; Machens, H.-G.; Schilling, A.F. Hydrogels for Engineering of Perfusable Vascular Networks. *Int. J. Mol. Sci.* **2015**, *16*, 15997–16016. [CrossRef] [PubMed]
145. Sergeeva, A.S. Porous Alginate Scaffolds: Design and Loading/Release Opportunities. Ph.D. Thesis, Technical University Berlin, Berlin, Germany, 2017.
146. Sergeeva, A.S.; Volkova, E.K.; Bratashov, D.N.; Shishkin, M.I.; Atkin, V.S.; Markin, A.V.; Skaptsov, A.A.; Volodkin, D.V.; Gorin, D.A. Layer-by-layer assembled highly absorbing hundred-layer films containing a phthalocyanine dye: Fabrication and photosensibilization by thermal treatment. *Thin Solid Film.* **2015**, *583*, 60–69. [CrossRef]
147. Bosio, V.E.; Cacicedo, M.L.; Calvignac, B.; León, I.; Beuvier, T.; Boury, F.; Castro, G.R. Synthesis and characterization of $CaCO_3$-biopolymer hybrid nanoporous microparticles for controlled release of doxorubicin. *Colloids Surf. B Biointerfaces* **2014**, *123*, 158–169. [CrossRef] [PubMed]
148. Hanafy, N.A.; El-Kemary, M.; Leporatti, S. Optimizing $CaCO_3$ Matrix Might Allow To Raise Their Potential Use In Biomedical Application. *J. Nanosci. Curr. Res.* **2018**, *3*, 124. [CrossRef]

Review

Biocatalysis by Transglutaminases: A Review of Biotechnological Applications

Maria Pia Savoca, Elisa Tonoli, Adeola G. Atobatele and Elisabetta A. M. Verderio *

School of Science and Technology, Interdisciplinary Biomedical Research Centre, Nottingham Trent University, Nottingham NG11 8NS, UK; maria.savoca2016@my.ntu.ac.uk (M.P.S.); elisa.tonoli2015@my.ntu.ac.uk (E.T.); adeola.atobatele022014@my.ntu.ac.uk (A.G.A.)

* Correspondence: elisabetta.verderio-edwards@ntu.ac.uk; Tel.: +44-(0)-115-848-6628

Received: 10 October 2018; Accepted: 23 October 2018; Published: 31 October 2018

Abstract: The biocatalytic activity of transglutaminases (TGs) leads to the synthesis of new covalent isopeptide bonds (crosslinks) between peptide-bound glutamine and lysine residues, but also the transamidation of primary amines to glutamine residues, which ultimately can result into protein polymerisation. Operating with a cysteine/histidine/aspartic acid (Cys/His/Asp) catalytic triad, TGs induce the post-translational modification of proteins at both physiological and pathological conditions (e.g., accumulation of matrices in tissue fibrosis). Because of the disparate biotechnological applications, this large family of protein-remodelling enzymes have stimulated an escalation of interest. In the past 50 years, both mammalian and microbial TGs polymerising activity has been exploited in the food industry for the improvement of aliments' quality, texture, and nutritive value, other than to enhance the food appearance and increased marketability. At the same time, the ability of TGs to crosslink extracellular matrix proteins, like collagen, as well as synthetic biopolymers, has led to multiple applications in biomedicine, such as the production of biocompatible scaffolds and hydrogels for tissue engineering and drug delivery, or DNA-protein bio-conjugation and antibody functionalisation. Here, we summarise the most recent advances in the field, focusing on the utilisation of TGs-mediated protein multimerisation in biotechnological and bioengineering applications.

Keywords: transglutaminases; crosslinking; polymerisation; food industry; biomedicine

1. Transglutaminases: Enzymatic Activity and Regulation

Mammalian transglutaminases (TGs) have been extensively characterised in the past 60 years since their discovery by Heinrich Waelsch in 1957 [1]. They constitute a family of eight catalytically active acyl-transferases (TG1-7 and factor XIIIa), plus the inactive erythrocyte protein band 4.2 (EPB4.2) [2]. TGs are mostly known for the ability to catalyse the formation of intra- and inter-molecular covalent bonds between proteins, also referred to as crosslinking activity [3]. The presence of ε-(γ-glutamyl)lysine crosslinks was first reported in human [4] and bovine [5] fibrin polymerised by FXIIIa. Over the years, it has been confirmed that also the other members of the TG family are capable of protein crosslinking, with the exception of EPB4.2, a catalytically inactive form [6–8].

The crosslinking reaction occurs in two consecutive steps: At first, an intermediate thioester is formed through the attack of an acyl donor (γ-carboxamide group of a peptide-bound glutamine residue) by the nucleophilic active thiolate (cysteine residue), with consequent release of ammonia. Secondly, the thiolate is restored by nucleophilic attack of an acyl acceptor substrate (ε-amino group of a peptide-bound lysine residue) (Figure 1a). This leads to the formation of a covalent inter-molecular ε-(γ-glutamyl)lysine isopeptide bond, which is resistant to physical and chemical degradation [3,9–11]. A similar reaction leads to the incorporation of primary amines, including polyamines, into the γ-carboxamide group of peptide-bound glutamine (Gln) residues (Figure 1b,c). Both reactions are calcium-dependent and together are referred to as protein transamidation. Notably, the transamidation reaction, which leads to amine incorporation, was actually the first to be identified by Waelsch and colleagues in 1957, by the detection of radiolabelled transamidated polyamines in guinea pig liver protein extracts [1,12].

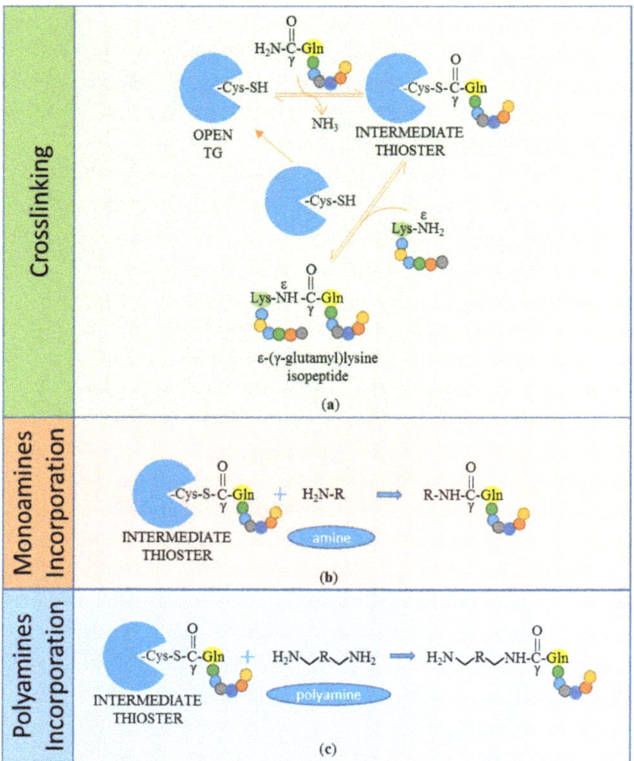

Figure 1. Transamidation reactions catalysed by transglutaminases (TGs). In response to triggering Ca^{2+} concentration and in appropriate redox conditions, TG conformation is open and the catalytic cysteine (Cys) thiol (SH) group is prone to bind the γ-carboxamide group of a peptide-bound glutamine residue (Gln). Therefore, a thioester bond is created between the TG's Cys and the Gln of a peptide target, with consequent ammonia release (**a**). TG catalyses the transfer of the acyl intermediate product to a nucleophilic substrate, like an ε-amino group of a peptide-bound lysine residue (Lys), leading to the formation of ε-(γ-glutamyl)lysine isopeptide bond, also called a crosslink (**a**). TG catalyses the incorporation of monoamines (**b**) or polyamines (**c**), acting as acyl-acceptors in a reaction similar to the crosslinking.

Among mammalian TGs, tissue transglutaminase (tTG) or transglutaminase 2 (TG2) (NM_004613.2) has been by far the most studied, mainly because of its diverse proprieties and involvement in multiple physiological and pathological processes. TG2 is composed by 687 amino acids (aa), it is ubiquitously expressed in several different cell types and, like the other active members, is defined by its calcium-dependent transamidating activity [13,14]. TG2 structure consists of four globular domains (Figure 2a). The core domain (aa 140–460), key for the transamidation activity, is characterised by the catalytic triad, cysteine-histidine-aspartic acid (Cys277-His335-Asp358) [15,16], plus two tryptophan residues (W241 and W332), which stabilise the reaction intermediate product [17,18]. The N-terminal β-sandwich domain (aa 1–139) includes the binding site for fibronectin (FN) [19–21], while the two C-terminal β-barrel domains (aa 461–586 and

587–687) are involved with the TG2 ability to bind and hydrolyse guanosine/adenosine triphosphate (GTP/ATP) [16,22–24]. TG2 undergoes an allosteric activation fostered by calcium (Ca^{2+}) with a dissociation constant (K_d) of 90 μmol·L^{-1} [25]. Seeing that the TG2-Ca^{2+} bound X-ray structure is not available, the crystal structures of other TGs (i.e., TG3 and FXIIIa) and computational homology-based three-dimensional models of TG2 have been used to study the Ca^{2+} binding sites [26,27]. Out of the six Ca^{2+} binding sites that have been identified, five influence enzymatic activity and act in a cooperative manner. In physiological conditions (Ca^{2+}: 0.1 μmol·L^{-1}/GTP: 100–150 μmol·L^{-1}), intracellular TG2 is completely inhibited. It is believed that TG2 inhibition is mainly accomplished by guanine nucleotides, i.e., GTP, guanosine diphosphate (GDP), guanosine monophosphate (GMP), and adenine nucleotides (ATP) [22,25,28]. When Ca^{2+} levels are sufficiently increased (0.5–1.5 mmol·L^{-1}), GTP inhibitory capability is significantly reduced, likely due to the conformational changes caused by Ca^{2+} binding [16,28–30]. Additionally, other molecules, such as heparan sulfate moieties of proteoglycans (HSPG), may influence TG conformation [31]. Besides TG2, also other members of the TGs family (TG3, 5 and 6) have been reported to be inhibited by purine nucleotides, with different responsiveness levels [22,28,32,33].

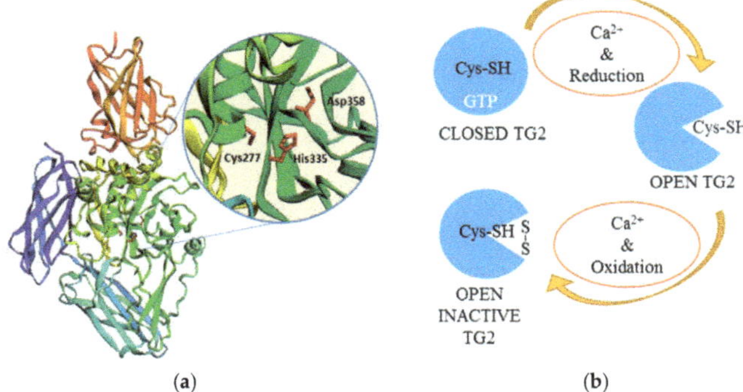

Figure 2. Human transglutaminase 2 (hTG2) structure and regulation. (**a**) hTG2 structure is shown in closed conformation. The catalytic site is composed by the triad, Cys277, His335, and Asp358. In this conformation, the protein is inactive, since the two β-barrels hide the catalytic pocket. Three-dimensional structure (PDB: 4PYG) was produced with the molecule modelling software, "EzMol" (version 1.22) [34]. (**b**) Effect of redox regulation on TG2 conformations. TG2 is locked in closed conformation when bound to guanosine triphosphate (GTP) and calcium concentration is low. Conversely, it assumes an open conformation after calcium binding, which can either be inactive when in oxidising conditions or active in reducing conditions [35].

Calcium and purine nucleotides are not the only regulators of TGs. In particular, the redox state affects the accessibility of the Cys active site and it is also essential for TGs' crosslinking activity [35–38]. Recent knowledge suggests that TG2 can assume three conformations: An inactive form bound to GTP, an inactive one bound to Ca^{2+}, but oxidised, and a reduced one activated by Ca^{2+} [38] (Figure 2b). Under reducing conditions, Ca^{2+} binding decreasing TGs' affinity for GTP/GDP leads to an enzymatically active "open" conformation [38–40], while GTP binding causes the "closed" conformation, blocking substrate access to the catalytic pocket [39,40]. Experimental data confirm that transamidation is not only dependent on the Ca^{2+}/GTP ratio. In fact, extracellular TG2 is mostly inactive even when the low GTP/Ca^{2+} ratio would theoretically promote activation, at least until induced by a chemical or physical injury [41]. This might be explained by TG2 being predominantly locked in closed conformation, possibly due to the redox state of the extracellular environment; however, other molecules may also further modulate TG conformation (e.g., HSPG, integrins) [31,42]. The formation of protein disulphide bridges between Cys370-Cys371 and Cys370-Cys230 in oxidising conditions is in fact sufficient to inactivate TG2 enzymatic activity [35]; conversely, reducing events cause its activation [35,38].

Besides protein transamidation, TGs are characterised by numerous other enzymatic activities. In the presence of water, TG2 is also able to hydrolyse target glutamine residues, thereby converting them into glutamic acid residues (deamidation) [43]. By deamidating gluten peptides and generating immunogenic epitopes, TG2 is responsible for the gluten-induced enteropathy celiac disease (CD) [44,45]. Additional TGs functions, such as GTPase and ATPase activity [24,28,46], protein kinase activity [47–50], and protein disulphide isomerase activity [51,52], have also been reported.

Research on TGs has led to the identification of TG homologous proteins in several species, from microorganisms to plants and animals [1,53,54]. In silico studies have allowed the identification of multiple conserved motifs in the TGs catalytic core in archaea, bacteria, and eukaryotes [55]. Conversely, the highest variability among these domains is present in the insert regions localised between the conserved motifs [55]. These studies confirmed the theory that genes codifying for TGs are derived from a unique ancestor gene expressing a cysteine protease, which then gave rise to two lineages through successive gene duplication events [55–57]. Specifically, one lineage includes orthologue genes from the majority of mammal TGs (TG2, TG3, TG5, TG6, TG7, and erythrocyte band 4.2), while the second one comprises the genes from invertebrates TGs, mammal TG1, and factor XIIIA [56,57].

Among the bacterial TGs, the most relevant is microbial transglutaminase (mTG), which was first isolated from the culture medium of *Streptomyces mobarensis* and characterised by Ando and colleagues in 1989 [58,59]. mTG is a monomeric protein of about 38 kDa, consisting of 331 aa [53] and, differently from eukaryotic TGs, it is characterised by a Ca^{2+}-independent crosslinking activity [58]. The overall sequence data and crystal structure indicate that mTG catalytic activity is dependent on a cysteine residue (Cys64), which, together with the adjacent Asp255 and His274 residues, overlaps well with the catalytic triad, "Cys-His-Asp", that characterises cysteine proteases and factor XIII-like TGs [60] (Figure 3). Regulation of mTG crosslinking activity is quite different from that of mammalian TGs. For instance, mTG is not dependent on Ca^{2+}, while it presents sensitivity to other cations, such as Cu^{2+}, Zn^{2+}, Pb^{2+}, and Li^{+} [58,59].

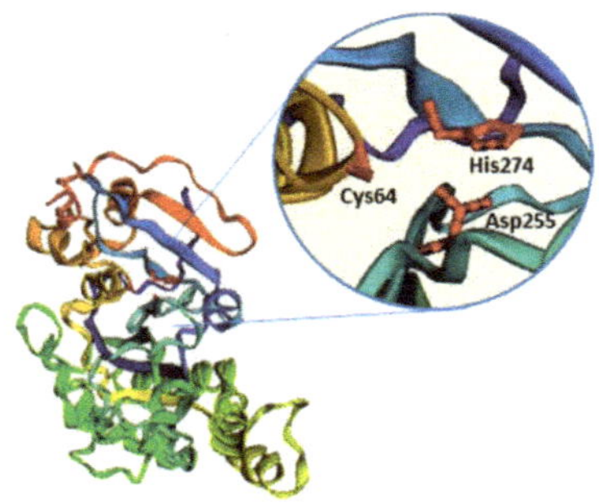

Figure 3. Microbial TG structure. mTG is composed by a single, compact domain. The amino acids of the active site (Cys64, Asp255, and His274) constitute the mTG catalytic triad. The modelling software, "EzMol", was used to generate the structure (PDB: 1IU4) [34].

2. TG2-Mediated Polymerisation of Extracellular Matrix Proteins

Among the molecules that are most likely target of TG2-induced multimerisation there are proteins found in the extracellular matrix (ECM). TG2 modifies these proteins through crosslinking, with an impact on overall matrix stabilisation/stiffness. The increased complexity of the ECM leads to increased cell-matrix interactions and changes in cell adhesion and migration [20,61–64]. FN is also a well-known target of TG2 crosslinking activity and, together with other ECM proteins (osteonectin, osteopontin, laminin, vitronectin, fibrinogen, and collagen), has been shown to be polymerised by extracellular TG2 in vitro and in various cell systems [63–71]. In vivo, TG2 transamidation of ECM proteins leads to their stabilisation and to the accumulation of polymeric complexes rich in isopeptide bonds, which are resistant to degradation by matrix metalloproteinases [9,72–74]. This results in the generation of a pathological matrix typical of fibrotic conditions, such as in kidney [72,75–77], lung [78–81], liver [82–84], and heart fibrosis [85–87]. Notably, skeletal phenotyping in TGs double knockout mice (Tgm2−/− and F13a1−/−) revealed that both TGs possess a synergistic function in maintaining bone mass, as their absence increases osteoclastogenesis and bone resorption. The authors also reported an increased expression of TG1 during osteoclastogenesis in both wild type and double null mice bone marrow MSCs, suggestive of a role of TG1 in osteoclast formation [88].

Hitomi's group has developed important probes for the identification of specific substrates of TG family members, with applications in liver and kidney disease [84,89]. The mechanism of externalisation of TG2 from cells to reach the ECM is an unconventional pathway that has fascinated many research groups. Different theories have been proposed, including TG2 loading into recycling endosomes [90], TG2 secretion via purinergic P2X7 receptor-mediated vesicle shedding [91,92], cell-surface trafficking via HSPG [93–95], and secretion via exosomes [96–98].

Furthermore, TG2 itself has been shown to act as a structural adhesive protein. For example, by interacting directly with FN through a specific binding site localised in the N-terminal β-sandwich domain, TG2 forms adhesive complexes and induces cell adhesion via cell surface HSPG (syndecan-4) independently from the classic RGD-dependent cell adhesion to integrin receptors [99,100]. At the same time, it has been reported to act as an integrin co-receptor reinforcing integrin-dependent cell adhesion [42]. Several research groups have shown an interest in studying the novel TG2-HS interaction [31,94,101]. Teesalu and colleagues initially identified two sites localised in TG2 catalytic domain (aa 202–215 and aa 261–274) possibly involved in heparin binding, also suggesting that the second one could theoretically compete with FN for the binding of cell surface HS proteoglycans [101]. Soon after, Wang et al. suggested that Lys205 and Arg209 residues, which are accessible on the TG2 surface, are essential for heparin binding [94]. Two positively charged clusters (aa 262–265 and aa 598–602) have been determined by Lortat-Jacob et al. as crucial for HS binding [31]. Interestingly, these distant clusters were shown to be in spatial proximity only when TG2 is in closed conformation, which is necessary for the formation of the high affinity heparin binding domain [31].

Gaining a deeper understanding of the TG2/ECM interplay has been extremely useful for the development of several practical applications, such as the production of crosslinked matrices or for tissue engineering, which will be addressed in the following sections.

3. Substrate Specificity of TGs Isozymes: mTG, TG2, and FXIIIa

Since their identification in 1957, TG2 and the other members of the TGs family have shown to catalyse the modification of a variety of substrates both in vivo and in vitro. Furthermore, they have displayed a preference for the recognition of their target proteins, revealing that the transamidating reactions may be restricted to specific consensus sequences. In search for a clarification of the physiological and pathological significance of the TG transamidation, many groups have focused their research on the identification of each isozyme substrates' specificities.

Cousson et al. have proposed the minimal requirements for TG2-dependent modification of a putative Gln side chain in a substrate protein: (i) The residue should be accessible, either by being exposed to the solvent or located in a highly flexible region of the protein; and (ii) the amino acid sequence around the Gln should allow the correct interaction with TG, which is mainly dependent on the amino acids' charge. Specifically, it has been shown that positively charged residues on the C-terminal side of the target Gln discourage TG catalysis [102]. Therefore, the localisation of the target amino acids is one of the main limiting factors for TG interaction, other than the protein conformation.

Seminal work by Hitomi's group screened potential TGs peptide substrates by creating a random peptide library by M13 phage-display. In particular, they analysed TG2 and Factor XIIIa substrates by the incorporation of biotin-labelled primary amine on phage clones expressed peptides. Among these, the following specific amino acid sequences were highlighted: QxPϕD(P), QxPϕ, and QxxϕDP for TG2; and QxxϕxWP for FXIIIa (where "x" stands for any amino acid and "ϕ" for any hydrophobic amino acid) [103]. Furthermore, a phage-display based study by Fesus' group identified Gln-donor substrates from a random heptapeptide library by binding to recombinant TG2 and consecutive elution with a synthetic amine-donor substrate. Among these, twenty-six substrates were successfully transamidated by TG2, especially the peptides GQQQTPY, GLQQASV, and WQTPMNS. This study also confirmed pQX(P,T,S)l (where "p" stands for any polar amino acid and "l" for any aliphatic amino acid) as a consensus sequence recognised by TG2 [104]. Recently, Malešević et al. used a fluorescence-based array of tripeptides and determined that mTG could specifically recognise X-Q-Q and L-Q-X peptides (where "X" is any amino acid), with a higher preference for Y-Q-R. They also analysed mTG substrate preference in relation to amino acids adjacent to the target, Gln, highlighting a relevance for hydrophobic residues at Gln+1 and Gln-1 positions. They identified other preferred amino acids, such as tyrosine and proline in position Gln-1, but not Gln+1, while arginine presence in the tripeptides gave the opposite effect, making them poor substrates for mTG transamidation [105]. Finally, using a small focused synthetic peptide library, the tetrapeptide, "TQGA", was identified as a novel highly specific substrate of mTG [106]. The necessity to open an active network on this topic led to the creation of TRANSDAB wiki (http://genomics.dote.hu/wiki/index.php/Main_Page/), a database that lists about 350 substrates for six human transglutaminases and mTG, along with additional interaction partners [107]. However, with the event of genomics and proteomics, we expect the number of substrates to have increased since the last update of the database was in 2010. In parallel, the database, TRANSIT (http://bioinformatica.isa.cnr.it/TRANSIT/), was generated to assess possible substrates by analysing their amino acid sequence [108]. Research in this field is ongoing, as a better knowledge of TGs specific substrates would further clarify their role in both physiology and disease. Furthermore, multiple TGs transamidation substrates have been exploited in a variety of applications, such as assay systems for in situ visualisation of TG activity [109], identification of endogenous targets of TGs in cells and tissues [84,89,110], and TGs-mediated bio-conjugation of proteins. Hence, a strong interest in finding novel and more specific TGs substrates is still alive in this area of research.

4. TGs Crosslinking Activity in Biotechnological Applications

Currently, the use of enzymes in biotechnological production processes is highly preferred by many biotech industries because of their wide variety and competitiveness in terms of production time and costs. The enzyme-driven crosslinking mechanism leads to changes in proteins' hydrophobicity, thus interfering with their solubility and other properties, such as gelation, emulsification, foaming, viscosity, and water-holding capacity [111–114]. In this context, TGs have become a very popular tool for the development of different applications, an overview of which is given in Table 1.

4.1. Applications in Food Industry

Since the 1980s, to improve the quality and nutritive value of food, research has focused on the potential use of TGs in food processing, exploiting their ability to catalyse intermolecular isopeptide bonds and polymerise proteins [115–117]. Hence, different groups begun investigating the best substrates for TGs activity in this area and many were identified, such as dairy proteins (e.g., caseins and whey proteins) [118–120], soybean globulins [121], wheat (gluten) [122,123], myosins [124,125], egg [126], and seafood proteins [127,128].

One of the most widespread applications for TGs crosslinking activity in the food industry is the restructuring of meat and seafood by treatment of chopped muscle pieces with mTG and the polymerisation of muscle proteins (myofibrillar protein and myosin) to increase the textural characteristics and quality of the products [129–134]. Notably, mTG is not only exploited in meat processing, but also for the manufacture of dairy and bakery products, because of its considerable potential in improving the firmness, flavour, colour, texture, viscosity, elasticity, and water-binding capacity of aliments. In particular, mTG treatment during yogurt preparation was shown to improve the gel-forming properties of caseins by intermolecular crosslinking, increasing the yogurt's breaking strength and texture, which, for example, has been applied in the production of low-fat yogurt [135–137]. Concerning bakery products, mTG and also guinea pig liver TG (gplTG) have been widely used to ameliorate bread and dough rheological proprieties (e.g., elasticity, stability, and volume) and shelf life [138–142].

A safety concern on the treatment of bakery products by TGs emerged because of the well-known involvement of TG2 and TG6 in CD [44,143]. Initial controversial studies have suggested that mTG-treated wheat and gluten-free breads increase IgA reactivity in few CD patients' sera [144], while more recent studies showed the opposite [145]. Moreover, Heil et al. demonstrated that standard concentrations of mTG in bakery preparations (2–8 mTG units/Kg of flour) have no impact on CD incidence, even though it is not possible to exclude that higher doses might be correlated with it [145,146].

mTG cross-linking activity has also been explored for other applications, such as the preparation of chitosan-whey proteins edible films [147,148] and fish gelatin films [149–152]. The enzyme action was shown to significantly increase the films' mechanical properties and improve other characteristics, like deformability and biodegradability [147,151]. Moreover, gplTG has been used to perform an uncommon TG-mediated glycosylation between fish gelatin hydrolysates and glucosamine, and the resulting glycopeptides had enhanced bioactivity, with a significant potential as antioxidants and antimicrobial agents [153].

Therefore, the relevance of TGs biocatalysis in the food industry continues to be explored.

4.2. Applications in Science and Biomedicine

TGs-mediated biocatalysis for biomedical applications has grown over the years. In this respect, TGs have raised interest because they can substitute, as non-toxic crosslinkers, the chemical agents commonly used to produce scaffold biomaterials (e.g., glutaraldehyde and formaldehyde), which leave toxic residues that are difficult to remove.

4.2.1. Hydrogels and Scaffolds

One of the main applications of TGs as biocatalysts is in skin tissue engineering, for their ability to polymerise ECM proteins, especially collagens, for the production of hydrogels and scaffolds.

Collagens are structural proteins naturally present in all vertebrates, with the important physiological function of maintaining the mechanical proprieties and integrity of connective tissues. Their characteristics of high tensile strength, high water solubility, low antigenicity, and good compatibility make them optimal to be used as biomaterials in the skin engineering sector [154]. Moreover, collagen-made matrices are known to support cell proliferation and infiltration [155,156], and thus are ideal for applications in wound-healing. However, collagens are also prone to rapid enzymatic degradation in vivo by collagenases and lack mechanical strength at high temperatures or when solubilised in aqueous media, hence they need to be stabilised by covalent crosslinks to be able to form stable structures. Various chemical crosslinking agents have been investigated, e.g., glutaraldehyde [157,158], which, however, have been shown to significantly reduce the solubility, antigenicity, and biodegradation of the collagen matrices in vitro and in vivo, presenting also some cytotoxic effects [159,160]. Several studies have been carried out to determine the chemical and physical proprieties of TG-derived collagen hydrogels as an alternative to the chemically produced ones [161–163]. TGs are known to induce intermolecular crosslinks in collagen fibrils [70] and to covalently bind collagen to other ECM proteins, such as FN [164]. Stachel and colleagues have shown that mTG is able to potentially create up to 5.4 crosslinks per monomer of type I collagen under denaturing conditions. When collagen is in its native conformation, half of the target Gln residues are hidden within the triple helix region of the protein, and thus are not accessible to TG catalysis, explaining why the crosslinks are efficiently created only after collagen denaturation at high temperatures [165]. In addition, TG2 and mTG biocatalysis has been exploited to incorporate polyamines and crosslink different collagens to form matrices and scaffolds [166–168], increasing their denaturation temperature, resistance to proteolysis, and biocompatibility, other than presenting the advantage of avoiding toxic leftovers [73,74,169,170]. Collagen-based matrices are able to enhance cell attachment, spreading, differentiation, and proliferation, as demonstrated in dermal fibroblasts and also mesenchymal stem cells (MSCs), with relevance to tissue and cartilage bioengineering [73,167]. Notably, TG2 overexpression in mammalian cell lines (human osteoblasts, endothelial cells, and mouse fibroblasts) was shown to enhance biological recognition of polymers, such as poly(DL lactide co-glycolide) (PLG), poly(e-caprolactone) (PCL), and poly(L lactide) (PLA), consequently, increasing cell attachment and spreading [171].

More recently, the suitable concentration of mTG (40 units/g) has been established for the creation of new collagen-based hydrogels, with a focus on collagenases' degradation time [172]. Moreover, in vitro cell attachment together with in vivo biodegradability and biocompatibility assays have confirmed the high potentiality of these biomaterials in tissue engineering [172,173].

Furthermore, the use of guinea pig TG2 to crosslink amniotic membrane (AM), a scaffold employed in regenerative medicine, was shown to improve the mechanical properties of the membrane without altering the visual transparency and biocompatibility. These are fundamental features for ocular surface reconstruction applications [174]. Crosslinked AM showed higher interconnectivity among collagen fibres and promoted in vitro cell growth and angiogenesis, without eliciting an immune response [174].

TGs crosslinking has also been used in bone grafting, for example, to enhance interfacial adhesion of collagen/osteopontin on mineral substrates, increase the fracture toughness of bone [175,176], and produce collagen/nano-hydroxyapatite/chondroitin sulfate scaffolds, with possible clinical applications for spinal fusion surgery [177]. Collagen-based biomaterials crosslinked by TG2 have shown to increase cellular response in bone healing, by promoting the expression of integrins in human osteoblasts [178].

Similarly, other natural biopolymers, such as cellulose, fibrin, alginate, and hyaluronic acid (HA), have been used for the production of both tissue engineering scaffolds and drug delivery matrices [179–184] and, in the last 20 years, also synthetic peptide-based biopolymers have been exploited. In particular, elastin-like polypeptides (ELPs) hydrogels have been applied in multiple medical procedures, like cartilage and intervertebral disc tissue repair, vascular grafts, stem cell matrices, and post-surgical wound treatment to mention a few [185–190]. McHale and colleagues designed Lys- and Gln-containing ELPs by substituting the residue in position X of the ELP repeat sequence, VPGXG(VPGVG)$_6$, with Lys and Gln, respectively (Figure 4). These ELPs can be crosslinked by TG2 in a biocompatible process and form hydrogels able to encapsulate chondrocytes, leading to an increased ECM deposition and mechanical integrity [191].

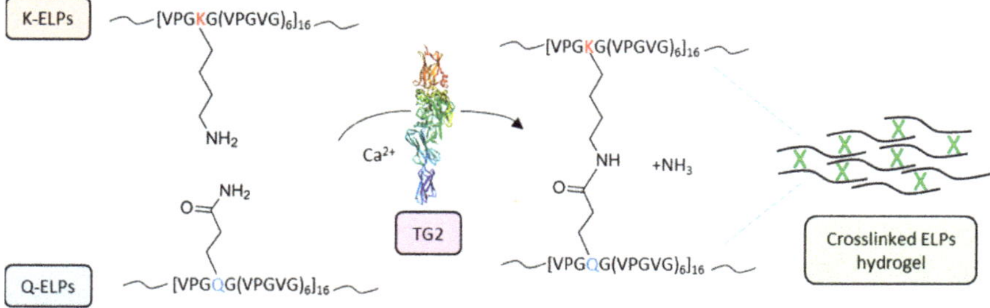

Figure 4. Schematic representation of TG2-crosslinked ELPs hydrogels. The repeat sequence [VPGXG(VPGVG)$_6$]$_{16}$ of ELPs was genetically designed to produce two libraries by substituting the X residue with Lys (K) and Gln (Q), thus generating K-ELPs and Q-ELPs, respectively. TG2 mediates the formation of a crosslink between the Gln and Lys residues of ELPs [191]. The Q and K side chains are superimposed to visualize the TG reaction. TG2 three-dimensional structure (PDB: 2Q3Z) was produced with the molecule modelling software, "EzMol" [34].

Interestingly, TG2 had already been tested as a biocompatible glue used to join two articular cartilage pieces in 1997 by Jürgensen, who demonstrated that TG2 treatment increased the adhesive strength by 40%, displaying a better performance compared to a commercial tissue sealant [192].

Recently, TG2 has been used to substitute the chemical agent, tris(hydroxylmethyl)phosphine (THP), for the production of resilin-based (RZ) protein gels, with applications in tissue engineering. TG-produced RZ matrices were more suitable for long-term cell attachment compared to those formed with THP. Moreover, as TG-derived matrices mimic the subendothelial environment more successfully compared to hard glass surfaces, they provide a more suitable environment for endothelial differentiation in vitro [193].

TGs have also shown promise for the fabrication of biopolymer microgels, small particles composed by crosslinked polymers that form three-dimensional (3D) structures filled with water, and are thus especially suitable for the delivery of nutrients and bioactive molecules [194–197]. One application of TGs-produced 3D hydrogels is in neuro tissue-engineering. In this context, the most commonly used hydrogels are made by chemical crosslinking of high molecular weight HA, which is an essential component of the central nervous system's ECM, with anti-inflammatory and anti-fibrotic properties, using agents, such as 1,4-butanediol diglycidyl ether (BDDE), adipic dihydrazide (ADH), or ethyl N, N-dimethylaminopropyl carbodiimide (EDC) [198,199]. The use of FXIIIa, able to crosslink HA modified by the addition of TG substrate peptides providing a reactive Gln or Lys residue (HA-TG), has been shown to allow the formation of better HA-based hydrogels compared to the chemically produced ones [200]. In fact, these hydrogels present higher chemical stability, more specific crosslinking, and features, such as tuneable gelation speed and stiffness, cytocompatibility, and injectability. Furthermore, HA-TG hydrogels can create covalent crosslinks with fibrin and other proteins, and can be a target of enzymatic degradation, which facilitates bioresorption [200].

Poly-ethyl-glycol (PEG) polymers are among the most commonly used molecules for the development of drug-delivery systems, as they are biocompatible and easily modified to form hydrogels. Indeed, these hydrogels can be produced by TGs crosslinking when functionalised with peptides, then mixed with therapeutic agents and potentially injected in the body. For example, TG2 was able to form highly elastic hydrogels in less than two minutes by creating crosslinks between the Gln residue of a PEG containing Ac-GQQQLG-NH_2 and the Lys residue of a PEG containing DOPA-FKG-NH_2 [201,202]. These hydrogels were tested as tissue glues on both guinea pig skin and collagen membranes, showing similar and higher adhesive strength, respectively, compared to fibrin tissue sealants [202]. Besides this application, PEGs have also been largely used in bio-conjugation processes.

4.2.2. Bio-Conjugation

PEGs conjugation of therapeutic proteins, defined as PEGylation, has been used to decrease proteins' immunogenicity, which is relevant for the improvement of the pharmacological proprieties of drugs [203,204]. By exploiting TGs' requirements for sequence and structure specificity for the targeting of Gln residues (amine acceptor site) [205], different groups have shown that classical random conjugation of PEG on proteins, which produces heterogeneous results, can be replaced by the more efficient TGs-dependent PEGylation (by mTG and FXIIIa), which allows the production of single site-specific conjugate isomers and, at the same time, the preservation of the proteins' bioactivities [206–210]. Interestingly, Sato et al. showed that the mTG-mediated incorporation of site-specific alkylamine-PEG conjugates into recombinant human interleukin-2 (rhIL-2), acting as the Gln donor substrate, did not affect rhIL-2 bioactivity, as opposed to random derivatisation. Moreover, pharmacokinetics studies in rodents revealed that the conjugates presented an increased half-life (up to 6-fold) compared to unmodified rhIL-2 [206]. mTG has also been used in vitro for protein lipidation, in order to increase the protein-lipid conjugate amphiphilicity and thus control its localisation at natural or artificial membranes' interfaces [211].

An alternative TGs substrate used for bio-conjugation is benzyloxycarbonyl-l-glutaminylglycine (Z-QG), already well-known in TG activity assays [212]. A specific application of Z-QG tags is the creation of protein–oligonucleotide (DNA) conjugates, which are useful tools in molecular biology, in particular for the production of protein microarrays. In order to overcome some issues related to the chemical manipulation commonly used for DNA-directed immobilisation, mTG has been successfully used to induce site-specific and covalent conjugation of DNA to peptide tags [213]. Specifically, mTG was able to mediate the labelling of K6-tagged recombinant proteins (K6 = MKHKGS), i.e., alkaline phosphatase (AP) and enhanced green fluorescent protein (EGFP), to an aminated DNA coupled with Z-QG, forming a protein-DNA conjugate (Z-QG-DNA tagged proteins) (Figure 5) [213].

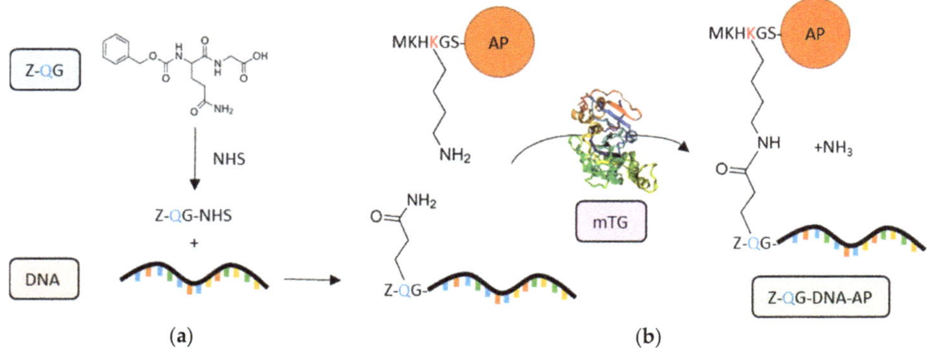

Figure 5. Protein-DNA conjugation. By exploiting benzyloxycarbonyl-L-glutaminylglycine (Z-QG) as a TG substrate, mTG induces site-specific and covalent conjugation of DNA to proteins, such as alkaline phosphatase (AP). (**a**) Chemical activation of Z-QG carboxylate with N-hydroxysuccinimide (NHS) and formation of the activated Z-QG-NHS, which is then modified by addition of an aminated oligodeoxynucleotide (DNA). (**b**) mTG mediates the formation of a crosslink between the Gln residue of Z-QG-DNA and the Lys residue of the MKHKGS peptide tag fused to AP. The Q and K side chains are superimposed to visualize the TG reaction [213]. The modelling software, "EzMol", was used to generate mTG structure (PDB: 1IU4) [34].

An improvement of the same approach was explored for the sensitive and cost-effective preparation of Z-QG-DNA–AP conjugates for filter and in situ hybridisation assays [214]. Moreover, this procedure was used to functionalise RNA (Z-QG-RNA-AP conjugates) and tested in tissue sections by in situ hybridisation [215]. The Z-QG conjugation approach has been broadened to the production of DNA aptamer-(protein)$_n$ conjugates for cell imaging through a two-step reaction mediated by terminal deoxynucleotidyl transferase (TdT) and mTG [216,217]. These biocompatible mTG-derived constructs offer novel opportunities for the development of non-invasive in vivo imaging [217].

The successful use of TGs for the creation of bio-conjugates has led to the application of this procedure for protein fluorescent labelling and immobilisation. Keillor and colleagues demonstrated that mTG can be used for the site-specific labelling of proteins genetically modified with encodable high-affinity Gln-substrates ('Q-tags') through incorporation of propargylamine into the Gln residues (propargylation) [218,219]. This strategy showed high potential for the conjugation of a wide range of azide derivatives for fluorescence labelling, with possible applications in living cells [219–221].

Notably, antibodies are among the proteins that can be functionalised by TGs for diagnostic and therapeutic purposes (Figure 6). Josten and colleagues proposed an original enzymatic biotinylation method useful for the production of low-biotinylated proteins, based on mTG-mediated incorporation of amino-modified derivatives on IgG Gln residues [222]. More recently, antibody functionalisation by TGs has been applied in radio immunodiagnosis and therapy antibody [223–225]. Both mTG and, to a lesser extent, human TG2 have shown the ability to perform the selective modification of antibodies heavy chains (IgGs), without interfering with their biological activities, such as antigen affinity and cell internalisation, as tested both in vitro and in vivo [223,225]. Interestingly, a new multi-loading approach that improves the drug-to-antibody ratio has been tested, with promising applications in targeted therapy [224]. Specifically, mTG conjugation of branched linkers on the heavy chain of an anti-HER2 monoclonal antibody have shown to increase the drug cytotoxicity against a HER2-expressing breast cancer cell line in vitro, compared to conjugates carrying the classic linear linkers [224].

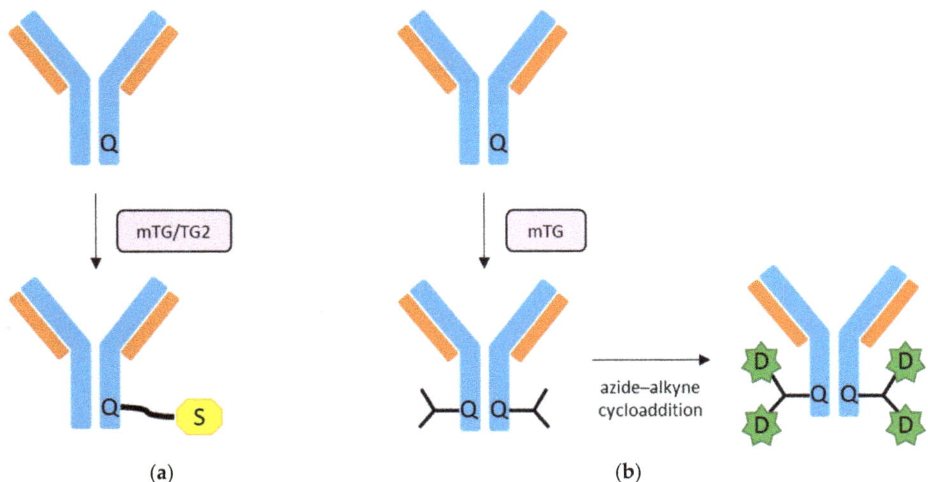

Figure 6. Schematic representation of TG-mediated antibody conjugation. (**a**) TGs can transamidate modified substrates (e.g., aminated or Lys carrying derivatives), such as biotin, fluorophores, or radioisotopes (schematically shown with the yellow "S"), to a carboxamide group of a Q residue on the antibody heavy chain peptide sequence [222,223,225]. (**b**) Branched linkers, conjugated by mTG to the heavy chain of an antibody (e.g., anti-HER2 monoclonal antibody), can be coupled by azide-alkyne cycloaddition to an antimitotic drug (shown as a green "D") [224].

Table 1. List of applications involving TGs activity.

Applications	References
Food Industry	
Dairy products (caseins and whey proteins)	[118,120,135–137]
Soybean proteins	[121]
Bakery products (gluten)	[122,123,138–142]
Meat (myosins and myofibrillar protein)	[124,125,129–134]
Eggs (ovalbumin)	[126]
Seafood and edible films	[127,128,147–153]
Biomedicine	
Collagen-based scaffolds and hydrogels	[70,73,74,161–163,166,167,169–178]
Other natural biopolymers-based hydrogels and microgels (Fibrin, gelatin, alginate, hyaluronic acid, casein)	[179–184,194,195,197–200]
Synthetic biopolymers-based scaffolds and 3D microgels (ELPs, RZ, PEG)	[189–191,193,196,201,202]
PEGylation/Lipidation	[206–211]
Protein-DNA conjugation	[213,214,216,217]
Protein fluorescent labelling	[218–221]
Antibodies functionalization	[223–225]

5. Conclusions

There is no question that TG-biocatalysis is instrumental in determining protein multimerisation, and that several are the natural substrates of TG, which can be permanently modified by transamidation. Although research has documented numerous physio-pathological conditions in which TG family members are involved, there is still much to be learnt about the way the catalytic activity of these enzymes is controlled in vivo. In parallel, there is a great interest in the application of transglutaminases, especially mTG and TG2, as a tool to catalyse the formation of amide bonds between peptide or protein bound glutamines and lysines, or to change protein proprieties via incorporation of polyamines in a variety of uses of particular relevance to the biomedical and food industry. Even though substrate specificity remains a challenge, the versatility and biocompatibility of transglutaminases continue to make them attractive for a wide range of biotechnological applications.

Author Contributions: M.P.S. and E.T. equally contributed to the main body of the manuscript and writing; A.G.A. contributed to specific paragraphs related to the "Hydrogels and Scaffolds" section; E.A.M.V. developed and coordinated the ideas and revised the manuscript.

Funding: This research was funded by NTU VC's Scholarship, John Turland bursary and UCB Pharmaceutical.

Conflicts of Interest: The authors declare no conflict of interest.

References

1. Sarkar, N.K.; Clarke, D.D.; Waelsch, H. An Enzymically Catalyzed Incorporation of Amines into Proteins. *Biochim. Biophys. Acta* **1957**, *25*, 451–452. [CrossRef]
2. Demeny, M.A.; Korponay-Szabo, I.; Fesus, L. Structure of Transglutaminases: Unique Features Serve Diverse Functions. In *Transglutaminases: Multiple Functional Modifiers and Targets for New Drug Discovery*; Hitomi, K., Kojima, S., Fesus, L., Eds.; Springer: Tokyo, Japan, 2015; pp. 1–2.
3. Folk, J.E. Mechanism of Action of Guinea Pig Liver Transglutaminase. VI. Order of Substrate Addition. *J. Biol. Chem.* **1969**, *244*, 3707–3713. [PubMed]
4. Pisano, J.J.; Finlayson, J.S.; Peyton, M.P. Cross-Link in Fibrin Polymerized by Factor 13: Epsilon-(Gamma-Glutamyl)Lysine. *Science* **1968**, *160*, 892–893. [CrossRef] [PubMed]
5. Matacic, S.; Loewy, A.G. The Identification of Isopeptide Crosslinks in Insoluble Fibrin. *Biochem. Biophys. Res. Commun.* **1968**, *30*, 356–362. [CrossRef]
6. Folk, J.E.; Park, M.H.; Chung, S.I.; Schrode, J.; Lester, E.P.; Cooper, H.L. Polyamines as Physiological Substrates for Transglutaminases. *J. Biol. Chem.* **1980**, *255*, 3695–3700. [PubMed]
7. Korsgren, C.; Lawler, J.; Lambert, S.; Speicher, D.; Cohen, C.M. Complete Amino Acid Sequence and Homologies of Human Erythrocyte Membrane Protein Band 4.2. *Proc. Natl. Acad. Sci. USA* **1990**, *87*, 613–617. [CrossRef] [PubMed]
8. Grenard, P.; Bates, M.K.; Aeschlimann, D. Evolution of Transglutaminase Genes: Identification of a Transglutaminase Gene Cluster on Human Chromosome 15q15. Structure of the Gene Encoding Transglutaminase X and a Novel Gene Family Member, Transglutaminase Z. *J. Biol. Chem.* **2001**, *276*, 33066–33078. [CrossRef] [PubMed]
9. Folk, J.; Finlayson, J. The ε-(Γ-Glutamyl) Lysine Crosslink and the Catalytic Role of Transglutaminases. *Adv. Protein Chem.* **1977**, *31*, 1–133. [PubMed]
10. Greenberg, C.S.; Birckbichler, P.J.; Rice, R.H. Transglutaminases: Multifunctional Cross-Linking Enzymes that Stabilize Tissues. *FASEB J.* **1991**, *5*, 3071–3077. [CrossRef] [PubMed]
11. Keillor, J.W.; Clouthier, C.M.; Apperley, K.Y.; Akbar, A.; Mulani, A. Acyl Transfer Mechanisms of Tissue Transglutaminase. *Bioorg. Chem.* **2014**, *57*, 186–197. [CrossRef] [PubMed]
12. Clarke, D.; Mycek, M.; Neidle, A.; Waelsch, H. The Incorporation of Amines into Protein. *Arch. Biochem. Biophys.* **1959**, *79*, 338–354. [CrossRef]
13. Mycek, M.J.; Clarke, D.D.; Neidle, A.; Waelsch, H. Amine Incorporation into Insulin as Catalyzed by Transglutaminase. *Arch. Biochem. Biophys.* **1959**, *84*, 528–540. [CrossRef]
14. Nurminskaya, M.V.; Belkin, A.M. Cellular Functions of Tissue Transglutaminase. *Int. Rev. Cell. Mol. Biol.* **2012**. [CrossRef]

15. Yee, V.C.; Pedersen, L.C.; Le Trong, I.; Bishop, P.D.; Stenkamp, R.E.; Teller, D.C. Three-Dimensional Structure of a Transglutaminase: Human Blood Coagulation Factor XIII. *Proc. Natl. Acad. Sci. USA* **1994**, *91*, 7296–7300. [CrossRef] [PubMed]
16. Liu, S.; Cerione, R.A.; Clardy, J. Structural Basis for the Guanine Nucleotide-Binding Activity of Tissue Transglutaminase and its Regulation of Transamidation Activity. *Proc. Natl. Acad. Sci. USA* **2002**, *99*, 2743–2747. [CrossRef] [PubMed]
17. Murthy, S.N.; Iismaa, S.; Begg, G.; Freymann, D.M.; Graham, R.M.; Lorand, L. Conserved Tryptophan in the Core Domain of Transglutaminase is Essential for Catalytic Activity. *Proc. Natl. Acad. Sci. USA* **2002**, *99*, 2738–2742. [CrossRef] [PubMed]
18. Iismaa, S.E.; Holman, S.; Wouters, M.A.; Lorand, L.; Graham, R.M.; Husain, A. Evolutionary Specialization of a Tryptophan Indole Group for Transition-State Stabilization by Eukaryotic Transglutaminases. *Proc. Natl. Acad. Sci. USA* **2003**, *100*, 12636–12641. [CrossRef] [PubMed]
19. Jeong, J.M.; Murthy, S.N.; Radek, J.T.; Lorand, L. The Fibronectin-Binding Domain of Transglutaminase. *J. Biol. Chem.* **1995**, *270*, 5654–5658. [CrossRef] [PubMed]
20. Gaudry, C.A.; Verderio, E.; Aeschlimann, D.; Cox, A.; Smith, C.; Griffin, M. Cell Surface Localization of Tissue Transglutaminase is Dependent on a Fibronectin-Binding Site in its N-Terminal Beta-Sandwich Domain. *J. Biol. Chem.* **1999**, *274*, 30707–30714. [CrossRef] [PubMed]
21. Hang, J.; Zemskov, E.A.; Lorand, L.; Belkin, A.M. Identification of a Novel Recognition Sequence for Fibronectin within the NH2-Terminal Beta-Sandwich Domain of Tissue Transglutaminase. *J. Biol. Chem.* **2005**, *280*, 23675–23683. [CrossRef] [PubMed]
22. Achyuthan, K.E.; Greenberg, C.S. Identification of a Guanosine Triphosphate-Binding Site on Guinea Pig Liver Transglutaminase. Role of GTP and Calcium Ions in Modulating Activity. *J. Biol. Chem.* **1987**, *262*, 1901–1906. [PubMed]
23. Bergamini, C.M.; Signorini, M.; Poltronieri, L. Inhibition of Erythrocyte Transglutaminase by GTP. *Biochim. Biophys. Acta* **1987**, *916*, 149–151. [CrossRef]
24. Nakaoka, H.; Perez, D.M.; Baek, K.J.; Das, T.; Husain, A.; Misono, K.; Im, M.J.; Graham, R.M. Gh: A GTP-Binding Protein with Transglutaminase Activity and Receptor Signaling Function. *Science* **1994**, *264*, 1593–1596. [CrossRef] [PubMed]
25. Bergamini, C.M. GTP Modulates Calcium Binding and Cation-Induced Conformational Changes in Erythrocyte Transglutaminase. *FEBS Lett.* **1988**, *239*, 255–258. [CrossRef]
26. Casadio, R.; Polverini, E.; Mariani, P.; Spinozzi, F.; Carsughi, F.; Fontana, A.; Polverino de Laureto, P.; Matteucci, G.; Bergamini, C.M. The Structural Basis for the Regulation of Tissue Transglutaminase by Calcium Ions. *Eur. J. Biochem.* **1999**, *262*, 672–679. [CrossRef] [PubMed]
27. Kiraly, R.; Csosz, E.; Kurtan, T.; Antus, S.; Szigeti, K.; Simon-Vecsei, Z.; Korponay-Szabo, I.R.; Keresztessy, Z.; Fesus, L. Functional Significance of Five Noncanonical Ca^{2+}-Binding Sites of Human Transglutaminase 2 Characterized by Site-Directed Mutagenesis. *FEBS J.* **2009**, *276*, 7083–7096. [CrossRef] [PubMed]
28. Candi, E.; Paradisi, A.; Terrinoni, A.; Pietroni, V.; Oddi, S.; Cadot, B.; Jogini, V.; Meiyappan, M.; Clardy, J.; Finazzi-Agro, A.; et al. Transglutaminase 5 is regulated by Guanine-Adenine Nucleotides. *Biochem. J.* **2004**, *381*, 313–319. [CrossRef] [PubMed]
29. Lai, T.S.; Slaughter, T.F.; Peoples, K.A.; Hettasch, J.M.; Greenberg, C.S. Regulation of Human Tissue Transglutaminase Function by Magnesium-Nucleotide Complexes. Identification of Distinct Binding Sites for mg-GTP and mg-ATP. *J. Biol. Chem.* **1998**, *273*, 1776–1781. [CrossRef] [PubMed]
30. Di Venere, A.; Rossi, A.; De Matteis, F.; Rosato, N.; Agro, A.F.; Mei, G. Opposite Effects of $Ca^{(2+)}$ and GTP Binding on Tissue Transglutaminase Tertiary Structure. *J. Biol. Chem.* **2000**, *275*, 3915–3921. [CrossRef] [PubMed]
31. Lortat-Jacob, H.; Burhan, I.; Scarpellini, A.; Thomas, A.; Imberty, A.; Vives, R.R.; Johnson, T.; Gutierrez, A.; Verderio, E.A. Transglutaminase-2 Interaction with Heparin: Identification of a Heparin Binding Site that Regulates Cell Adhesion to Fibronectin-Transglutaminase-2 Matrix. *J. Biol. Chem.* **2012**, *287*, 18005–18017. [CrossRef] [PubMed]
32. Boeshans, K.M.; Mueser, T.C.; Ahvazi, B. A Three-Dimensional Model of the Human Transglutaminase 1: Insights into the Understanding of Lamellar Ichthyosis. *J. Mol. Model.* **2007**, *13*, 233–246. [CrossRef] [PubMed]

33. Thomas, H.; Beck, K.; Adamczyk, M.; Aeschlimann, P.; Langley, M.; Oita, R.C.; Thiebach, L.; Hils, M.; Aeschlimann, D. Transglutaminase 6: A Protein Associated with Central Nervous System Development and Motor Function. *Amino Acids* **2013**, *44*, 161–177. [CrossRef] [PubMed]
34. Reynolds, C.R.; Islam, S.A.; Sternberg, M.J.E. EzMol: A Web Server Wizard for the Rapid Visualization and Image Production of Protein and Nucleic Acid Structures. *J. Mol. Biol.* **2018**, *430*, 2244–2248. [CrossRef] [PubMed]
35. Stamnaes, J.; Pinkas, D.M.; Fleckenstein, B.; Khosla, C.; Sollid, L.M. Redox Regulation of Transglutaminase 2 Activity. *J. Biol. Chem.* **2010**, *285*, 25402–25409. [CrossRef] [PubMed]
36. Boothe, R.L.; Folk, J.E. A Reversible, Calcium-Dependent, Copper-Catalyzed Inactivation of Guinea Pig Liver Transglutaminase. *J. Biol. Chem.* **1969**, *244*, 399–405. [PubMed]
37. Connellan, J.M.; Folk, J.E. Mechanism of the Inactivation of Guinea Pig Liver Transglutaminase by 5,5′-Dithiobis-(2-Nitrobenzoic Acid). *J. Biol. Chem.* **1969**, *244*, 3173–3181. [PubMed]
38. Jin, X.; Stamnaes, J.; Klock, C.; DiRaimondo, T.R.; Sollid, L.M.; Khosla, C. Activation of Extracellular Transglutaminase 2 by Thioredoxin. *J. Biol. Chem.* **2011**, *286*, 37866–37873. [CrossRef] [PubMed]
39. Begg, G.E.; Carrington, L.; Stokes, P.H.; Matthews, J.M.; Wouters, M.A.; Husain, A.; Lorand, L.; Iismaa, S.E.; Graham, R.M. Mechanism of Allosteric Regulation of Transglutaminase 2 by GTP. *Proc. Natl. Acad. Sci. USA* **2006**, *103*, 19683–19688. [CrossRef] [PubMed]
40. Pinkas, D.M.; Strop, P.; Brunger, A.T.; Khosla, C. Transglutaminase 2 Undergoes a Large Conformational Change upon Activation. *PLoS Biol.* **2007**, *5*, e327. [CrossRef] [PubMed]
41. Siegel, M.; Strnad, P.; Watts, R.E.; Choi, K.; Jabri, B.; Omary, M.B.; Khosla, C. Extracellular Transglutaminase 2 is Catalytically Inactive, but is Transiently Activated upon Tissue Injury. *PLoS ONE* **2008**, *3*, e1861. [CrossRef] [PubMed]
42. Akimov, S.S.; Krylov, D.; Fleischman, L.F.; Belkin, A.M. Tissue Transglutaminase is an Integrin-Binding Adhesion Coreceptor for Fibronectin. *J. Cell Biol.* **2000**, *148*, 825–838. [CrossRef] [PubMed]
43. Mycek, M.J.; Waelsch, H. The Enzymatic Deamidation of Proteins. *J. Biol. Chem.* **1960**, *235*, 3513–3517. [PubMed]
44. Dieterich, W.; Ehnis, T.; Bauer, M.; Donner, P.; Volta, U.; Riecken, E.O.; Schuppan, D. Identification of Tissue Transglutaminase as the Autoantigen of Celiac Disease. *Nat. Med.* **1997**, *3*, 797–801. [CrossRef] [PubMed]
45. Molberg, O.; Mcadam, S.N.; Korner, R.; Quarsten, H.; Kristiansen, C.; Madsen, L.; Fugger, L.; Scott, H.; Noren, O.; Roepstorff, P.; et al. Tissue Transglutaminase Selectively Modifies Gliadin Peptides that are Recognized by Gut-Derived T Cells in Celiac Disease. *Nat. Med.* **1998**, *4*, 713–717. [CrossRef] [PubMed]
46. Singh, U.S.; Erickson, J.W.; Cerione, R.A. Identification and Biochemical Characterization of an 80 Kilodalton GTP-Binding/Transglutaminase from Rabbit Liver Nuclei. *Biochemistry* **1995**, *34*, 15863–15871. [CrossRef] [PubMed]
47. Mishra, S.; Murphy, L.J. Tissue Transglutaminase has Intrinsic Kinase Activity: Identification of Transglutaminase 2 as an Insulin-Like Growth Factor-Binding Protein-3 Kinase. *J. Biol. Chem.* **2004**, *279*, 23863–23868. [CrossRef] [PubMed]
48. Mishra, S.; Saleh, A.; Espino, P.S.; Davie, J.R.; Murphy, L.J. Phosphorylation of Histones by Tissue Transglutaminase. *J. Biol. Chem.* **2006**, *281*, 5532–5538. [CrossRef] [PubMed]
49. Mishra, S.; Murphy, L.J. The p53 Oncoprotein is a Substrate for Tissue Transglutaminase Kinase Activity. *Biochem. Biophys. Res. Commun.* **2006**, *339*, 726–730. [CrossRef] [PubMed]
50. Mishra, S.; Melino, G.; Murphy, L.J. Transglutaminase 2 Kinase Activity Facilitates Protein Kinase A-Induced Phosphorylation of Retinoblastoma Protein. *J. Biol. Chem.* **2007**, *282*, 18108–18115. [CrossRef] [PubMed]
51. Hasegawa, G.; Suwa, M.; Ichikawa, Y.; Ohtsuka, T.; Kumagai, S.; Kikuchi, M.; Sato, Y.; Saito, Y. A Novel Function of Tissue-Type Transglutaminase: Protein Disulphide Isomerase. *Biochem. J.* **2003**, *373*, 793–803. [CrossRef] [PubMed]
52. Mastroberardino, P.G.; Farrace, M.G.; Viti, I.; Pavone, F.; Fimia, G.M.; Melino, G.; Rodolfo, C.; Piacentini, M. "Tissue" Transglutaminase Contributes to the Formation of Disulphide Bridges in Proteins of Mitochondrial Respiratory Complexes. *Biochim. Biophys. Acta* **2006**, *1757*, 1357–1365. [CrossRef] [PubMed]
53. Kanaji, T.; Ozaki, H.; Takao, T.; Kawajiri, H.; Ide, H.; Motoki, M.; Shimonishi, Y. Primary Structure of Microbial Transglutaminase from Streptoverticillium Sp. Strain s-8112. *J. Biol. Chem.* **1993**, *268*, 11565–11572. [PubMed]

54. Del Duca, S.; Beninati, S.; Serafini-Fracassini, D. Polyamines in Chloroplasts: Identification of their Glutamyl and Acetyl Derivatives. *Biochem. J.* **1995**, *305 Pt 1*, 233–237. [CrossRef]
55. Makarova, K.S.; Aravind, L.; Galperin, M.Y.; Grishin, N.V.; Tatusov, R.L.; Wolf, Y.I.; Koonin, E.V. Comparative Genomics of the Archaea (Euryarchaeota): Evolution of Conserved Protein Families, the Stable Core, and the Variable Shell. *Genome Res.* **1999**, *9*, 608–628. [PubMed]
56. Polakowska, R.R.; Eickbush, T.; Falciano, V.; Razvi, F.; Goldsmith, L.A. Organization and Evolution of the Human Epidermal Keratinocyte Transglutaminase I Gene. *Proc. Natl. Acad. Sci. USA* **1992**, *89*, 4476–4480. [CrossRef] [PubMed]
57. Tokunaga, F.; Muta, T.; Iwanaga, S.; Ichinose, A.; Davie, E.W.; Kuma, K.; Miyata, T. Limulus Hemocyte Transglutaminase. cDNA Cloning, Amino Acid Sequence, and Tissue Localization. *J. Biol. Chem.* **1993**, *268*, 262–268. [PubMed]
58. Ando, H.; Adachi, M.; Umeda, K.; Matsuura, A.; Nonaka, M.; Uchio, R.; Tanaka, H.; Motoki, M. Purification and Characteristics of a Novel Transglutaminase Derived from Microorganisms. *Agric. Biol. Chem.* **1989**, *53*, 2613–2617.
59. Nonaka, M.; Tanaka, H.; Okiyama, A.; Motoki, M.; Ando, H.; Umeda, K.; Matsuura, A. Polymerization of several Proteins by Ca2 -Independent Transglutaminase Derived from Microorganisms. *Agric. Biol. Chem.* **1989**, *53*, 2619–2623.
60. Kashiwagi, T.; Yokoyama, K.; Ishikawa, K.; Ono, K.; Ejima, D.; Matsui, H.; Suzuki, E. Crystal Structure of Microbial Transglutaminase from Streptoverticillium Mobaraense. *J. Biol. Chem.* **2002**, *277*, 44252–44260. [CrossRef] [PubMed]
61. Fesus, L.; Metsis, M.L.; Muszbek, L.; Koteliansky, V.E. Transglutaminase-Sensitive Glutamine Residues of Human Plasma Fibronectin Revealed by Studying its Proteolytic Fragments. *Eur. J. Biochem.* **1986**, *154*, 371–374. [CrossRef] [PubMed]
62. Upchurch, H.F.; Conway, E.; Patterson, M., Jr.; Maxwell, M.D. Localization of Cellular Transglutaminase on the Extracellular Matrix after Wounding: Characteristics of the Matrix Bound Enzyme. *J. Cell. Physiol.* **1991**, *149*, 375–382. [CrossRef] [PubMed]
63. Jones, R.A.; Nicholas, B.; Mian, S.; Davies, P.J.; Griffin, M. Reduced Expression of Tissue Transglutaminase in a Human Endothelial Cell Line Leads to Changes in Cell Spreading, Cell Adhesion and Reduced Polymerisation of Fibronectin. *J. Cell Sci.* **1997**, *110*, 2461–2472. [PubMed]
64. Verderio, E.; Nicholas, B.; Gross, S.; Griffin, M. Regulated Expression of Tissue Transglutaminase in Swiss 3T3 Fibroblasts: Effects on the Processing of Fibronectin, Cell Attachment, and Cell Death. *Exp. Cell Res.* **1998**, *239*, 119–138. [CrossRef] [PubMed]
65. Sane, D.C.; Moser, T.L.; Pippen, A.M.; Parker, C.J.; Achyuthan, K.E.; Greenberg, C.S. Vitronectin is a Substrate for Transglutaminases. *Biochem. Biophys. Res. Commun.* **1988**, *157*, 115–120. [CrossRef]
66. Martinez, J.; Rich, E.; Barsigian, C. Transglutaminase-Mediated Cross-Linking of Fibrinogen by Human Umbilical Vein Endothelial Cells. *J. Biol. Chem.* **1989**, *264*, 20502–20508. [PubMed]
67. Aeschlimann, D.; Paulsson, M. Cross-Linking of Laminin-Nidogen Complexes by Tissue Transglutaminase. A Novel Mechanism for Basement Membrane Stabilization. *J. Biol. Chem.* **1991**, *266*, 15308–15317. [PubMed]
68. Barsigian, C.; Stern, A.M.; Martinez, J. Tissue (Type II) Transglutaminase Covalently Incorporates itself, Fibrinogen, Or Fibronectin into High Molecular Weight Complexes on the Extracellular Surface of Isolated Hepatocytes. Use of 2-[(2-Oxopropyl)Thio] Imidazolium Derivatives as Cellular Transglutaminase Inactivators. *J. Biol. Chem.* **1991**, *266*, 22501–22509. [PubMed]
69. Aeschlimann, D.; Kaupp, O.; Paulsson, M. Transglutaminase-Catalyzed Matrix Cross-Linking in Differentiating Cartilage: Identification of Osteonectin as a Major Glutaminyl Substrate. *J. Cell Biol.* **1995**, *129*, 881–892. [CrossRef] [PubMed]
70. Kleman, J.P.; Aeschlimann, D.; Paulsson, M.; van der Rest, M. Transglutaminase-Catalyzed Cross-Linking of Fibrils of Collagen V/XI in A204 Rhabdomyosarcoma Cells. *Biochemistry* **1995**, *34*, 13768–13775. [CrossRef] [PubMed]
71. Kaartinen, M.T.; Pirhonen, A.; Linnala-Kankkunen, A.; Maenpaa, P.H. Transglutaminase-Catalyzed Cross-Linking of Osteopontin is inhibited by Osteocalcin. *J. Biol. Chem.* **1997**, *272*, 22736–22741. [CrossRef] [PubMed]

72. Johnson, T.S.; Skill, N.J.; El Nahas, A.M.; Oldroyd, S.D.; Thomas, G.L.; Douthwaite, J.A.; Haylor, J.L.; Griffin, M. Transglutaminase Transcription and Antigen Translocation in Experimental Renal Scarring. *J. Am. Soc. Nephrol.* **1999**, *10*, 2146–2157. [PubMed]
73. Chau, D.Y.; Collighan, R.J.; Verderio, E.A.; Addy, V.L.; Griffin, M. The Cellular Response to Transglutaminase-Cross-Linked Collagen. *Biomaterials* **2005**, *26*, 6518–6529. [CrossRef] [PubMed]
74. Jones, R.A.; Kotsakis, P.; Johnson, T.S.; Chau, D.Y.; Ali, S.; Melino, G.; Griffin, M. Matrix Changes Induced by Transglutaminase 2 Lead to Inhibition of Angiogenesis and Tumor Growth. *Cell Death Differ.* **2006**, *13*, 1442–1453. [CrossRef] [PubMed]
75. Johnson, T.S.; Griffin, M.; Thomas, G.L.; Skill, J.; Cox, A.; Yang, B.; Nicholas, B.; Birckbichler, P.J.; Muchaneta-Kubara, C.; Meguid El Nahas, A. The Role of Transglutaminase in the Rat Subtotal Nephrectomy Model of Renal Fibrosis. *J. Clin. Investig.* **1997**, *99*, 2950–2960. [CrossRef] [PubMed]
76. Johnson, T.S.; El-Koraie, A.F.; Skill, N.J.; Baddour, N.M.; El Nahas, A.M.; Njloma, M.; Adam, A.G.; Griffin, M. Tissue Transglutaminase and the Progression of Human Renal Scarring. *J. Am. Soc. Nephrol.* **2003**, *14*, 2052–2062. [CrossRef] [PubMed]
77. Burhan, I.; Furini, G.; Lortat-Jacob, H.; Atobatele, A.G.; Scarpellini, A.; Schroeder, N.; Atkinson, J.; Maamra, M.; Nutter, F.H.; Watson, P.; et al. Interplay between Transglutaminases and Heparan Sulphate in Progressive Renal Scarring. *Sci. Rep.* **2016**, *6*, 31343. [CrossRef] [PubMed]
78. Griffin, M.; Smith, L.L.; Wynne, J. Changes in Transglutaminase Activity in an Experimental Model of Pulmonary Fibrosis Induced by Paraquat. *Br. J. Exp. Pathol.* **1979**, *60*, 653–661. [PubMed]
79. Richards, R.J.; Masek, L.C.; Brown, R.F. Biochemical and Cellular Mechanisms of Pulmonary Fibrosis. *Toxicol. Pathol.* **1991**, *19*, 526–539. [PubMed]
80. Oh, K.; Park, H.B.; Byoun, O.J.; Shin, D.M.; Jeong, E.M.; Kim, Y.W.; Kim, Y.S.; Melino, G.; Kim, I.G.; Lee, D.S. Epithelial Transglutaminase 2 is Needed for T Cell Interleukin-17 Production and Subsequent Pulmonary Inflammation and Fibrosis in Bleomycin-Treated Mice. *J. Exp. Med.* **2011**, *208*, 1707–1719. [CrossRef] [PubMed]
81. Olsen, K.C.; Sapinoro, R.E.; Kottmann, R.M.; Kulkarni, A.A.; Iismaa, S.E.; Johnson, G.V.; Thatcher, T.H.; Phipps, R.P.; Sime, P.J. Transglutaminase 2 and its Role in Pulmonary Fibrosis. *Am. J. Respir. Crit. Care Med.* **2011**, *184*, 699–707. [CrossRef] [PubMed]
82. Mirza, A.; Liu, S.L.; Frizell, E.; Zhu, J.; Maddukuri, S.; Martinez, J.; Davies, P.; Schwarting, R.; Norton, P.; Zern, M.A. A Role for Tissue Transglutaminase in Hepatic Injury and Fibrogenesis, and its Regulation by NF-κB. *Am. J. Physiol.* **1997**. [CrossRef]
83. Grenard, P.; Bresson-Hadni, S.; El Alaoui, S.; Chevallier, M.; Vuitton, D.A.; Ricard-Blum, S. Transglutaminase-Mediated Cross-Linking is Involved in the Stabilization of Extracellular Matrix in Human Liver Fibrosis. *J. Hepatol.* **2001**, *35*, 367–375. [CrossRef]
84. Tatsukawa, H.; Tani, Y.; Otsu, R.; Nakagawa, H.; Hitomi, K. Global Identification and Analysis of Isozyme-Specific Possible Substrates Crosslinked by Transglutaminases using Substrate Peptides in Mouse Liver Fibrosis. *Sci. Rep.* **2017**, *7*, 45049. [CrossRef] [PubMed]
85. Small, K.; Feng, J.F.; Lorenz, J.; Donnelly, E.T.; Yu, A.; Im, M.J.; Dorn, G.W., 2nd; Liggett, S.B. Cardiac Specific Overexpression of Transglutaminase II (G(H)) Results in a Unique Hypertrophy Phenotype Independent of Phospholipase C Activation. *J. Biol. Chem.* **1999**, *274*, 21291–21296. [CrossRef] [PubMed]
86. Shinde, A.V.; Dobaczewski, M.; de Haan, J.J.; Saxena, A.; Lee, K.K.; Xia, Y.; Chen, W.; Su, Y.; Hanif, W.; Kaur Madahar, I.; et al. Tissue Transglutaminase Induction in the Pressure-Overloaded Myocardium Regulates Matrix Remodelling. *Cardiovasc. Res.* **2017**, *113*, 892–905. [CrossRef] [PubMed]
87. Wang, Z.; Stuckey, D.J.; Murdoch, C.E.; Camelliti, P.; Lip, G.Y.H.; Griffin, M. Cardiac Fibrosis can be Attenuated by Blocking the Activity of Transglutaminase 2 using a Selective Small-Molecule Inhibitor. *Cell Death Dis.* **2018**. [CrossRef] [PubMed]
88. Mousa, A.; Cui, C.; Song, A.; Myneni, V.D.; Sun, H.; Li, J.J.; Murshed, M.; Melino, G.; Kaartinen, M.T. Transglutaminases Factor XIII-A and TG2 Regulate Resorption, Adipogenesis and Plasma Fibronectin Homeostasis in Bone and Bone Marrow. *Cell Death Differ.* **2017**, *24*, 844–854. [CrossRef] [PubMed]
89. Tatsukawa, H.; Otsu, R.; Tani, Y.; Wakita, R.; Hitomi, K. Isozyme-Specific Comprehensive Characterization of Transglutaminase-Crosslinked Substrates in Kidney Fibrosis. *Sci. Rep.* **2018**, *8*, 7306. [CrossRef] [PubMed]

90. Zemskov, E.A.; Mikhailenko, I.; Hsia, R.C.; Zaritskaya, L.; Belkin, A.M. Unconventional Secretion of Tissue Transglutaminase Involves Phospholipid-Dependent Delivery into Recycling Endosomes. *PLoS ONE* **2011**, *6*, e19414. [CrossRef] [PubMed]
91. Adamczyk, M.; Griffiths, R.; Dewitt, S.; Knauper, V.; Aeschlimann, D. P2X7 Receptor Activation Regulates Rapid Unconventional Export of Transglutaminase-2. *J. Cell Sci.* **2015**, *128*, 4615–4628. [CrossRef] [PubMed]
92. Aeschlimann, D.; Knauper, V. P2X7 Receptor-Mediated TG2 Externalization: A Link to Inflammatory Arthritis? *Amino Acids* **2017**, *49*, 453–460. [CrossRef] [PubMed]
93. Scarpellini, A.; Germack, R.; Lortat-Jacob, H.; Muramatsu, T.; Billett, E.; Johnson, T.; Verderio, E.A. Heparan Sulfate Proteoglycans are Receptors for the Cell-Surface Trafficking and Biological Activity of Transglutaminase-2. *J. Biol. Chem.* **2009**, *284*, 18411–18423. [CrossRef] [PubMed]
94. Wang, Z.; Collighan, R.J.; Pytel, K.; Rathbone, D.L.; Li, X.; Griffin, M. Characterization of Heparin-Binding Site of Tissue Transglutaminase: Its Importance in Cell Surface Targeting, Matrix Deposition, and Cell Signaling. *J. Biol. Chem.* **2012**, *287*, 13063–13083. [CrossRef] [PubMed]
95. Scarpellini, A.; Huang, L.; Burhan, I.; Schroeder, N.; Funck, M.; Johnson, T.S.; Verderio, E.A. Syndecan-4 Knockout Leads to Reduced Extracellular Transglutaminase-2 and Protects Against Tubulointerstitial Fibrosis. *J. Am. Soc. Nephrol.* **2014**, *25*, 1013–1027. [CrossRef] [PubMed]
96. Antonyak, M.A.; Li, B.; Boroughs, L.K.; Johnson, J.L.; Druso, J.E.; Bryant, K.L.; Holowka, D.A.; Cerione, R.A. Cancer Cell-Derived Microvesicles Induce Transformation by Transferring Tissue Transglutaminase and Fibronectin to Recipient Cells. *Proc. Natl. Acad. Sci. USA* **2011**, *108*, 4852–4857. [CrossRef] [PubMed]
97. Diaz-Hidalgo, L.; Altuntas, S.; Rossin, F.; D'Eletto, M.; Marsella, C.; Farrace, M.G.; Falasca, L.; Antonioli, M.; Fimia, G.M.; Piacentini, M. Transglutaminase Type 2-Dependent Selective Recruitment of Proteins into Exosomes Under Stressful Cellular Conditions. *Biochim. Biophys. Acta* **2016**, *1863*, 2084–2092. [CrossRef] [PubMed]
98. Furini, G.; Schroeder, N.; Huang, L.; Boocock, D.; Scarpellini, A.; Coveney, C.; Tonoli, E.; Ramaswamy, R.; Ball, G.; Verderio, C.; et al. Proteomic Profiling Reveals the Transglutaminase-2 Externalization Pathway in Kidneys After Unilateral Ureteric Obstruction. *J. Am. Soc. Nephrol.* **2018**, *29*, 880–905. [CrossRef] [PubMed]
99. Verderio, E.A.; Telci, D.; Okoye, A.; Melino, G.; Griffin, M. A Novel RGD-Independent Cel Adhesion Pathway Mediated by Fibronectin-Bound Tissue Transglutaminase Rescues Cells from Anoikis. *J. Biol. Chem.* **2003**, *278*, 42604–42614. [CrossRef] [PubMed]
100. Telci, D.; Wang, Z.; Li, X.; Verderio, E.A.; Humphries, M.J.; Baccarini, M.; Basaga, H.; Griffin, M. Fibronectin-Tissue Transglutaminase Matrix Rescues RGD-Impaired Cell Adhesion through Syndecan-4 and Beta1 Integrin Co-Signaling. *J. Biol. Chem.* **2008**, *283*, 20937–20947. [CrossRef] [PubMed]
101. Teesalu, K.; Uibo, O.; Uibo, R.; Utt, M. Kinetic and Functional Characterisation of the Heparin-binding Peptides from Human Transglutaminase 2. *J. Pept. Sci.* **2012**, *18*, 350–356. [CrossRef] [PubMed]
102. Coussons, P.J.; Price, N.C.; Kelly, S.M.; Smith, B.; Sawyer, L. Factors that Govern the Specificity of Transglutaminase-Catalysed Modification of Proteins and Peptides. *Biochem. J.* **1992**, *282 Pt 3*, 929–930. [CrossRef]
103. Sugimura, Y.; Hosono, M.; Wada, F.; Yoshimura, T.; Maki, M.; Hitomi, K. Screening for the Preferred Substrate Sequence of Transglutaminase using a Phage-Displayed Peptide Library: Identification of Peptide Substrates for TGASE 2 and Factor XIIIA. *J. Biol. Chem.* **2006**, *281*, 17699–17706. [CrossRef] [PubMed]
104. Keresztessy, Z.; Csosz, E.; Harsfalvi, J.; Csomos, K.; Gray, J.; Lightowlers, R.N.; Lakey, J.H.; Balajthy, Z.; Fesus, L. Phage Display Selection of Efficient Glutamine-Donor Substrate Peptides for Transglutaminase 2. *Protein Sci.* **2006**, *15*, 2466–2480. [CrossRef] [PubMed]
105. Malesevic, M.; Migge, A.; Hertel, T.C.; Pietzsch, M. A Fluorescence-Based Array Screen for Transglutaminase Substrates. *Chembiochem* **2015**, *16*, 1169–1174. [CrossRef] [PubMed]
106. Caporale, A.; Selis, F.; Sandomenico, A.; Jotti, G.S.; Tonon, G.; Ruvo, M. The LQSP Tetrapeptide is a New Highly Efficient Substrate of Microbial Transglutaminase for the Site-Specific Derivatization of Peptides and Proteins. *Biotechnol. J.* **2015**, *10*, 154–161. [CrossRef] [PubMed]
107. Csosz, E.; Mesko, B.; Fesus, L. Transdab Wiki: The Interactive Transglutaminase Substrate Database on Web 2.0 Surface. *Amino Acids* **2009**, *36*, 615–617. [CrossRef] [PubMed]
108. Facchiano, A.M.; Facchiano, A.; Facchiano, F. Active Sequences Collection (ASC) Database: A New Tool to Assign Functions to Protein Sequences. *Nucleic Acids Res.* **2003**, *31*, 379–382. [CrossRef] [PubMed]

109. Itoh, M.; Kawamoto, T.; Tatsukawa, H.; Kojima, S.; Yamanishi, K.; Hitomi, K. In Situ Detection of Active Transglutaminases for Keratinocyte Type (TGase 1) and Tissue Type (TGase 2) using Fluorescence-Labeled Highly Reactive Substrate Peptides. *J. Histochem. Cytochem.* **2011**, *59*, 180–187. [CrossRef] [PubMed]
110. Watanabe, K.; Tsunoda, K.; Itoh, M.; Fukui, M.; Mori, H.; Hitomi, K. Transglutaminase 2 and Factor XIII Catalyze Distinct Substrates in Differentiating Osteoblastic Cell Line: Utility of Highly Reactive Substrate Peptides. *Amino Acids* **2013**, *44*, 209–214. [CrossRef] [PubMed]
111. Feeney, R.E.; Whitaker, J.R. Food Proteins. Lmprovement through Chemical and Enzymatic Modification. *Chem. Inform.* **1977**, *8*, 95.
112. Motoki, M.; Nio, N.; Takinami, K. Functional Properties of Food Proteins Polymerized by Transglutaminase. *Agric. Biol. Chem.* **1984**, *48*, 1257–1261.
113. Nio, N.; Motoki, M.; Takinami, K. Gelation of Casein and Soybean Globulins by Transglutaminase. *Agric. Biol. Chem.* **1985**, *49*, 2283–2286.
114. Nonaka, M.; Sakamoto, H.; Toiguchi, S.; Kawajiri, H.; Soeda, T.; Motoki, M. Sodium Caseinate and Skim Milk Gels Formed by Incubation with Microbial Transglutaminase. *J. Food Sci.* **1992**, *57*, 1214–1241. [CrossRef]
115. Ikura, K.; Yoshikawa, M.; Sasaki, R.; Chiba, H. Incorporation of Amino Acids into Food Proteins by Transglutaminase. *Agric. Biol. Chem.* **1981**, *45*, 2587–2592.
116. Kurth, L.; Rogers, P. Transglutaminase Catalyzed Cross-linking of Myosin to Soya Protein, Casein and Gluten. *J. Food Sci.* **1984**, *49*, 573–576. [CrossRef]
117. Aboumahmoud, R.; Savello, P. Crosslinking of Whey Protein by Transglutaminase. *J. Dairy Sci.* **1990**, *73*, 256–263. [CrossRef]
118. Ikura, K.; Kometani, T.; Yoshikawa, M.; Sasaki, R.; Chiba, H. Crosslinking of Casein Components by Transglutaminase. *Agric. Biol. Chem.* **1980**, *44*, 1567–1573.
119. Motoki, M.; Nio, N. Crosslinking between Different Food Proteins by Transglutaminase. *J. Food Sci.* **1983**, *48*, 561–566. [CrossRef]
120. Bercovici, D.; Gaertner, H.F.; Puigserver, A.J. Transglutaminase-Catalyzed Incorporation of Lysine Oligomers into Casein. *J. Agric. Food Chem.* **1987**, *35*, 301–304. [CrossRef]
121. Ikura, K.; Kometani, T.; Sasaki, R.; Chiba, H. Crosslinking of Soybean 7S and 11S Proteins by Transglutaminase. *Agric. Biol. Chem.* **1980**, *44*, 2979–2984.
122. Alexandre, M.C.; Popineau, Y.; Viroben, G.; Chiarello, M.; Lelion, A.; Gueguen, J. Wheat. Gamma. Gliadin as Substrate for Bovine Plasma Factor XIII. *J. Agric. Food Chem.* **1993**, *41*, 2208–2214. [CrossRef]
123. Gerrard, J.; Fayle, S.; Wilson, A.; Newberry, M.; Ross, M.; Kavale, S. Dough Properties and Crumb Strength of White Pan Bread as Affected by Microbial Transglutaminase. *J. Food Sci.* **1998**, *63*, 472–475. [CrossRef]
124. Cohen, I.; Young-Bandala, L.; Blankenberg, T.A.; Siefring, G.E., Jr.; Bruner-Lorand, J. Fibrinoligase-Catalyzed Cross-Linking of Myosin from Platelet and Skeletal Muscle. *Arch. Biochem. Biophys.* **1979**, *192*, 100–111. [CrossRef]
125. Kahn, D.R.; Cohen, I. Factor XIIIa-Catalyzed Coupling of Structural Proteins. *Biochim. Biophys. Acta* **1981**, *668*, 490–494. [CrossRef]
126. Giosafatto, C.; Rigby, N.; Wellner, N.; Ridout, M.; Husband, F.; Mackie, A. Microbial Transglutaminase-Mediated Modification of Ovalbumin. *Food Hydrocoll.* **2012**, *26*, 261–267. [CrossRef]
127. Seki, N.; Uno, H.; Lee, N.H.; Kimura, I.; Toyoda, K.; Fujita, T.; Arai, K. Transglutaminase Activity in Alaska Pollack Muscle and Surimi [Minced Fish Meat], and Its Reaction with Myosin B [Purified from Carp]. In *Bulletin of the Japanese Society of Scientific Fisheries*; JSFS: Tokyo, Japan, 1990.
128. Maruyama, N.; Nozawa, H.; Kimura, I.; Satake, M.; Seki, N. Transglutaminase-Induced Polymerization of a Mixture of Diffrent Fish Myosins. *Fish. Sci.* **1995**, *61*, 495–500. [CrossRef]
129. Kuraishi, C.; Sakamoto, J.; Yamazaki, K.; Susa, Y.; Kuhara, C.; Soeda, T. Production of Restructured Meat using Microbial Transglutaminase without Salt Or Cooking. *J. Food Sci.* **1997**, *62*, 488–490. [CrossRef]
130. Ahhmed, A.M.; Kuroda, R.; Kawahara, S.; Ohta, K.; Nakade, K.; Aoki, T.; Muguruma, M. Dependence of Microbial Transglutaminase on Meat Type in Myofibrillar Proteins Cross-Linking. *Food Chem.* **2009**, *112*, 354–361. [CrossRef]
131. Canto, A.C.; Lima, B.R.; Suman, S.P.; Lazaro, C.A.; Monteiro, M.L.; Conte-Junior, C.A.; Freitas, M.Q.; Cruz, A.G.; Santos, E.B.; Silva, T.J. Physico-Chemical and Sensory Attributes of Low-Sodium Restructured Caiman Steaks Containing Microbial Transglutaminase and Salt Replacers. *Meat Sci.* **2014**, *96*, 623–632. [CrossRef] [PubMed]

132. Lesiow, T.; Rentfrow, G.K.; Xiong, Y.L. Polyphosphate and Myofibrillar Protein Extract Promote Transglutaminase-Mediated Enhancements of Rheological and Textural Properties of PSE Pork Meat Batters. *Meat Sci.* **2017**, *128*, 40–46. [CrossRef] [PubMed]
133. Wang, X.; Xiong, Y.L.; Sato, H. Rheological Enhancement of Pork Myofibrillar Protein–Lipid Emulsion Composite Gels Via Glucose Oxidase Oxidation/Transglutaminase Cross-Linking Pathway. *J. Agric. Food Chem.* **2017**, *65*, 8451–8458. [CrossRef] [PubMed]
134. Sorapukdee, S.; Tangwatcharin, P. Quality of Steak Restructured from Beef Trimmings Containing Microbial Transglutaminase and Impacted by Freezing and Grading by Fat Level. *Asian-Aust. J. Anim. Sci.* **2018**, *31*, 129–137. [CrossRef] [PubMed]
135. Lauber, S.; Henle, T.; Klostermeyer, H. Relationship between the Crosslinking of Caseins by Transglutaminase and the Gel Strength of Yoghurt. *Eur. Food Res. Technol.* **2000**, *210*, 305–309. [CrossRef]
136. Abou-Soliman, N.H.I.; Sakr, S.S.; Awad, S. Physico-Chemical, Microstructural and Rheological Properties of Camel-Milk Yogurt as Enhanced by Microbial Transglutaminase. *J. Food Sci. Technol.* **2017**, *54*, 1616–1627. [CrossRef] [PubMed]
137. Garcia-Gomez, B.; Romero-Rodriguez, A.; Vazquez-Oderiz, L.; Munoz-Ferreiro, N.; Vazquez, M. Physicochemical Evaluation of Low-Fat Yoghurt Produced with Microbial Transglutaminase. *J. Sci. Food Agric.* **2018**. [CrossRef] [PubMed]
138. Caballero, P.A.; Gómez, M.; Rosell, C.M. Improvement of Dough Rheology, Bread Quality and Bread Shelf-Life by Enzymes Combination. *J. Food Eng.* **2007**, *81*, 42–53. [CrossRef]
139. Marco, C.; Rosell, C.M. Breadmaking Performance of Protein Enriched, Gluten-Free Breads. *Eur. Food Res. Technol.* **2008**, *227*, 1205–1213. [CrossRef]
140. Huang, W.; Li, L.; Wang, F.; Wan, J.; Tilley, M.; Ren, C.; Wu, S. Effects of Transglutaminase on the Rheological and Mixolab Thermomechanical Characteristics of Oat Dough. *Food Chem.* **2010**, *121*, 934–939. [CrossRef]
141. Grossmann, I.; Doring, C.; Jekle, M.; Becker, T.; Koehler, P. Compositional Changes and Baking Performance of Rye Dough as Affected by Microbial Transglutaminase and Xylanase. *J. Agric. Food Chem.* **2016**, *64*, 5751–5758. [CrossRef] [PubMed]
142. Niu, M.; Hou, G.G.; Kindelspire, J.; Krishnan, P.; Zhao, S. Microstructural, Textural, and Sensory Properties of Whole-Wheat Noodle Modified by Enzymes and Emulsifiers. *Food Chem.* **2017**, *223*, 16–24. [CrossRef] [PubMed]
143. Hadjivassiliou, M.; Aeschlimann, P.; Strigun, A.; Sanders, D.S.; Woodroofe, N.; Aeschlimann, D. Autoantibodies in Gluten Ataxia Recognize a Novel Neuronal Transglutaminase. *Ann. Neurol.* **2008**, *64*, 332–343. [CrossRef] [PubMed]
144. Cabrera-Chavez, F.; Rouzaud-Sandez, O.; Sotelo-Cruz, N.; Calderon de la Barca, A.M. Transglutaminase Treatment of Wheat and Maize Prolamins of Bread Increases the Serum IgA Reactivity of Celiac Disease Patients. *J. Agric. Food Chem.* **2008**, *56*, 1387–1391. [CrossRef] [PubMed]
145. Ruh, T.; Ohsam, J.; Pasternack, R.; Yokoyama, K.; Kumazawa, Y.; Hils, M. Microbial Transglutaminase Treatment in Pasta-Production does Not Affect the Immunoreactivity of Gliadin with Celiac Disease Patients' Sera. *J. Agric. Food Chem.* **2014**, *62*, 7604–7611. [CrossRef] [PubMed]
146. Heil, A.; Ohsam, J.; van Genugten, B.; Diez, O.; Yokoyama, K.; Kumazawa, Y.; Pasternack, R.; Hils, M. Microbial Transglutaminase used in Bread Preparation at Standard Bakery Concentrations does Not Increase Immunodetectable Amounts of Deamidated Gliadin. *J. Agric. Food Chem.* **2017**, *65*, 6982–6990. [CrossRef] [PubMed]
147. Di Pierro, P.; Chico, B.; Villalonga, R.; Mariniello, L.; Damiao, A.; Masi, P.; Porta, R. Chitosan-Whey Protein Edible Films Produced in the Absence Or Presence of Transglutaminase: Analysis of their Mechanical and Barrier Properties. *Biomacromolecules* **2006**, *7*, 744–749. [CrossRef] [PubMed]
148. Di Pierro, P.; Sorrentino, A.; Mariniello, L.; Giosafatto, C.V.L.; Porta, R. Chitosan/Whey Protein Film as Active Coating to Extend Ricotta Cheese Shelf-Life. *LWT Food Sci. Technol.* **2011**, *44*, 2324–2327. [CrossRef]
149. Lee, H.; Lanier, T.; Hamann, D.; Knopp, J. Transglutaminase Effects on Low Temperature Gelation of Fish Protein Sols. *J. Food Sci.* **1997**, *62*, 20–24. [CrossRef]
150. Fernandez-Dıaz, M.; Montero, P.; Gomez-Guillen, M. Gel Properties of Collagens from Skins of Cod (Gadus Morhua) and Hake (Merluccius Merluccius) and their Modification by the Coenhancers Magnesium Sulphate, Glycerol and Transglutaminase. *Food Chem.* **2001**, *74*, 161–167. [CrossRef]

151. Al-Hassan, A.; Norziah, M. Effect of Transglutaminase Induced Crosslinking on the Properties of Starch/Gelatin Films. *Food Packag. Shelf Life* **2017**, *13*, 15–19. [CrossRef]
152. Huang, T.; Tu, Z.C.; Shangguan, X.; Wang, H.; Zhang, N.; Zhang, L.; Sha, X. Gelation Kinetics and Characterization of Enzymatically Enhanced Fish Scale Gelatin-Pectin Coacervate. *J. Sci. Food Agric.* **2018**, *98*, 1024–1032. [CrossRef] [PubMed]
153. Hong, P.K.; Gottardi, D.; Ndagijimana, M.; Betti, M. Glycation and Transglutaminase Mediated Glycosylation of Fish Gelatin Peptides with Glucosamine Enhance Bioactivity. *Food Chem.* **2014**, *142*, 285–293. [CrossRef] [PubMed]
154. Miyata, T.; Taira, T.; Noishiki, Y. Collagen Engineering for Biomaterial Use. *Clin. Mater.* **1992**, *9*, 139–148. [CrossRef]
155. Rath, N.C.; Reddi, A.H. Collagenous Bone Matrix is a Local Mitogen. *Nature* **1979**, *278*, 855–857. [CrossRef] [PubMed]
156. Kleinman, H.K.; Klebe, R.J.; Martin, G.R. Role of Collagenous Matrices in the Adhesion and Growth of Cells. *J. Cell Biol.* **1981**, *88*, 473–485. [CrossRef] [PubMed]
157. Nimni, M.E. A Defect in the Intramolecular and Intermolecular Cross-Linking of Collagen Caused by Penicillamine. I. Metabolic and Functional Abnormalities in Soft Tissues. *J. Biol. Chem.* **1968**, *243*, 1457–1466. [PubMed]
158. Damink, L.O.; Dijkstra, P.J.; Van Luyn, M.; Van Wachem, P.; Nieuwenhuis, P.; Feijen, J. Glutaraldehyde as a Crosslinking Agent for Collagen-Based Biomaterials. *J. Mater. Sci. Mater. Med.* **1995**, *6*, 460–472. [CrossRef]
159. Speer, D.P.; Chvapil, M.; Eskelson, C.D.; Ulreich, J. Biological Effects of Residual Glutaraldehyde in Glutaraldehyde-Tanned Collagen Biomaterials. *J. Biomed. Mater. Res.* **1980**, *14*, 753–764. [CrossRef] [PubMed]
160. Huang-Lee, L.L.; Cheung, D.T.; Nimni, M.E. Biochemical Changes and Cytotoxicity Associated with the Degradation of Polymeric Glutaraldehyde Derived Crosslinks. *J. Biomed. Mater. Res.* **1990**, *24*, 1185–1201. [CrossRef] [PubMed]
161. Crescenzi, V.; Francescangeli, A.; Taglienti, A. New Gelatin-Based Hydrogels via Enzymatic Networking. *Biomacromolecules* **2002**, *3*, 1384–1391. [CrossRef] [PubMed]
162. Liu, L.; Wen, H.; Rao, Z.; Zhu, C.; Liu, M.; Min, L.; Fan, L.; Tao, S. Preparation and Characterization of Chitosan—Collagen Peptide/Oxidized Konjac Glucomannan Hydrogel. *Int. J. Biol. Macromol.* **2018**, *108*, 376–382. [CrossRef] [PubMed]
163. Valero, C.; Amaveda, H.; Mora, M.; Garcia-Aznar, J.M. Combined Experimental and Computational Characterization of Crosslinked Collagen-Based Hydrogels. *PLoS ONE* **2018**, *13*, e0195820. [CrossRef] [PubMed]
164. Mosher, D.F.; Schad, P.E. Cross-Linking of Fibronectin to Collagen by Blood Coagulation Factor XIIIa. *J. Clin. Investig.* **1979**, *64*, 781–787. [CrossRef] [PubMed]
165. Stachel, I.; Schwarzenbolz, U.; Henle, T.; Meyer, M. Cross-Linking of Type I Collagen with Microbial Transglutaminase: Identification of Cross-Linking Sites. *Biomacromolecules* **2010**, *11*, 698–705. [CrossRef] [PubMed]
166. Bowness, J.M.; Folk, J.E.; Timpl, R. Identification of a Substrate Site for Liver Transglutaminase on the Aminopropeptide of Type III Collagen. *J. Biol. Chem.* **1987**, *262*, 1022–1024. [PubMed]
167. Shanmugasundaram, S.; Logan-Mauney, S.; Burgos, K.; Nurminskaya, M. Tissue Transglutaminase Regulates Chondrogenesis in Mesenchymal Stem Cells on Collagen Type XI Matrices. *Amino Acids* **2012**, *42*, 1045–1053. [CrossRef] [PubMed]
168. Halloran, D.M.; Collighan, R.J.; Griffin, M.; Pandit, A.S. Characterization of a Microbial Transglutaminase Cross-Linked Type II Collagen Scaffold. *Tissue Eng.* **2006**, *12*, 1467–1474. [CrossRef] [PubMed]
169. Orban, J.M.; Wilson, L.B.; Kofroth, J.A.; El-Kurdi, M.S.; Maul, T.M.; Vorp, D.A. Crosslinking of Collagen Gels by Transglutaminase. *J. Biomed. Mater. Res. Part A* **2004**, *68*, 756–762. [CrossRef] [PubMed]
170. Yang, G.; Xiao, Z.; Long, H.; Ma, K.; Zhang, J.; Ren, X.; Zhang, J. Assessment of the Characteristics and Biocompatibility of Gelatin Sponge Scaffolds Prepared by various Crosslinking Methods. *Sci. Rep.* **2018**, *8*, 1616. [CrossRef] [PubMed]
171. Verderio, E.; Coombes, A.; Jones, R.A.; Li, X.; Heath, D.; Downes, S.; Griffin, M. Role of the Cross-Linking Enzyme Tissue Transglutaminase in the Biological Recognition of Synthetic Biodegradable Polymers. *J. Biomed. Mater. Res.* **2001**, *54*, 294–304. [CrossRef]

172. Zhao, L.; Li, X.; Zhao, J.; Ma, S.; Ma, X.; Fan, D.; Zhu, C.; Liu, Y. A Novel Smart Injectable Hydrogel Prepared by Microbial Transglutaminase and Human-Like Collagen: Its Characterization and Biocompatibility. *Mater. Sci. Eng. C Mater. Biol. Appl.* **2016**, *68*, 317–326. [CrossRef] [PubMed]
173. Chen, R.N.; Ho, H.O.; Sheu, M.T. Characterization of Collagen Matrices Crosslinked using Microbial Transglutaminase. *Biomaterials* **2005**, *26*, 4229–4235. [CrossRef] [PubMed]
174. Chau, D.Y.; Brown, S.V.; Mather, M.L.; Hutter, V.; Tint, N.L.; Dua, H.S.; Rose, F.R.; Ghaemmaghami, A.M. Tissue Transglutaminase (TG-2) Modified Amniotic Membrane: A Novel Scaffold for Biomedical Applications. *Biomed. Mater.* **2012**. [CrossRef] [PubMed]
175. Hoac, B.; Nelea, V.; Jiang, W.; Kaartinen, M.T.; McKee, M.D. Mineralization-Inhibiting Effects of Transglutaminase-Crosslinked Polymeric Osteopontin. *Bone* **2017**, *101*, 37–48. [CrossRef] [PubMed]
176. Cavelier, S.; Dastjerdi, A.K.; McKee, M.D.; Barthelat, F. Bone Toughness at the Molecular Scale: A Model for Fracture Toughness using Crosslinked Osteopontin on Synthetic and Biogenic Mineral Substrates. *Bone* **2018**, *110*, 304–311. [CrossRef] [PubMed]
177. Sharma, A.; Brand, D.; Fairbank, J.; Ye, H.; Lavy, C.; Czernuszka, J. A Self-Organising Biomimetic Collagen/Nano-Hydroxyapatite-Glycosaminoglycan Scaffold for Spinal Fusion. *J. Mater. Sci.* **2017**, *52*, 12574–12592. [CrossRef] [PubMed]
178. Fortunati, D.; Chau, D.Y.; Wang, Z.; Collighan, R.J.; Griffin, M. Cross-Linking of Collagen I by Tissue Transglutaminase Provides a Promising Biomaterial for Promoting Bone Healing. *Amino Acids* **2014**, *46*, 1751–1761. [CrossRef] [PubMed]
179. Paige, K.T.; Cima, L.G.; Yaremchuk, M.J.; Vacanti, J.P.; Vacanti, C.A. Injectable Cartilage. *Plast. Reconstr. Surg.* **1995**, *96*, 1390–1398. [CrossRef] [PubMed]
180. Smeds, K.A.; Pfister-Serres, A.; Miki, D.; Dastgheib, K.; Inoue, M.; Hatchell, D.L.; Grinstaff, M.W. Photocrosslinkable Polysaccharides for in Situ Hydrogel Formation. *J. Biomed. Mater. Res.* **2001**, *54*, 115–121. [CrossRef]
181. Nettles, D.L.; Vail, T.P.; Morgan, M.T.; Grinstaff, M.W.; Setton, L.A. Photocrosslinkable Hyaluronan as a Scaffold for Articular Cartilage Repair. *Ann. Biomed. Eng.* **2004**, *32*, 391–397. [CrossRef] [PubMed]
182. Ehrbar, M.; Metters, A.; Zammaretti, P.; Hubbell, J.A.; Zisch, A.H. Endothelial Cell Proliferation and Progenitor Maturation by Fibrin-Bound VEGF Variants with Differential Susceptibilities to Local Cellular Activity. *J. Control. Release* **2005**, *101*, 93–109. [CrossRef] [PubMed]
183. Yung, C.W.; Wu, L.Q.; Tullman, J.A.; Payne, G.F.; Bentley, W.E.; Barbari, T.A. Transglutaminase Crosslinked Gelatin as a Tissue Engineering Scaffold. *J. Biomed. Mater. Res. Part A* **2007**, *83*, 1039–1046. [CrossRef] [PubMed]
184. Ranga, A.; Lutolf, M.P.; Hilborn, J.; Ossipov, D.A. Hyaluronic Acid Hydrogels Formed in Situ by Transglutaminase-Catalyzed Reaction. *Biomacromolecules* **2016**, *17*, 1553–1560. [CrossRef] [PubMed]
185. Panitch, A.; Yamaoka, T.; Fournier, M.J.; Mason, T.L.; Tirrell, D.A. Design and Biosynthesis of Elastin-Like Artificial Extracellular Matrix Proteins Containing Periodically Spaced Fibronectin CS5 Domains. *Macromolecules* **1999**, *32*, 1701–1703. [CrossRef]
186. Welsh, E.R.; Tirrell, D.A. Engineering the Extracellular Matrix: A Novel Approach to Polymeric Biomaterials. I. Control of the Physical Properties of Artificial Protein Matrices Designed to Support Adhesion of Vascular Endothelial Cells. *Biomacromolecules* **2000**, *1*, 23–30. [CrossRef] [PubMed]
187. Betre, H.; Setton, L.A.; Meyer, D.E.; Chilkoti, A. Characterization of a Genetically Engineered Elastin-Like Polypeptide for Cartilaginous Tissue Repair. *Biomacromolecules* **2002**, *3*, 910–916. [CrossRef] [PubMed]
188. Heilshorn, S.C.; DiZio, K.A.; Welsh, E.R.; Tirrell, D.A. Endothelial Cell Adhesion to the Fibronectin CS5 Domain in Artificial Extracellular Matrix Proteins. *Biomaterials* **2003**, *24*, 4245–4252. [CrossRef]
189. Bandiera, A. Transglutaminase-Catalyzed Preparation of Human Elastin-Like Polypeptide-Based Three-Dimensional Matrices for Cell Encapsulation. *Enzym. Microb. Technol.* **2011**, *49*, 347–352. [CrossRef] [PubMed]
190. Bozzini, S.; Giuliano, L.; Altomare, L.; Petrini, P.; Bandiera, A.; Conconi, M.T.; Fare, S.; Tanzi, M.C. Enzymatic Cross-Linking of Human Recombinant Elastin (HELP) as Biomimetic Approach in Vascular Tissue Engineering. *J. Mater. Sci. Mater. Med.* **2011**, *22*, 2641–2650. [CrossRef] [PubMed]
191. McHale, M.K.; Setton, L.A.; Chilkoti, A. Synthesis and in Vitro Evaluation of Enzymatically Cross-Linked Elastin-Like Polypeptide Gels for Cartilaginous Tissue Repair. *Tissue Eng.* **2005**, *11*, 1768–1779. [CrossRef] [PubMed]

192. Jurgensen, K.; Aeschlimann, D.; Cavin, V.; Genge, M.; Hunziker, E. A New Biological Glue for Cartilage-Cartilage Interfaces: Tissue Transglutaminase. *JBJS* **1997**, *79*, 185–193. [CrossRef]
193. Kim, Y.; Gill, E.E.; Liu, J.C. Enzymatic Cross-Linking of Resilin-Based Proteins for Vascular Tissue Engineering Applications. *Biomacromolecules* **2016**, *17*, 2530–2539. [CrossRef] [PubMed]
194. Huppertz, T.; Smiddy, M.A.; de Kruif, C.G. Biocompatible Micro-Gel Particles from Cross-Linked Casein Micelles. *Biomacromolecules* **2007**, *8*, 1300–1305. [CrossRef] [PubMed]
195. Huppertz, T.; de Kruif, C.G. Structure and Stability of Nanogel Particles Prepared by Internal Cross-Linking of Casein Micelles. *Int. Dairy J.* **2008**, *18*, 556–565. [CrossRef]
196. Zhang, Z.; Decker, E.A.; McClements, D.J. Encapsulation, Protection, and Release of Polyunsaturated Lipids using Biopolymer-Based Hydrogel Particles. *Food Res. Int.* **2014**, *64*, 520–526. [CrossRef] [PubMed]
197. McClements, D.J. Designing Biopolymer Microgels to Encapsulate, Protect and Deliver Bioactive Components: Physicochemical Aspects. *Adv. Colloid Interface Sci.* **2017**, *240*, 31–59. [CrossRef] [PubMed]
198. Tian, W.M.; Hou, S.P.; Ma, J.; Zhang, C.L.; Xu, Q.Y.; Lee, I.S.; Li, H.D.; Spector, M.; Cui, F.Z. Hyaluronic Acid-Poly-D-Lysine-Based Three-Dimensional Hydrogel for Traumatic Brain Injury. *Tissue Eng.* **2005**, *11*, 513–525. [CrossRef] [PubMed]
199. Ma, J.; Tian, W.M.; Hou, S.P.; Xu, Q.Y.; Spector, M.; Cui, F.Z. An Experimental Test of Stroke Recovery by Implanting a Hyaluronic Acid Hydrogel Carrying a Nogo Receptor Antibody in a Rat Model. *Biomed. Mater.* **2007**, *2*, 233–240. [CrossRef] [PubMed]
200. Broguiere, N.; Isenmann, L.; Zenobi-Wong, M. Novel Enzymatically Cross-Linked Hyaluronan Hydrogels Support the Formation of 3D Neuronal Networks. *Biomaterials* **2016**, *99*, 47–55. [CrossRef] [PubMed]
201. Hu, B.H.; Messersmith, P.B. Rational Design of Transglutaminase Substrate Peptides for Rapid Enzymatic Formation of Hydrogels. *J. Am. Chem. Soc.* **2003**, *125*, 14298–14299. [CrossRef] [PubMed]
202. Hu, B.H.; Messersmith, P.B. Enzymatically Cross-Linked Hydrogels and their Adhesive Strength to Biosurfaces. *Orthod. Craniofac. Res.* **2005**, *8*, 145–149. [CrossRef] [PubMed]
203. Abuchowski, A.; van Es, T.; Palczuk, N.C.; Davis, F.F. Alteration of Immunological Properties of Bovine Serum Albumin by Covalent Attachment of Polyethylene Glycol. *J. Biol. Chem.* **1977**, *252*, 3578–3581. [PubMed]
204. Abuchowski, A.; McCoy, J.R.; Palczuk, N.C.; van Es, T.; Davis, F.F. Effect of Covalent Attachment of Polyethylene Glycol on Immunogenicity and Circulating Life of Bovine Liver Catalase. *J. Biol. Chem.* **1977**, *252*, 3582–3586. [PubMed]
205. Gorman, J.J.; Folk, J.E. Structural Features of Glutamine Substrates for Transglutaminases. Role of Extended Interactions in the Specificity of Human Plasma Factor XIIIa and of the Guinea Pig Liver Enzyme. *J. Biol. Chem.* **1984**, *259*, 9007–9010. [PubMed]
206. Sato, H. Enzymatic Procedure for Site-Specific Pegylation of Proteins. *Adv. Drug Deliv. Rev.* **2002**, *54*, 487–504. [CrossRef]
207. Mero, A.; Schiavon, M.; Veronese, F.M.; Pasut, G. A New Method to Increase Selectivity of Transglutaminase Mediated PEGylation of Salmon Calcitonin and Human Growth Hormone. *J. Control. Release* **2011**, *154*, 27–34. [CrossRef] [PubMed]
208. Khameneh, B.; Jaafari, M.R.; Hassanzadeh-Khayyat, M.; Varasteh, A.; Chamani, J.; Iranshahi, M.; Mohammadpanah, H.; Abnous, K.; Saberi, M.R. Preparation, Characterization and Molecular Modeling of PEGylated Human Growth Hormone with Agonist Activity. *Int. J. Biol. Macromol.* **2015**, *80*, 400–409. [CrossRef] [PubMed]
209. Spolaore, B.; Raboni, S.; Satwekar, A.A.; Grigoletto, A.; Mero, A.; Montagner, I.M.; Rosato, A.; Pasut, G.; Fontana, A. Site-Specific Transglutaminase-Mediated Conjugation of Interferon Alpha-2b at Glutamine Or Lysine Residues. *Bioconjug. Chem.* **2016**, *27*, 2695–2706. [CrossRef] [PubMed]
210. Braun, A.C.; Gutmann, M.; Mueller, T.D.; Luhmann, T.; Meinel, L. Bioresponsive Release of Insulin-Like Growth Factor-I from its PEGylated Conjugate. *J. Control. Release* **2018**, *279*, 17–28. [CrossRef] [PubMed]
211. Abe, H.; Goto, M.; Kamiya, N. Protein Lipidation Catalyzed by Microbial Transglutaminase. *Chemistry* **2011**, *17*, 14004–14008. [CrossRef] [PubMed]
212. Folk, J.E.; Cole, P.W. Mechanism of Action of Guinea Pig Liver Transglutaminase. I. Purification and Properties of the Enzyme: Identification of a Functional Cysteine Essential for Activity. *J. Biol. Chem.* **1966**, *241*, 5518–5525. [PubMed]

213. Tominaga, J.; Kemori, Y.; Tanaka, Y.; Maruyama, T.; Kamiya, N.; Goto, M. An Enzymatic Method for Site-Specific Labeling of Recombinant Proteins with Oligonucleotides. *Chem. Commun.* **2007**, *4*, 401–403. [CrossRef] [PubMed]
214. Kitaoka, M.; Tsuruda, Y.; Tanaka, Y.; Goto, M.; Mitsumori, M.; Hayashi, K.; Hiraishi, Y.; Miyawaki, K.; Noji, S.; Kamiya, N. Transglutaminase-Mediated Synthesis of a DNA-(Enzyme)N Probe for Highly Sensitive DNA Detection. *Chemistry* **2011**, *17*, 5387–5392. [CrossRef] [PubMed]
215. Kitaoka, M.; Mitsumori, M.; Hayashi, K.; Hiraishi, Y.; Yoshinaga, H.; Nakano, K.; Miyawaki, K.; Noji, S.; Goto, M.; Kamiya, N. Transglutaminase-Mediated in Situ Hybridization (TransISH) System: A New Methodology for Simplified mRNA Detection. *Anal. Chem.* **2012**, *84*, 5885–5891. [CrossRef] [PubMed]
216. Takahara, M.; Hayashi, K.; Goto, M.; Kamiya, N. Tailing DNA Aptamers with a Functional Protein by Two-Step Enzymatic Reaction. *J. Biosci. Bioeng.* **2013**, *116*, 660–665. [CrossRef] [PubMed]
217. Takahara, M.; Wakabayashi, R.; Minamihata, K.; Goto, M.; Kamiya, N. Primary Amine-Clustered DNA Aptamer for DNA-Protein Conjugation Catalyzed by Microbial Transglutaminase. *Bioconjug. Chem.* **2017**, *28*, 2954–2961. [CrossRef] [PubMed]
218. Gnaccarini, C.; Ben-Tahar, W.; Mulani, A.; Roy, I.; Lubell, W.D.; Pelletier, J.N.; Keillor, J.W. Site-Specific Protein Propargylation using Tissue Transglutaminase. *Org. Biomol. Chem.* **2012**, *10*, 5258–5265. [CrossRef] [PubMed]
219. Oteng-Pabi, S.K.; Pardin, C.; Stoica, M.; Keillor, J.W. Site-Specific Protein Labelling and Immobilization Mediated by Microbial Transglutaminase. *Chem. Commun.* **2014**, *50*, 6604–6606. [CrossRef] [PubMed]
220. Guy, J.; Castonguay, R.; Campos-Reales Pineda, N.B.; Jacquier, V.; Caron, K.; Michnick, S.W.; Keillor, J.W. De Novo Helical Peptides as Target Sequences for a Specific, Fluorogenic Protein Labelling Strategy. *Mol. Biosyst.* **2010**, *6*, 976–987. [CrossRef] [PubMed]
221. Oteng-Pabi, S.K.; Clouthier, C.M.; Keillor, J.W. Design of a Glutamine Substrate Tag Enabling Protein Labelling Mediated by Bacillus Subtilis Transglutaminase. *PLoS ONE* **2018**, *13*, e0197956. [CrossRef] [PubMed]
222. Josten, A.; Haalck, L.; Spener, F.; Meusel, M. Use of Microbial Transglutaminase for the Enzymatic Biotinylation of Antibodies. *J. Immunol. Methods* **2000**, *240*, 47–54. [CrossRef]
223. Jeger, S.; Zimmermann, K.; Blanc, A.; Grunberg, J.; Honer, M.; Hunziker, P.; Struthers, H.; Schibli, R. Site-Specific and Stoichiometric Modification of Antibodies by Bacterial Transglutaminase. *Angew. Chem. Int. Ed.* **2010**, *49*, 9995–9997. [CrossRef] [PubMed]
224. Anami, Y.; Xiong, W.; Gui, X.; Deng, M.; Zhang, C.C.; Zhang, N.; An, Z.; Tsuchikama, K. Enzymatic Conjugation using Branched Linkers for Constructing Homogeneous Antibody-Drug Conjugates with High Potency. *Org. Biomol. Chem.* **2017**, *15*, 5635–5642. [CrossRef] [PubMed]
225. Mindt, T.L.; Jungi, V.; Wyss, S.; Friedli, A.; Pla, G.; Novak-Hofer, I.; Grunberg, J.; Schibli, R. Modification of Different IgG1 Antibodies Via Glutamine and Lysine using Bacterial and Human Tissue Transglutaminase. *Bioconjug. Chem.* **2008**, *19*, 271–278. [CrossRef] [PubMed]

Review

Progress in Photo-Responsive Polypeptide Derived Nano-Assemblies

Lu Yang [1], Houliang Tang [2] and Hao Sun [1,*]

[1] Department of Chemistry, University of Florida, Gainesville, FL 32611, USA; yanglulucia@chem.ufl.edu
[2] Department of Chemistry, Southern Methodist University, Dallas, TX 75275, USA; houliangt@smu.edu
* Correspondence: chrisun@ufl.edu; Tel.: +1-352-281-4799

Received: 25 May 2018; Accepted: 11 June 2018; Published: 13 June 2018

Abstract: Stimuli-responsive polymeric materials have attracted significant attention in a variety of high-value-added and industrial applications during the past decade. Among various stimuli, light is of particular interest as a stimulus because of its unique advantages, such as precisely spatiotemporal control, mild conditions, ease of use, and tunability. In recent years, a lot of effort towards the synthesis of a biocompatible and biodegradable polypeptide has resulted in many examples of photo-responsive nanoparticles. Depending on the specific photochemistry, those polypeptide derived nano-assemblies are capable of crosslinking, disassembling, or morphing into other shapes upon light irradiation. In this review, we aim to assess the current state of photo-responsive polypeptide based nanomaterials. Firstly, those 'smart' nanomaterials will be categorized by their photo-triggered events (i.e., crosslinking, degradation, and isomerization), which are inherently governed by photo-sensitive functionalities, including *O*-nitrobenzyl, coumarin, azobenzene, cinnamyl, and spiropyran. In addition, the properties and applications of those polypeptide nanomaterials will be highlighted as well. Finally, the current challenges and future directions of this subject will be evaluated.

Keywords: stimuli-responsive polymers; synthetic polypeptide; photo-sensitive; self-assembly; morphological transformation

1. Introduction

Stimuli-responsive or 'Smart' polymers are capable of changing their physical and/or chemical properties upon receiving external triggers, such as temperature, pH, redox, mechanical forces, and light [1–9]. These tailor-made polymers are receiving significant interest in the fields of drug delivery, biosensor, tissue engineering, coatings, and self-healing materials [10–14]. In particular, light has recently garnered tremendous attention as a stimulus, as it can be not only triggered remotely but also provides spatiotemporal control [15–23]. Moreover, irradiation parameters, including wavelength, power, and time, can be easily tuned to fit the system (e.g., on-demand and controllable drug release rate) [24–27]. Typically, the ability of smart polymers to respond to light stems from the incorporation of photo-sensitive chemical structures [28–30]. Those moieties can be classified into three general categories based on their specific photo-chemistry (Scheme 1A–C). In the first category, represented by cinnamyl and coumarin, photo-induced dimerization of those groups takes place upon irradiation at a certain wavelength, while the dimer can undergo a reversal reaction at another wavelength with a higher energy (i.e., shorter wavelength). The second (e.g., *O*-nitrobenzyl) involves irreversible photo-triggered degradation, which can liberate the unprotected functionality, leading to a dramatic change in solvability. The last subset includes functional groups, such as azobenzene and spiropyran, which are capable of reversibly isomerizing under different wavelengths. By taking advantage of the above-mentioned photo-chemistry, various light-triggered morphological transformations (e.g., crosslinking, dissociation, and shape change) of polypeptide nano-assemblies have been achieved (Scheme 1D).

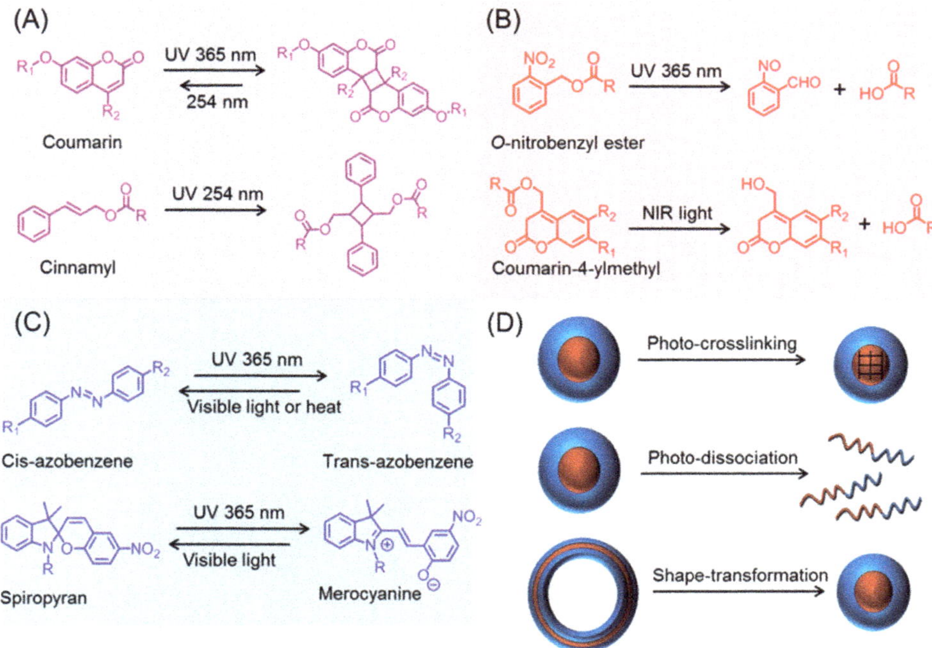

Scheme 1. Photo-chemistry of various functional groups. (**A**) Ultraviolet (UV)-induced dimerization; (**B**) UV or near infrared (NIR) promoted cleavage; (**C**) reversible isomerization by UV and visible light; and (**D**) photo-triggered metamorphosis of polypeptide derived nano-objects.

Inspired by natural protein, synthetic polypeptides or poly(amino acids) based nanomaterials are receiving increasing interest in the field of polymer science because of their inherent biocompatibility and biodegradability [31,32]. Furthermore, synthetic polypeptides have exhibited their unique ability to form higher order secondary structures, including α-helix, β-sheet, and β-turn, thanks to non-covalent interactions (i.e., hydrogen bonds, pi–pi stacking, and hydrophobic interaction) between amino acids side chains [33,34]. Those non-covalent interactions are highly sensitive to local environments, such as temperature, pH, the presence and concentration of reducing agent, ionic strength, and even light. A small change in the local environment could have a noticeable impact on non-covalent interactions, resulting in the transformation of secondary structures and concomitant change in bio-activity and function of polypeptides [35].

The rapid development of polymerization methodology has empowered polymer chemists with the ability to easily prepare unique polypeptides with diverse architecture and functionalities [36–39]. Numerous polypeptides have been successfully prepared via various living polymerization approaches, such as ring-opening polymerization of *N*-carboxyanhydrides (NCA) [40–43], reversible addition-fragmentation polymerization [38], atom transfer radical polymerization [44,45], and ring-opening metathesis polymerization [46–50]. In a typical case, living or controlled polymerization techniques are capable of producing polymers with precise chain lengths, excellent functionalities tolerance, and narrow polydispersity [51–55]. In addition, complicated architectures, such as block, cyclic, brush, and star, which were previously inaccessible, can now easily be made via living polymerization techniques [56–64]. Owing to those features arising from the living polymerizations (vide supra), one can design and tune the hydrophobic to hydrophilic balance, which dictates the critical packing parameters and give rise to nanoparticles with predictable morphologies (Scheme 2) [65].

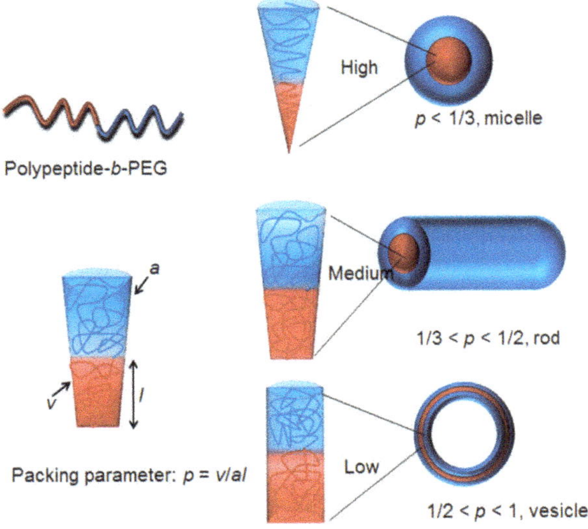

Scheme 2. The shapes of polypeptide derived nano-objects are dictated by critical packing parameters [65]. Diblock copolymer polypeptide-*b*-PEG exemplifies amphiphilic polypeptide based copolymers.

With the recent success of light-responsive amphiphilic polypeptides in nanotechnology and nanomedicine, we believe it is necessary to assess the current state of those smart nanomaterials. In this review, the main focus will be placed on the photo-chemistry of various light-sensitive functional groups that are incorporated into the polypeptide nanoparticles. Furthermore, we will discuss the influence on the size or morphologies of nano-assemblies as a consequence of light treatment and how this may assist the prediction of potential applications of those materials. Recent examples of photo-responsive polypeptide deriving nanoparticles are summarized in Table 1. Finally, we believe it is crucial to evaluate the current challenges and future directions of this field.

Table 1. Summary of photo-responsive polypeptide nano-assemblies.

Polypeptide	Synthetic Method	Photo-Responsive Moiety	Light-Triggered Events	Application	Ref.
PEG-*b*-P(LGA/CLG)	ROP [a]	Cinamyl	Micellar Core-Crosslink	Drug Delivery	[66]
PEG-*b*-PCLG	ROP	Cinamyl	Micellar Core-Crosslink	Drug Delivery	[67]
PEG-$CA_4LS_4Co_4$	Solution-PPS [b]	Coumarin	Micellar Core-Crosslink	Drug Delivery	[68]
PNBC-*b*-PEO	ROP	*O*-nitrobenzyl	Micellar Disassembly	Drug Delivery	[69]
PNBC-*b*-PEO	ROP	*O*-nitrobenzyl	Vesicle to Micelle	N/A	[70]
PNBC-*b*-PEO	ROP	*O*-nitrobenzyl	Composite Nanoparticle	Drug Delivery	[13]
PNBL-*b*-PLL	ROP	*O*-nitrobenzyl	Sol-Gel Transition	N/A	[71]
PNBL-*b*-PEO	ROP	*O*-nitrobenzyl	Sol-Gel Transition	N/A	[72]
PEtOx-*b*-P[MetNB][Br]	ROP	*O*-nitrobenzyl	Composite Nanoparticle	Gene Delivery	[73]
PEO-*b*-P(LGA-*co*-COU)	ROP	6-Bromo-7-hydroxyl-coumarine	Micellar Disassembly	Drug Delivery	[74]
Fmoc-Phe-pazoDbg	Solution-PPS	Azobenzene	Micelle to Fiber	N/A	[75]
PBLG-azobenzene	ROP	Azobenzene	Vesicle Disruption	Cargo Release	[76]
P(OEG-Azo)	ROP	Azobenzene	Sol-Gel Transition	N/A	[40]
Azo-GFGH	Solid-PPS	Azobenzene	Nanofiber Assembly	Catalysis	[77]
PLGASP-*b*-PEO	ROP	Spiropyran	Micellar Disassembly	Drug Delivery	[78]

Note: [a] ROP—ring-opening polymerization; [b] PPS—phase peptide synthesis.

2. Photo-Chemistry of Light-Responsive Polypeptide Nanoparticles

2.1. Photo-Crosslinkable Nanoparticles

Photo-dimerizable or crosslinkable groups including cinnamic, coumarin, and anthracene can undergo a crosslinking reaction via [2+2] cycloaddition of the carbon-carbon double bonds after UV-irradiation [79–81]. They have been mainly utilized for the photo-crosslinking of micelles, leading to micelles or nanogels with enhanced colloidal stability, even in very dilute condition [82–84]. Compared with traditional crosslinking methods such as 'click' chemistry and carbodiimide coupling, the photo-crosslinking approach is relatively inexpensive, rapid, and highly efficient at room temperature. Furthermore, no byproduct is generated during the photo-dimerization process, rendering a final product with a high purity [85].

Chen et al. demonstrated the first example of photo-crosslinkable polypeptide based micelle [66]. In their work, diblock copolymer poly(ethylene glycol)-*b*-poly(L-glutamic acid) was synthesized by ring-opening polymerization (ROP) of L-glutamate-NCA monomer in the presence of PEG-amine macroinitiator. The resulting diblock copolymer further underwent deprotection and subsequent modification with cinnamyl alcohol, yielding amphiphilic PEG-*b*-polypeptide, containing pendent cinnamyl functionalities. A core-shell micellar structure was formed by the self-assembly of the PEG-*b*-polypeptide into water. Moreover, UV-irradiation at 254 nm led to photo-crosslinking of the micellar core, which was directly proven by dynamic light scattering (DLS), showing a decreased size of the nanoparticle after core-crosslinking. Jing and coworkers reported the synthesis and ROP of a functional NCA monomer bearing a cinnamyl moiety [67]. Water soluble PEG-amine macroinitiator was utilized during the polymerization process, leading to well-defined PEG-*b*-polypeptide copolymers, which possess cinnamyl groups in the side chains of hydrophobic polypeptide (Figure 1A). The block copolymer was capable of self-assembling into micelles that could be core-crosslinked under UV light (Figure 1B,C). It is noteworthy to mention that Jing's direct polymerization approach achieved full functionalization of cinnamyl in the repeating units of the polypeptide chain. However, in the case of Chen's post-modification method, only a partial functionalization of repeating units with cinnamyl could be realized because of the low efficiency of esterification under steric environment of the polypeptide.

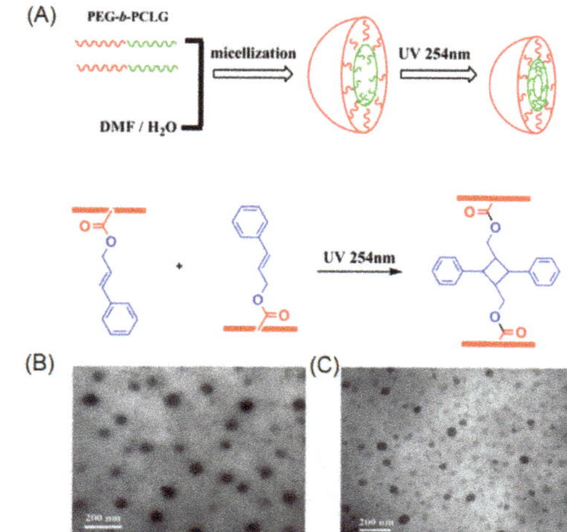

Figure 1. (**A**) Self-assembly of PEG-*b*-polypeptide and subsequent photo-induced core-crosslinking; (**B**) Transmission electron microscopy (TEM) image before UV irradiation; and (**C**) TEM image after UV irradiation. Reproduced with permission from [67].

Beyond cinnamyl photo-chemistry, coumarin based reversible dimerization has also been illustrated in the fabrication of photo-crosslinkable polypeptide micelles. In a pioneering work by Luo, solution-phase peptide synthesis (PPS) was employed to obtain a linear-dendritic block copolymer composed of hydrophilic PEG (5 KDa) and hydrophobic branched polylysine containing peripheral coumarin groups (Figure 2A) [68]. As the block copolymer was amphiphilic, micellar nanoparticles were observed in water as a result of self-assembly (Figure 2B). When long-wavelength UV irradiation (>310 nm) was applied to micelle solutions, the core-crosslinking event was rapidly completed within 400 seconds, as indicated by UV-Vis spectra. More interestingly, photo-induced decrosslinking occurred upon exposure to a short wavelength UV light (254 nm), elucidating the reversibility of this process. Notably, the decrosslinking reaction of coumarin dimer underwent a significantly slower kinetics (over 100 min) compared with that of crosslinking process.

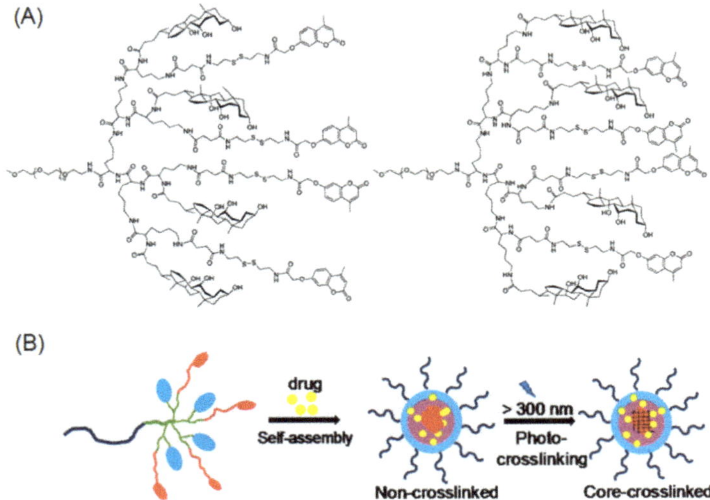

Figure 2. (**A**) Chemical structures of linear-dendritic PEG-*b*-polylysine containing coumarin moieties at their periphery; (**B**) light-triggered photo crosslinking of drug-loaded micelles. Reproduced with permission from [68].

2.2. Photo-Cleavable Nano-Objects

While polypeptide derived diblock copolymer micelles can acquire enhanced stability through the photo-crosslinking process, the concern regarding the lack of degradability still remains, especially in biomedical applications [86,87]. In view of this, photo-cleavage chemistry has emerged as an alternative approach to photo-sensitive polypeptide nanoparticles [13,69–72,74,88]. More importantly, photo-cleavage reactions are typically accompanied by a dramatic increase in the water-solubility of the hydrophobic segment, which could promote either a disassembly or morphological transformation of nano-objects.

Several illustrative examples of photo-cleavable polypeptide nanoparticles involving *O*-nitrobenzyl groups have been reported by Dong et al. In their first work, a photo-sensitive *S*-(*O*-nitrobenzyl)-L-cysteine NCA monomer (abbreviated as NBC) was designed and polymerized with PEG-amine as macroinitiator, giving rise to a diblock copolymer PNBC-*b*-PEG [69]. Since the NBC repeating units are hydrophobic because of the presense of *O*-nitrobenzyl moieties in the side chains, the amphiphilic block copolymer is able to form micelles with a size of 79 nm. This approach conferred photo-degradability to the micelles, because the hydrophobic core consists of numerous UV-labile *O*-nitrobenzyl groups. Transmission electron microscopy (TEM) and dynamic light scattering demonstrated that the block copolymer micelles were capable of dissociating into smaller nanoparticles (44 nm) upon UV irradiation at 365 nm. The reduction in particle size is because of the photo-cleavage of *O*-nitrobenzyl groups, producing free thiols with enhanced water solubility. A later report described the photo-induced shape transformation of polypeptide-containing vesicles (Figure 3A) [70]. In that work, $PNBC_{56}$-*b*-PEG_{114} (the subscript stands for the number of repeating units) was synthesized and used for constructing a vesicle morphology in an aqueous solution. The vesicle solution was subsequently exposed to 365 nm UV light, promoting the cleavage of *O*-nitrobenzyl groups and a concomitant increment in hydrophilicity of PNBC block. As the ratio of hydrophilicity to hydrophobicity increased, the critical packing parameters of the nano-assemblies decreased, inducing a morphological transition from vesicle to micelles (Figure 3B–D). In addition, the free thiol inside the micellar core can be further oxidized in the presence of an oxidizer (i.e., hydrogen peroxide), resulting in formation of disulfide linkages, which prompt the aggregation of the micelles.

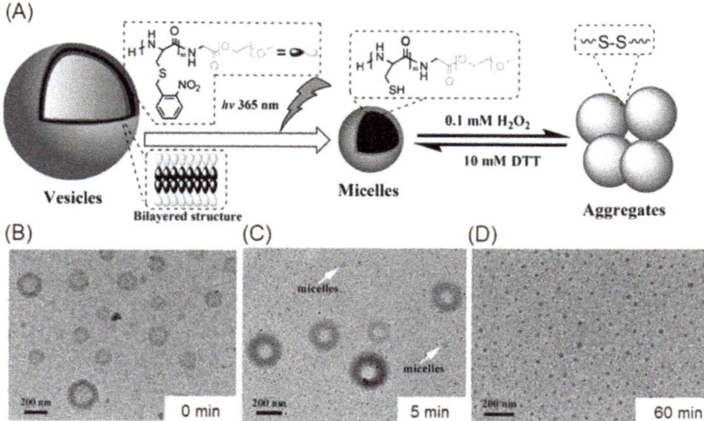

Figure 3. (**A**) Shape programming of polypeptide based nanoparticles through photo-regulation and redox process; (**B**) vesicular structures before UV treatment; (**C**) a mixture of vesicles and micelles after UV irradiation for 5 mintes; and (**D**) after UV-irradiation for 1 h, the vesicles were fully transformed to micelles. Reproduced with permission from [70].

Very recently, the same group invented NIR-responsive PNBC-*b*-PEG upconversion composite micelles (Figure 4) [13]. During the block copolymer self-assembly process, upconversion nanoparticles (UCNP) were encapsulated inside the PNBC core. The composite micelles were capable of disassembling with the help of UCNP, converting NIR light (980 nm) to UV light (365 nm). Moreover, Zhao and coworkers reported a novel NIR light-sensitive micellar system based on a diblock copolymer, consisting of PEG and poly(L-glutamic acid) bearing pendent 6-bromo-7-hydroxycoumarin-4-ylmethyl groups, an efficient NIR two-photon-absorbing chromophore (Figure 5) [74]. Upon irradiation with 794 nm of NIR light, the chromophores were gradually removed from the polypeptide chain, shifting the hydrophilic–hydrophobic balance toward a disassembly of micelles in water. Notably, nearly 200 min of irradiation was needed to fully cleave the side chain groups, demonstrating the potential of controlled release kinetics.

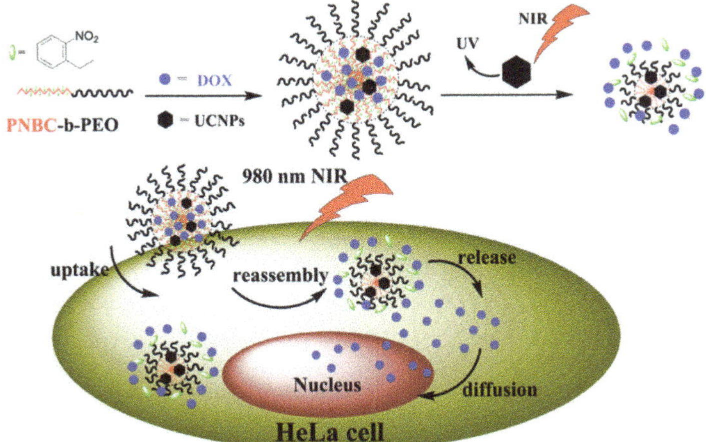

Figure 4. Fabrication of NIR-responsive polypeptide micelles via encapsulation of upconversion nanoparticles (UCNP) into block copolymer micellar core. Upon NIR irradiation, UCNP converts NIR into UV light, which further cleaves *O*-nitrobenzyl moieties and induces disassembly of micelles. Reproduced with permission from [13].

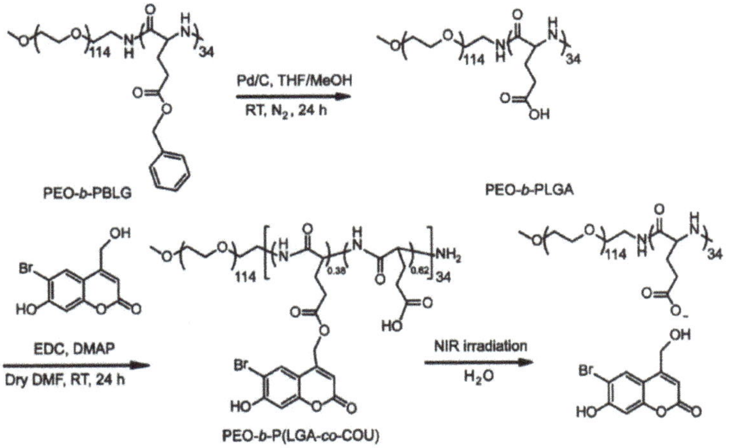

Figure 5. Synthetic route to NIR-responsive diblock copolymer consisting of PEG and polypeptide bearing coumarin groups. Reproduced with permission from [74].

2.3. Photo-Isomerizable Nano-Assemblies

According to the properties of the aforementioned photo-crosslinkable and photo-cleavable polypeptide nanoparticles (vide supra), we can easily draw the conclusion that the photo-induced shape transformation or micellar disruption based on those functionalities are non-reversible under common conditions. While the de-crosslinking reaction of coumarin dimer can be literally achieved, the condition (i.e., 254 nm) is harsh and the slow reaction could cause the decomposition of coumarin and lead to undesired side reactions [89]. In the case of *O*-nitrobenzyl, the UV-induced photo-redox cleavage would generate *O*-nitrosobenzaldehyde that cannot reform the original *O*-nitrobenzyl moiety. To further pursue efficient and reversible photo-responsiveness of polypeptide nano-assemblies, some research groups have designed smart nanoparticle systems, which rely on photo-isomerizable functionalities, such as azobenzene and spiropyran [40,75–78,90–92].

Azobenzene is capable of transitioning between two isomers (i.e., *cis* and *trans*) through manipulation of UV light (365 nm) and visible light. When UV light is present, a polar *cis*-isomer is favorably formed. On the other hand, visible light or heat can promote the shift of isomerization towards thermodynamically favored non-polar *trans*-isomer. To date, azobenzene derivatives have been extensively incorporated into many peptides, either in the side chains or in the backbone. In a report by Moretto, azobenzene served as a central linker for diblock poly(γ-benzyl-L-glutamate) (PBLG) (Figure 6A,B) [76]. Before UV irradiation, diblock PBLG *trans*-isomer vesicles were observed, as evidenced by TEM and SEM (Figure 6C,D). After exposure to UV light, a rapid and gradual collapse of those ordered vesicles was observed, probably owing to *trans*-to-*cis* azobenzene transformation, which induced change in 3D geometry of diblock polypeptide (Figure 6F–I). Lu and coworkers were able to synthesize photo-responsive polypeptides via ROP of NCA monomers that consisted of pendent azobenzene and oligoethylene glycol (OEG), affording P(OEG-Azo) [40]. Because of the presence of both hydrophobic azobenzene and hydrophilic OEG, P(PEG-Azo) can self-assemble into nanoparticles in an aqueous solution. Moreover, a α-helical conformation of polypeptide was observed in the case of azobenzene *trans*-isomer. Upon UV treatments, *trans-cis* isomerization occurred and forced the polypeptides to adopt a disordered conformation, as evidenced by the circular dichroism spectroscopy. Importantly, a reversible conformation switch was found when heating the UV-treated *cis*-polypeptides at 70 °C.

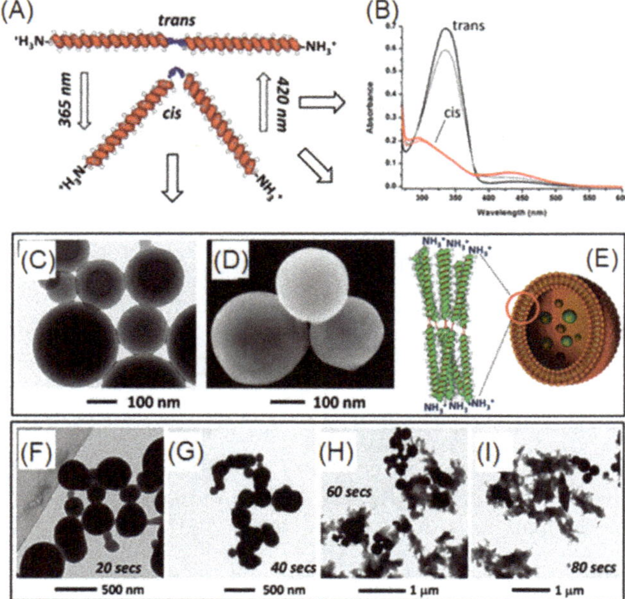

Figure 6. (**A**) Reversible geometry change of azobenzene-containing polypeptide via UV and visible light; (**B**) UV-Vis absorptions of *cis*- and *trans*-isomers; (**C**,**D**) TEM and SEM images of vesicles arising from self-assembly of *trans*-isomer of polypeptides; (**E**) cartoon representation of vesicular structure based on *trans*-polypeptide; and (**F**–**I**) time-dependent UV-induced degradation of vesicles. Reproduced with permission from [76].

Spiropyran (SP) is a widely-used photochromic molecule, thanks to light-induced spiropyran-to-merocyanine (SP–MC) isomerization [93,94]. Original SP derivatives, in their closed form, appear as colorless, nonpolar, and hydrophobic compounds. Isomerization toward MC (open form) occurred under UV treatment, leading to MC derivatives, which are colored, polar, and hydrophilic. Mezzenga et al. presented an excellent example of photo-reversible micelle system, based on spiropyran-containing polypeptide-*b*-PEG diblock copolymer (Figure 7A) [78]. Firstly, they performed a kinetic study of SP–MC and MC–SP isomerization, using UV-Vis spectroscopy. Before UV irradiation (365 nm), the solution was colorless, suggesting the absence of the MC form. After UV irradiation, the absorption peak at 544 nm progressively increased and reached maximum value within 5 min, indicative of fast and complete SP–MC isomerization. Nevertheless, MC–SP isomerization happened much slower and reached full conversion after 180 min in the presence of visible light (590 nm). After demonstrating the photo-regulated reversibility of SP–PC isomerization, the authors further utilized TEM to observe the reversible aggregation–dissolution–aggregation process of block copolymers in water. According to their results, original SP isomer containing polymers were capable of self-assembling into micelles (Figure 7B). UV irradiation fully disrupted the micellar structure after 5 min, because of the formation of hydrophilic MC moieties (Figure 7C). Interestingly, micelles were successfully recovered as a consequence of visible light treatment for 3 h (Figure 7D).

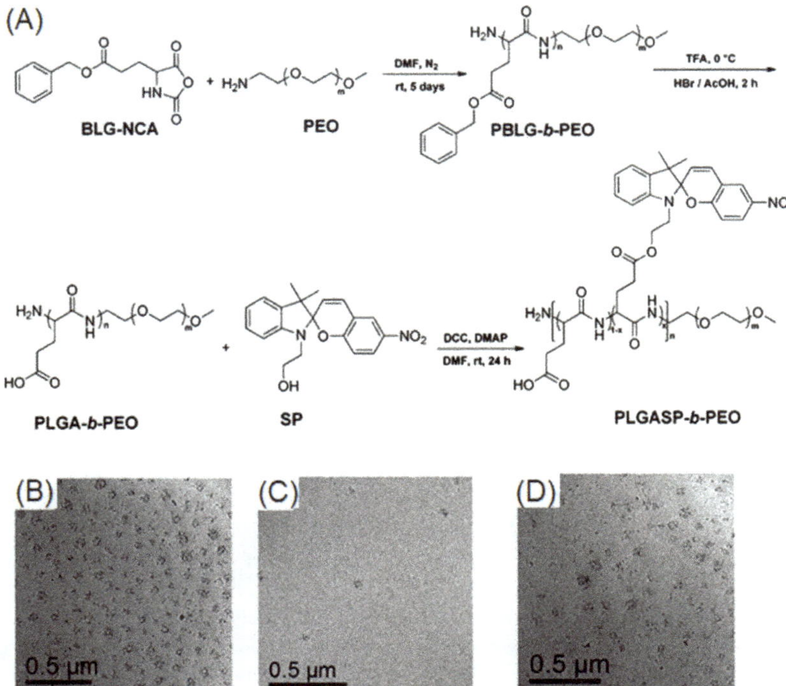

Figure 7. (**A**) Synthetic route to spiropyran-bearing polypeptide-*b*-PEG diblock copolymer; (**B**) TEM image of polymer nano-objects before UV treatment; (**C**) TEM image after UV irradiation; and (**D**) TEM image of regenerated micelles after applying visible light to UV-treated polymer solution. Reproduced with permission from [78].

3. Properties and Applications

Apparently, polypeptide derived nanoparticles hold great potential to serve as excellent drug delivery systems because of their biocompatibility and biodegradability. Moreover, the aforementioned photo-chemistry confers those nanoparticles with attractive properties, such as enhanced colloidal stability and on-demand drug release. Jing and coworkers investigated in vitro paclitaxel (PTX) release from two batches of PTX-loaded peptide micelles, with one batch treated with UV light [67]. According to their results, the drug release from the crosslinked micelles was significantly slower than that from the non-crosslinked micelle. For instance, only 20% of the drug was leaked from a crosslinked micelle during 55 h incubation in a phosphate buffered saline (PBS) buffer, while almost 100% of the drug was released from a non-crosslinked micelle under same condition (Figure 8). In Zhao's study, NIR-responsive Rifampicin-encapsulated polypeptide micelles showed a neglectable release after 55 h in the absence of NIR irradiation. When the NIR laser was turned on, a progressive drug release was observed, demonstrating the feasibility of this drug delivery system to achieve on-demand drug release.

Very recently, Mandal and coworkers employed ROP to prepare a cationic block copolymer consisting of poly(2-ethyl-2-oxazoline) and positively charged *O*-nitrobenzyl modified polymethionine (P[MetNB][Br]) [73]. This cationic polypeptide was capable of forming an electrostatic complex with negatively charged calf thymus DNA (ctDNA). When UV-light was applied to the polypeptide-DNA complex, a photo-driven cleavage of *O*-nitrobenzyl moieties occurred, resulting in neutral polypeptides, which had no binding affinity with ctDNA (Figure 9A). According to the gel electrophoresis, free ctDNA was rapidly released from the complex after irradiation with UV light. Those results demonstrated the potential of using photo-responsive cationic polypeptide as a DNA delivery platform (Figure 9B,C).

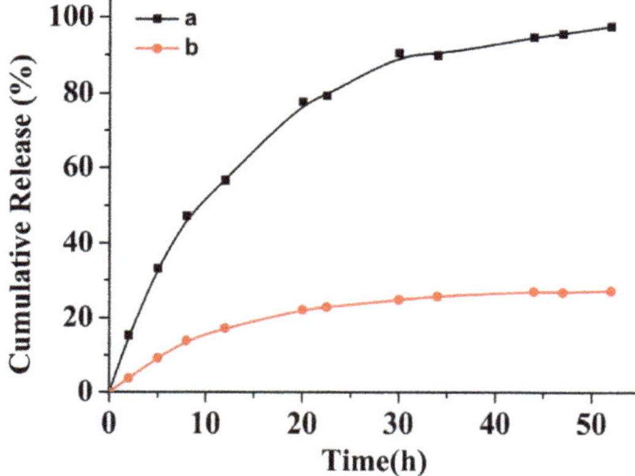

Figure 8. Cumulative release of drugs from non-crosslinked polypeptide micelle (black square) and photo-crosslinked polypeptide micelle (red dots). Reproduced with permission from [67].

In addition to biomedical applications, photo-responsive polypeptides have been used in the field of catalysis as well. He and coworkers designed a peptide-based artificial hydrolase, which consisted of a catalytic histidine residue and a photo-responsive azobenzene group in the peptide chain (Figure 10) [77]. Before UV irradiation, the peptide exhibited an antiparallel β-sheet conformation, enabling self-assembly into a peptide fibril. An enhanced catalytic activity on *p*-nitrophenyl acetate was observed, because of the hydrophobic environment of peptide fibril and proximity effect of histidine groups. However, a significant reduction in catalytic efficiency occurred upon exposure to UV-light, which caused a conformational conversion of peptide from β-sheet to random coil and thus disrupted the supramolecular fibril structure. Most importantly, the authors were able to demonstrate that the activity of the peptide-based artificial hydrolase could be reversibly controlled using visible and UV light.

Finally, the application of photo-sensitive polypeptides was successfully translated into macroscopic materials involving reversible sol-gel process, as described by Hu and Li (Figure 11) [40]. In their study, an organogel was formulated by dissolving azo-bearing polypeptide-b-PEG-b-polypeptide triblock copolymer in THF (Figure 11A). Interestingly, the gel was capable of switching physical states between gel and solution upon alternating the visible and UV treatment. According to the atom force microscope images, the gel revealed a densely crosslinked fibrous network, while the solution exhibited a much smaller degree of crosslinking after UV irradiation (Figure 11B,C).

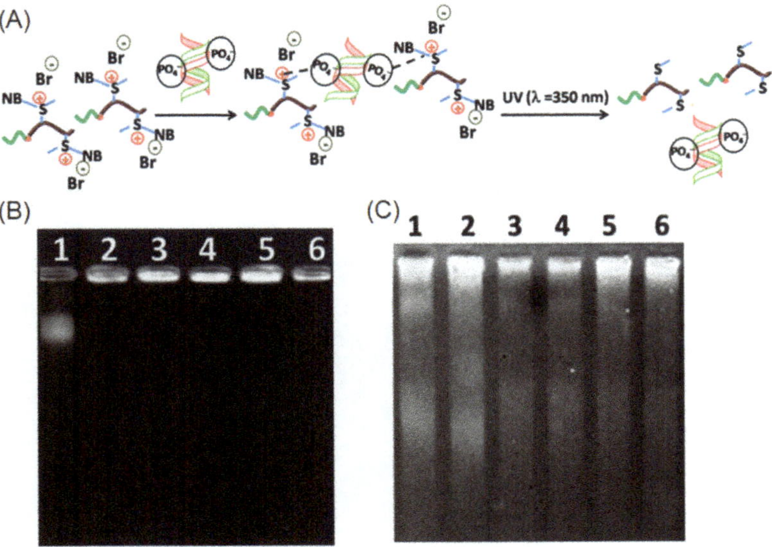

Figure 9. (**A**) Schematic illustration of polypeptide-DNA complexation and UV-induced release of DNA; (**B**) electrophoretic mobility of polypeptide-DNA conjugates at different concentrations of cationic polypeptide before UV irradiation; and (**C**) electrophoretic mobility of polypeptide-DNA conjugates upon UV treatment. Note: lanes 1–6 correspond to 12.7 mM ctDNA complexing with polypeptide at 0, 0.2, 0.4, 0.6, 0.8, and 1.0 mM, respectively. Reproduced with permission from [73].

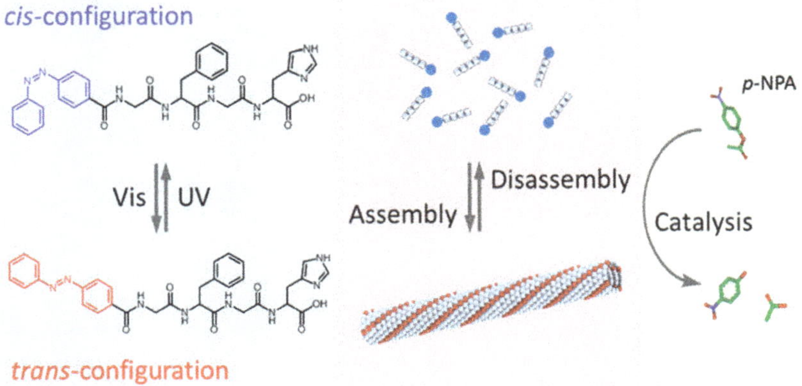

Figure 10. Molecular structures of azobenzene-terminated peptide and photo-switchable assembly and catalytic activities of the peptide, based artificial hydrolase. Reproduced with permission from [77].

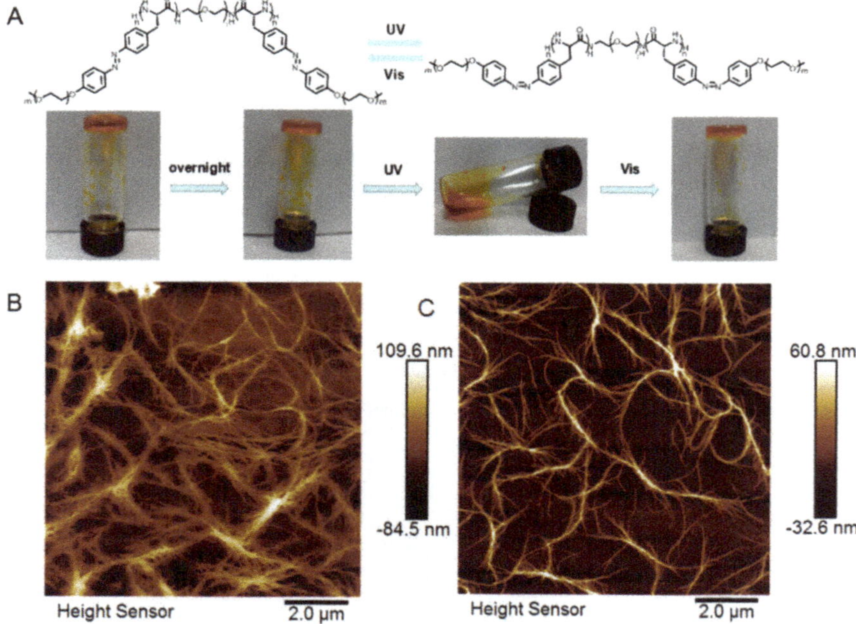

Figure 11. (**A**) Reversible sol-gel process by switching light wavelength between UV and visible; (**B**) atomic force microscopy (AFM) image of polymer solution before UV treatment; and (**C**) AFM image of polymer solution after UV irradiation. Reproduced with permission from [40].

4. Current Challenges and Prospective

Many relatively recent developments in photo-responsive polypeptides have greatly expanded the scope of smart nanomaterials, providing us with many new possibilities and opportunities in various applications, such as drug delivery, self-healing materials, and catalysis. Indeed, the marriage of polypeptide and photo-chemistry not only confer biocompatibility to the nanomaterials, but also facilitate the structural control of peptide chains or nano-assemblies because of the ease of using light. In view of photo-chemistry relying on different light-sensitive functionalities, a number of photo-sensitive peptide nanoparticles with distinct properties have been accomplished.

Despite the tremendous success that has been described above, many challenges still remain. One significant barrier is the translation of light-responsive polypeptide drug delivery system into clinical use. Indeed, the majority of examples in this review involve the use of UV light or visible light, which has a poor penetration depth into human tissue. Moreover, UV light has been shown to be detrimental to healthy cells and tissues [95–99]. Because of these downsides of using UV/Vis light, NIR-responsive polypeptide nanoparticles represent a more promising platform for nanomedicine [100]. However, the current NIR-responsive polypeptide derived drug delivery systems suffer from either slow drug release kinetics or an introduction of cytotoxic UCNP [13,74]. Therefore, more careful design and study are essential in order to translate those nanomaterials into biomedical applications. Moreover, photo-responsive polypeptide nano-objects have not yet been reported by means of controlled radical polymerization (CRP) and ring-opening metathesis polymerization (ROMP). Considering the robustness of CRP and ROMP techniques to prepare polymers with complex architectures and functions, we envision that one of the next directions for photo-responsive polypeptides will be updating the synthetic toolbox, in order to achieve more sophisticated polypeptide structures. According to the above-mentioned examples illustrated in

Table 1, photo-responsive polypeptide deriving nanomaterials have been overwhelmingly exploited for potential biomedicine use, such as drug delivery and gene release. However, only a few reports demonstrated the promise of those materials in other utilities, such as switchable catalysis and reversible macroscopic gel materials. Since light can be easily used to dictate when and where the photo-reaction happens, it can be anticipated that the significant attention on photo-responsive polypeptides will be shifted to some other applications, including self-healing materials, lithography or 3D-printing technology, which may exhibit excellent performance with the help of light-stimulus. Given the considerable success of traditional stimuli-responsive materials in biomedicine and manufacturing, we believe that photo-responsive polypeptide nanomaterials will take on more important roles to next generation of supramolecular peptide nanotechnology and material science.

Author Contributions: Conceptualization: H.S. and L.Y.; literature research: L.Y.; writing (original draft preparation): H.S.; writing (review and editing): H.S., L.Y., and H.T.; supervision: H.S.; and funding acquisition: H.S.

Funding: This research received no external funding.

Acknowledgments: H.S. gratefully acknowledges the fellowship support from EASTMAN Chemical Company and the Chinese government award for out-standing self-financed students abroad.

Conflicts of Interest: The authors declare no conflict of interest.

References

1. Sun, H.; Kabb, C.P.; Dai, Y.; Hill, M.R.; Ghiviriga, I.; Bapat, A.P.; Sumerlin, B.S. Macromolecular metamorphosis via stimulus-induced transformations of polymer architecture. *Nat. Chem.* **2017**, *9*, 817–823. [CrossRef] [PubMed]
2. Wei, M.; Gao, Y.; Li, X.; Serpe, M. Stimuli-responsive polymers and their applications. *Polym. Chem.* **2017**, *8*, 127–143. [CrossRef]
3. Herbert, K.M.; Schrettl, S.; Rowan, S.J.; Weder, C. 50th anniversary perspective: Solid-state multistimuli, multiresponsive polymeric materials. *Macromolecules* **2017**, *50*, 8845–8870. [CrossRef]
4. Rowan, S.; Cantrill, S.; Cousins, G.; Sanders, J.; Stoddart, J. Dynamic covalent chemistry. *Angew. Chem. Int. Ed.* **2002**, *41*, 898–952. [CrossRef]
5. Stuart, M.; Huck, W.; Genzer, J.; Muller, M.; Ober, C.; Stamm, M.; Sukhorukov, G.; Szleifer, I.; Tsukruk, V.; Urban, M.; et al. Emerging applications of stimuli-responsive polymer materials. *Nat. Mater.* **2010**, *9*, 101–113. [CrossRef] [PubMed]
6. Alfurhood, J.A.; Sun, H.; Kabb, C.P.; Tucker, B.S.; Matthews, J.H.; Luesch, H.; Sumerlin, B.S. Poly(*n*-(2-hydroxypropyl)-methacrylamide)-valproic acid conjugates as block copolymer nanocarriers. *Polym. Chem.* **2017**, *8*, 4983–4987. [CrossRef] [PubMed]
7. Tang, H.L.; Tsarevsky, N.V. Preparation and functionalization of linear and reductively degradable highly branched cyanoacrylate-based polymers. *J. Polym. Sci. Part A Polym. Chem.* **2016**, *54*, 3683–3693. [CrossRef]
8. Tang, H.L.; Tsarevsky, N.V. Lipoates as building blocks of sulfur-containing branched macromolecules. *Polym. Chem.* **2015**, *6*, 6936–6945. [CrossRef]
9. Wan, S.; Zhang, L.Q.; Wang, S.; Liu, Y.; Wu, C.C.; Cui, C.; Sun, H.; Shi, M.L.; Jiang, Y.; Li, L.; et al. Molecular recognition-based DNA nanoassemblies on the surfaces of nanosized exosomes. *J. Am. Chem. Soc.* **2017**, *139*, 5289–5292. [CrossRef] [PubMed]
10. Sun, H.; Dobbins, D.J.; Dai, Y.; Kabb, C.P.; Wu, S.; Alfurhood, J.A.; Rinaldi, C.; Sumerlin, B.S. Radical departure: Thermally-triggered degradation of azo-containing poly(β-thioester)s. *ACS Macro Lett.* **2016**, *5*, 688–693. [CrossRef]
11. Dai, Y.; Sun, H.; Pal, S.; Zhang, Y.; Park, S.; Kabb, C.; Wei, W.; Sumerlin, B. Near-ir-induced dissociation of thermally-sensitive star polymers. *Chem. Sci.* **2017**, *8*, 1815–1821. [CrossRef] [PubMed]
12. Alfurhood, J.A.; Sun, H.; Bachler, P.R.; Sumerlin, B.S. Hyperbranched poly(*n*-(2-hydroxypropyl)methacrylamide) via raft self-condensing vinyl polymerization. *Polym. Chem.* **2016**, *7*, 2099–2104. [CrossRef]

13. Liu, G.; Liu, N.; Zhou, L.Z.; Su, Y.; Dong, C.M. Nir-responsive polypeptide copolymer upconversion composite nanoparticles for triggered drug release and enhanced cytotoxicity. *Polym. Chem.* **2015**, *6*, 4030–4039. [CrossRef]
14. Liu, Y.; Hou, W.J.; Sun, H.; Cui, C.; Zhang, L.Q.; Jiang, Y.; Wu, Y.X.; Wang, Y.Y.; Li, J.; Sumerlin, B.S.; et al. Thiol-ene click chemistry: A biocompatible way for orthogonal bioconjugation of colloidal nanoparticles. *Chem. Sci.* **2017**, *8*, 6182–6187. [CrossRef] [PubMed]
15. Chen, M.; Zhong, M.; Johnson, J. Light-controlled radical polymerization: Mechanisms, methods, and applications. *Chem. Rev.* **2016**, *116*, 10167–10211. [CrossRef] [PubMed]
16. Tan, J.B.; Li, X.L.; Zeng, R.M.; Liu, D.D.; Xu, Q.; He, J.; Zhang, Y.X.; Dai, X.C.; Yu, L.L.; Zeng, Z.H.; et al. Expanding the scope of polymerization-induced self-assembly: Z-raft-mediated photoinitiated dispersion polymerization. *ACS Macro Lett.* **2018**, *7*, 255–262. [CrossRef]
17. Yeow, J.; Boyer, C. Photoinitiated polymerization-induced self-assembly (photo-pisa): New insights and opportunities. *Adv. Sci.* **2017**, *4*, 1700137. [CrossRef] [PubMed]
18. Tan, J.B.; Huang, C.D.; Liu, D.D.; Zhang, X.C.; Bai, Y.H.; Zhang, L. Alcoholic photoinitiated polymerization-induced self-assembly (photo-pisa): A fast route toward poly(isobornyl acrylate)-based diblock copolymer nano-objects. *ACS Macro Lett.* **2016**, *5*, 894–899. [CrossRef]
19. Tan, J.B.; Sun, H.; Yu, M.G.; Sumerlin, B.S.; Zhang, L. Photo-pisa: Shedding light on polymerization-induced self-assembly. *ACS Macro Lett.* **2015**, *4*, 1249–1253. [CrossRef]
20. Niu, J.; Lunn, D.J.; Pusuluri, A.; Yoo, J.I.; O'Malley, M.A.; Mitragotri, S.; Soh, H.T.; Hawker, C.J. Engineering live cell surfaces with functional polymers via cytocompatible controlled radical polymerization. *Nat. Chem.* **2017**, *9*, 537–545. [CrossRef] [PubMed]
21. Treat, N.; Sprafke, H.; Kramer, J.; Clark, P.; Barton, B.; de Alaniz, J.; Fors, B.; Hawker, C. Metal-free atom transfer radical polymerization. *J. Am. Chem. Soc.* **2014**, *136*, 16096–16101. [CrossRef] [PubMed]
22. Goetz, A.E.; Boydston, A.J. Metal-free preparation of linear and cross-linked polydicyclopentadiene. *J. Am. Chem. Soc.* **2015**, *137*, 7572–7575. [CrossRef] [PubMed]
23. Ogawa, K.A.; Goetz, A.E.; Boydston, A.J. Metal-free ring-opening metathesis polymerization. *J. Am. Chem. Soc.* **2015**, *137*, 1400–1403. [CrossRef] [PubMed]
24. Blasco, E.; Sims, M.B.; Goldmann, A.S.; Sumerlin, B.S.; Barner-Kowollik, C. 50th anniversary perspective: Polymer functionalization. *Macromolecules* **2017**, *50*, 5215–5252. [CrossRef]
25. Xu, J.; Shanmugam, S.; Fu, C.; Aguey-Zinsou, K.; Boyer, C. Selective photoactivation: From a single unit monomer insertion reaction to controlled polymer architectures. *J. Am. Chem. Soc.* **2016**, *138*, 3094–3106. [CrossRef] [PubMed]
26. Kottisch, V.; Michaudel, Q.; Fors, B.P. Photocontrolled interconversion of cationic and radical polymerizations. *J. Am. Chem. Soc.* **2017**, *139*, 10665–10668. [CrossRef] [PubMed]
27. Michaudel, Q.; Chauvire, T.; Kottisch, V.; Supej, M.J.; Stawiasz, K.J.; Shen, L.X.; Zipfel, W.R.; Abruna, H.D.; Freed, J.H.; Fors, B.P. Mechanistic insight into the photocontrolled cationic polymerization of vinyl ethers. *J. Am. Chem. Soc.* **2017**, *139*, 15530–15538. [CrossRef] [PubMed]
28. Rwei, A.Y.; Wang, W.P.; Kohane, D.S. Photoresponsive nanoparticles for drug delivery. *Nano Today* **2015**, *10*, 451–467. [CrossRef] [PubMed]
29. Zhao, Y.; Tremblay, L.; Zhao, Y. Doubly photoresponsive and water-soluble block copolymers: Synthesis and thermosensitivity. *J. Polym. Sci. Part A Polym. Chem.* **2010**, *48*, 4055–4066. [CrossRef]
30. Fomina, N.; McFearin, C.L.; Sermsakdi, M.; Morachis, J.M.; Almutairi, A. Low power, biologically benign nir light triggers polymer disassembly. *Macromolecules* **2011**, *44*, 8590–8597. [CrossRef] [PubMed]
31. Bauri, K.; Nandi, M.; De, P. Amino acid-derived stimuli-responsive polymers and their applications. *Polym. Chem.* **2018**, *9*, 1257–1287. [CrossRef]
32. Gradisar, H.; Bozic, S.; Doles, T.; Vengust, D.; Hafner-Bratkovic, I.; Mertelj, A.; Webb, B.; Sali, A.; Klavzar, S.; Jerala, R. Design of a single-chain polypeptide tetrahedron assembled from coiled-coil segments. *Nat. Chem. Biol.* **2013**, *9*, 362–366. [CrossRef] [PubMed]
33. Franco, L.; del Valle, L.J.; Puiggali, J. Smart systems related to polypeptide sequences. *AIMS Mater. Sci.* **2016**, *3*, 289–323. [CrossRef]
34. Ashkenasy, N.; Schneider, J. Functional peptide and protein nanostructures. *Isr. J. Chem.* **2015**, *55*, 621. [CrossRef]

35. Aemissegger, A.; Hilvert, D. Synthesis and application of an azobenzene amino acid as a light-switchable turn element in polypeptides. *Nat. Protoc.* **2007**, *2*, 161–167. [CrossRef] [PubMed]
36. Ungerleider, J.L.; Kammeyer, J.K.; Braden, R.L.; Christman, K.L.; Gianneschi, N.C. Enzyme-targeted nanoparticles for delivery to ischemic skeletal muscle. *Polym. Chem.* **2017**, *8*, 5212–5219. [CrossRef] [PubMed]
37. Carlini, A.S.; Adamiak, L.; Gianneschi, N.C. Biosynthetic polymers as functional materials. *Macromolecules* **2016**, *49*, 4379–4394. [CrossRef] [PubMed]
38. Chen, C.; Thang, S.H. Raft polymerization of a rgd peptide-based methacrylamide monomer for cell adhesion. *Polym. Chem.* **2018**, *9*, 1780–1786. [CrossRef]
39. Wang, Z.; Li, Y.W.; Huang, Y.R.; Thompson, M.P.; LeGuyader, C.L.M.; Sahu, S.; Gianneschi, N.C. Enzyme-regulated topology of a cyclic peptide brush polymer for tuning assembly. *Chem. Commun.* **2015**, *51*, 17108–17111. [CrossRef] [PubMed]
40. Xiong, W.; Fu, X.H.; Wan, Y.M.; Sun, Y.L.; Li, Z.B.; Lu, H. Synthesis and multimodal responsiveness of poly(alpha-amino acid)s bearing oegylated azobenzene side-chains. *Polym. Chem.* **2016**, *7*, 6375–6382. [CrossRef]
41. Deming, T.J. Synthesis of side-chain modified polypeptides. *Chem. Rev.* **2016**, *116*, 786–808. [CrossRef] [PubMed]
42. Zou, J.; Fan, J.W.; He, X.; Zhang, S.Y.; Wang, H.; Wooley, K.L. A facile glovebox-free strategy to significantly accelerate the syntheses of well-defined polypeptides by n-carboxyanhydride (nca) ring-opening polymerizations. *Macromolecules* **2013**, *46*, 4223–4226. [CrossRef] [PubMed]
43. Song, Z.Y.; Han, Z.Y.; Lv, S.X.; Chen, C.Y.; Chen, L.; Yin, L.C.; Cheng, J.J. Synthetic polypeptides: From polymer design to supramolecular assembly and biomedical application. *Chem. Soc. Rev.* **2017**, *46*, 6570–6599. [CrossRef] [PubMed]
44. Le Droumaguet, B.; Nicolas, J. Recent advances in the design of bioconjugates from controlled/living radical polymerization. *Polym. Chem.* **2010**, *1*, 563–598. [CrossRef]
45. Ayres, L.; Koch, K.; Adams, P.H.H.M.; van Hest, J.C.M. Stimulus responsive behavior of elastin-based side chain polymers. *Macromolecules* **2005**, *38*, 1699–1704. [CrossRef]
46. Blum, A.P.; Kammeyer, J.K.; Gianneschi, N.C. Activating peptides for cellular uptake via polymerization into high density brushes. *Chem. Sci.* **2016**, *7*, 989–994. [CrossRef] [PubMed]
47. Gianneschi, N. Peptide-polymer amphiphiles as programmable synthons for biologically-responsive nanomaterials. *Abstr. Pap. Am. Chem. Soc.* **2015**, *249*, 317.
48. Blum, A.P.; Kammeyer, J.K.; Rush, A.M.; Callmann, C.E.; Hahn, M.E.; Gianneschi, N.C. Stimuli-responsive nanomaterials for biomedical applications. *J. Am. Chem. Soc.* **2015**, *137*, 2140–2154. [CrossRef] [PubMed]
49. Baumgartner, R.; Fu, H.L.; Song, Z.Y.; Lin, Y.; Cheng, J.J. Cooperative polymerization of alpha-helices induced by macromolecular architecture. *Nat. Chem.* **2017**, *9*, 614–622. [CrossRef] [PubMed]
50. Fan, J.W.; Borguet, Y.P.; Su, L.; Nguyen, T.P.; Wang, H.; He, X.; Zou, J.; Wooley, K.L. Two-dimensional controlled syntheses of polypeptide molecular brushes via n-carboxyanhydride ring-opening polymerization and ring-opening metathesis polymerization. *ACS Macro Lett.* **2017**, *6*, 1031–1035. [CrossRef] [PubMed]
51. Sun, H.; Kabb, C.P.; Sumerlin, B.S. Thermally-labile segmented hyperbranched copolymers: Using reversible-covalent chemistry to investigate the mechanism of self-condensing vinyl copolymerization. *Chem. Sci.* **2014**, *5*, 4646–4655. [CrossRef]
52. Dong, P.; Sun, H.; Quan, D.P. Synthesis of poly(L-lactide-*co*-5-amino-5-methyl-1,3-dioxan-2-ones) [*p*(L-la-*co*-tac)] containing amino groups via organocatalysis and post-polymerization functionalization. *Polymer* **2016**, *97*, 614–622. [CrossRef]
53. Perrier, S. 50th anniversary perspective: Raft polymerization—A user guide. *Macromolecules* **2017**, *50*, 7433–7447. [CrossRef]
54. Grubbs, R.B.; Grubbs, R.H. 50th anniversary perspective: Living polymerization-emphasizing the molecule in macromolecules. *Macromolecules* **2017**, *50*, 6979–6997. [CrossRef]
55. Tan, J.B.; He, J.; Li, X.L.; Xu, Q.; Huang, C.D.; Liu, D.D.; Zhang, L. Rapid synthesis of well-defined all-acrylic diblock copolymer nano-objects via alcoholic photoinitiated polymerization-induced self-assembly (photo-pisa). *Polym. Chem.* **2017**, *8*, 6853–6864. [CrossRef]
56. Ren, J.; McKenzie, T.; Fu, Q.; Wong, E.; Xu, J.; An, Z.; Shanmugam, S.; Davis, T.; Boyer, C.; Qiao, G. Star polymers. *Chem. Rev.* **2016**, *116*, 6743–6836. [CrossRef] [PubMed]

57. Kerr, A.; Hartlieb, M.; Sanchis, J.; Smith, T.; Perrier, S. Complex multiblock bottle-brush architectures by raft polymerization. *Chem. Commun.* **2017**, *53*, 11901–11904. [CrossRef] [PubMed]
58. Matyjaszewski, K. Atom transfer radical polymerization (atrp): Current status and future perspectives. *Macromolecules* **2012**, *45*, 4015–4039. [CrossRef]
59. Matyjaszewski, K. Architecturally complex polymers with controlled heterogeneity. *Science* **2011**, *333*, 1104–1105. [CrossRef] [PubMed]
60. Bapat, A.; Roy, D.; Ray, J.; Savin, D.; Sumerlin, B. Dynamic-covalent macromolecular stars with boronic ester linkages. *J. Am. Chem. Soc.* **2011**, *133*, 19832–19838. [CrossRef] [PubMed]
61. Sumerlin, B.; Vogt, A. Macromolecular engineering through click chemistry and other efficient transformations. *Macromolecules* **2010**, *43*, 1–13. [CrossRef]
62. Aoki, D.; Aibara, G.; Uchida, S.; Takata, T. A rational entry to cyclic polymers via selective cyclization by self-assembly and topology transformation of linear polymers. *J. Am. Chem. Soc.* **2017**, *139*, 6791–6794. [CrossRef] [PubMed]
63. Yamamoto, T.; Yagyu, S.; Tezuka, Y. Light- and heat-triggered reversible linear-cyclic topological conversion of telechelic polymers with anthryl end groups. *J. Am. Chem. Soc.* **2016**, *138*, 3904–3911. [CrossRef] [PubMed]
64. Bielawski, C.; Benitez, D.; Grubbs, R. An "endless" route to cyclic polymers. *Science* **2002**, *297*, 2041–2044. [CrossRef] [PubMed]
65. Feng, H.B.; Lu, X.Y.; Wang, W.Y.; Kang, N.G.; Mays, J.W. Block copolymers: Synthesis, self-assembly, and applications. *Polymers* **2017**, *9*, 494. [CrossRef]
66. Ding, J.X.; Zhuang, X.L.; Xiao, C.S.; Cheng, Y.L.; Zhao, L.; He, C.L.; Tang, Z.H.; Chen, X.S. Preparation of photo-cross-linked ph-responsive polypeptide nanogels as potential carriers for controlled drug delivery. *J. Mater. Chem.* **2011**, *21*, 11383–11391. [CrossRef]
67. Yan, L.S.; Yang, L.X.; He, H.Y.; Hu, X.L.; Xie, Z.G.; Huang, Y.B.; Jing, X.B. Photo-cross-linked mpeg-poly(gamma-cinnamyl-L-glutamate) micelles as stable drug carriers. *Polym. Chem.* **2012**, *3*, 1300–1307. [CrossRef]
68. Shao, Y.; Shi, C.Y.; Xu, G.F.; Guo, D.D.; Luo, J.T. Photo and redox dual responsive reversibly cross-linked nanocarrier for efficient tumor-targeted drug delivery. *ACS Appl. Mater. Interfaces* **2014**, *6*, 10381–10392. [CrossRef] [PubMed]
69. Liu, G.; Dong, C.M. Photoresponsive poly(*s*-(*o*-nitrobenzyl)-L-cysteine)-*b*-peo from a L-cysteine *n*-carboxyanhydride monomer: Synthesis, self-assembly, and phototriggered drug release. *Biomacromolecules* **2012**, *13*, 1573–1583. [CrossRef] [PubMed]
70. Liu, G.; Zhou, L.Z.; Guan, Y.F.; Su, Y.; Dong, C.M. Multi-responsive polypeptidosome: Characterization, morphology transformation, and triggered drug delivery. *Macromol. Rapid Commun.* **2014**, *35*, 1673–1678. [CrossRef] [PubMed]
71. Negri, G.E.; Deming, T.J. Triggered copolypeptide hydrogel degradation using photolabile lysine protecting groups. *ACS Macro Lett.* **2016**, *5*, 1253–1256. [CrossRef]
72. Li, P.; Zhang, J.C.; Dong, C.M. Photosensitive poly(*o*-nitrobenzyloxycarbonyl-L-lysine)-*b*-peo polypeptide copolymers: Synthesis, multiple self-assembly behaviors, and the photo/ph-thermo-sensitive hydrogels. *Polym. Chem.* **2017**, *8*, 7033–7043. [CrossRef]
73. Jana, S.; Biswas, Y.; Mandal, T.K. Methionine-based cationic polypeptide/polypeptide block copolymer with triple-stimuli responsiveness: DNA polyplexation and phototriggered release. *Polym. Chem.* **2018**, *9*, 1869–1884. [CrossRef]
74. Kumar, S.; Allard, J.F.; Morris, D.; Dory, Y.L.; Lepage, M.; Zhao, Y. Near-infrared light sensitive polypeptide block copolymer micelles for drug delivery. *J. Mater. Chem.* **2012**, *22*, 7252–7257. [CrossRef]
75. Mba, M.; Mazzier, D.; Silvestrini, S.; Toniolo, C.; Fatas, P.; Jimenez, A.I.; Cativiela, C.; Moretto, A. Photocontrolled self-assembly of a bis-azobenzene-containing alpha-amino acid. *Chemistry* **2013**, *19*, 15841–15846. [CrossRef] [PubMed]
76. Mazzier, D.; Maran, M.; Perucchin, O.P.; Crisma, M.; Zerbetto, M.; Causin, V.; Toniolo, C.; Moretto, A. Photoresponsive supramolecular architectures based on polypeptide hybrids. *Macromolecules* **2014**, *47*, 7272–7283. [CrossRef]
77. Zhao, Y.A.; Lei, B.Q.; Wang, M.F.; Wu, S.T.; Qi, W.; Su, R.X.; He, Z.M. A supramolecular approach to construct a hydrolase mimic with photo-switchable catalytic activity. *J. Mater. Chem. B* **2018**, *6*, 2444–2449. [CrossRef]

78. Kotharangannagari, V.K.; Sanchez-Ferrer, A.; Ruokolainen, J.; Mezzenga, R. Photoresponsive reversible aggregation and dissolution of rod-coil polypeptide diblock copolymers. *Macromolecules* **2011**, *44*, 4569–4573. [CrossRef]
79. Bhola, R.; Payamyar, P.; Murray, D.J.; Kumar, B.; Teator, A.J.; Schmidt, M.U.; Hammer, S.M.; Saha, A.; Sakamoto, J.; Schluter, A.D.; et al. A two-dimensional polymer from the anthracene dimer and triptycene motifs. *J. Am. Chem. Soc.* **2013**, *135*, 14134–14141. [CrossRef] [PubMed]
80. He, J.; Tremblay, L.; Lacelle, S.; Zhao, Y. Preparation of polymer single chain nanoparticles using intramolecular photodimerization of coumarin. *Soft Matter* **2011**, *7*, 2380–2386. [CrossRef]
81. Inkinen, J.; Niskanen, J.; Talka, T.; Sahle, C.J.; Muller, H.; Khriachtchev, L.; Hashemi, J.; Akbari, A.; Hakala, M.; Huotari, S. X-ray induced dimerization of cinnamic acid: Time-resolved inelastic X-ray scattering study. *Sci. Rep.* **2015**, *5*, 15851. [CrossRef] [PubMed]
82. Wei, J.; Yu, Y.L. Photodeformable polymer gels and crosslinked liquid-crystalline polymers. *Soft Matter* **2012**, *8*, 8050–8059. [CrossRef]
83. Jamroz-Piegza, M.; Walach, W.; Dworak, A.; Trzebicka, B. Polyether nanoparticles from covalently crosslinked copolymer micelles. *J. Colloid Interface Sci.* **2008**, *325*, 141–148. [CrossRef] [PubMed]
84. Liu, Y.; Chang, H.; Jiang, J.Q.; Yan, X.Y.; Liu, Z.T.; Liu, Z.W. The photodimerization characteristics of anthracene pendants within amphiphilic polymer micelles in aqueous solution. *RSC Adv.* **2014**, *4*, 25912–25915. [CrossRef]
85. Jiang, J.Q.; Qi, B.; Lepage, M.; Zhao, Y. Polymer micelles stabilization on demand through reversible photo-cross-linking. *Macromolecules* **2007**, *40*, 790–792. [CrossRef]
86. Tardy, A.; Nicolas, J.; Gigmes, D.; Lefay, C.; Guillaneuf, Y. Radical ring-opening polymerization: Scope, limitations, and application to (bio)degradable materials. *Chem. Rev.* **2017**, *117*, 1319–1406. [CrossRef] [PubMed]
87. Yu, B.; Jiang, X.S.; Yin, J. Responsive fluorescent core-crosslinked polymer particles based on the anthracene-containing hyperbranched poly(ether amine) (hPEA-An). *Soft Matter* **2011**, *7*, 6853–6862. [CrossRef]
88. Xie, X.Y.; Yang, Y.F.; Yang, Y.; Zhang, H.; Li, Y.; Mei, X.G. A photo-responsive peptide- and asparagine-glycine-arginine (NGR) peptide-mediated liposomal delivery system. *Drug Deliv.* **2016**, *23*, 2445–2456. [CrossRef] [PubMed]
89. Chen, Y.; Chou, C.F. Reversible photodimerization of coumarin derivatives dispersed in poly(vinyl acetate). *J. Polym. Sci. Part A Polym. Chem.* **1995**, *33*, 2705–2714. [CrossRef]
90. Flint, D.G.; Kumita, J.R.; Smart, O.S.; Woolley, G.A. Using an azobenzene cross-linker to either increase or decrease peptide helix content upon *trans*-to-*cis* photoisomerization. *Chem. Biol.* **2002**, *9*, 391–397. [CrossRef]
91. Kinoshita, T.; Sato, M.; Takizawa, A.; Tsujita, Y. Photocontrol of polypeptide membrane functions by *cis-trans* isomerization in side-chain azobenzene groups. *Macromolecules* **1986**, *19*, 51–55. [CrossRef]
92. Li, Y.F.; Niu, Y.L.; Hu, D.; Song, Y.W.; He, J.W.; Liu, X.Y.; Xia, X.N.; Lu, Y.B.; Xu, W.J. Preparation of light-responsive polyester micelles via ring-opening polymerization of *o*-carboxyanhydride and azide-alkyne click chemistry. *Macromol. Chem. Phys.* **2015**, *216*, 77–84. [CrossRef]
93. Klajn, R. Spiropyran-based dynamic materials. *Chem. Soc. Rev.* **2014**, *43*, 148–184. [CrossRef] [PubMed]
94. Ciardelli, F.; Fabbri, D.; Pieroni, O.; Fissi, A. Photomodulation of polypeptide conformation by sunlight in spiropyran-containing poly(L-glutamic acid). *J. Am. Chem. Soc.* **1989**, *111*, 3470–3472. [CrossRef]
95. Yin, R.; Dai, T.H.; Avci, P.; Jorge, A.E.S.; de Melo, W.C.M.A.; Vecchio, D.; Huang, Y.Y.; Gupta, A.; Hamblin, M.R. Light based anti-infectives: Ultraviolet c irradiation, photodynamic therapy, blue light, and beyond. *Curr. Opin. Pharmacol.* **2013**, *13*, 731–762. [CrossRef] [PubMed]
96. Bertrand, O.; Gohy, J. Photo-responsive polymers: Synthesis and applications. *Polym. Chem.* **2017**, *8*, 52–73. [CrossRef]
97. Zhao, Y. Rational design of light-controllable polymer micelles. *Chem. Rec.* **2007**, *7*, 286–294. [CrossRef] [PubMed]
98. Jackson, J.; Chen, A.; Zhang, H.B.; Burt, H.; Chiao, M. Design and near-infrared actuation of a gold nanorod-polymer microelectromechanical device for on-demand drug delivery. *Micromachines* **2018**, *9*, 28. [CrossRef]

99. Qiu, M.; Wang, D.; Liang, W.Y.; Liu, L.P.; Zhang, Y.; Chen, X.; Sang, D.K.; Xing, C.Y.; Li, Z.J.; Dong, B.Q.; et al. Novel concept of the smart NIR-light-controlled drug release of black phosphorus nanostructure for cancer therapy. *Proc. Natl. Acad. Sci. USA* **2018**, *115*, 501–506. [CrossRef] [PubMed]
100. Linsley, C.S.; Wu, B.M. Recent advances in light-responsive on-demand drug- delivery systems. *Ther. Deliv.* **2017**, *8*, 89–107. [CrossRef] [PubMed]

 *micromachines*

Article

Marine Polysaccharide-Collagen Coatings on Ti6Al4V Alloy Formed by Self-Assembly

Karl Norris [1], Oksana I. Mishukova [2], Agata Zykwinska [3], Sylvia Colliec-Jouault [3], Corinne Sinquin [3], Andrei Koptioug [4], Stéphane Cuenot [5], Jemma G. Kerns [6], Maria A. Surmeneva [2], Roman A. Surmenev [2] and Timothy E.L. Douglas [1,7,*]

1 Engineering Department, Lancaster University, Lancaster LA1 4YW, UK; hwbkn3@gmail.com
2 Physical Materials Science and Composite Materials Centre, National research Tomsk Polytechnic University, Tomsk 634050, Russia; Oksana_mishukova@mail.ru (O.I.M.); surmenevamaria@mail.ru (M.A.S.); rsurmenev@mail.ru (R.A.S.)
3 IFREMER, Laboratoire Ecosystèmes Microbiens et Molécules Marines pour les Biotechnologies, F-44311 Nantes, France; agata.zykwinska@ifremer.fr (A.Z.); sylvia.colliec.jouault@ifremer.fr (S.C.-J.); corinne.sinquin@ifremer.fr (C.S.)
4 Sports Tech Research Centre, Mid-Sweden University, Akademigatan 1, 83125 Östersund, Sweden; andrey.koptyug@miun.se
5 Institut des Matériaux Jean Rouxel (IMN), Université de Nantes-CNRS, 44322 Nantes, France; Stephane.Cuenot@cnrs-imn.fr
6 Lancaster Medical School, Faculty of Health and Medicine, Lancaster University, Lancaster LA1 4YW, UK; j.kerns@lancaster.ac.uk
7 Materials Science Institute (MSI), Lancaster University, Lancaster LA1 4YW, UK
* Correspondence: t.douglas@lancaster.ac.uk; Tel.: +44-1524-594-450

Received: 30 November 2018; Accepted: 18 January 2019; Published: 19 January 2019

Abstract: Polysaccharides of marine origin are gaining interest as biomaterial components. Bacteria derived from deep-sea hydrothermal vents can produce sulfated exopolysaccharides (EPS), which can influence cell behavior. The use of such polysaccharides as components of organic, collagen fibril-based coatings on biomaterial surfaces remains unexplored. In this study, collagen fibril coatings enriched with HE800 and GY785 EPS derivatives were deposited on titanium alloy (Ti6Al4V) scaffolds produced by rapid prototyping and subjected to physicochemical and cell biological characterization. Coatings were formed by a self-assembly process whereby polysaccharides were added to acidic collagen molecule solution, followed by neutralization to induced self-assembly of collagen fibrils. Fibril formation resulted in collagen hydrogel formation. Hydrogels formed directly on Ti6Al4V surfaces, and fibrils adsorbed onto the surface. Scanning electron microscopy (SEM) analysis of collagen fibril coatings revealed association of polysaccharides with fibrils. Cell biological characterization revealed good cell adhesion and growth on bare Ti6Al4V surfaces, as well as coatings of collagen fibrils only and collagen fibrils enhanced with HE800 and GY785 EPS derivatives. Hence, the use of both EPS derivatives as coating components is feasible. Further work should focus on cell differentiation.

Keywords: marine exopolysaccharide; collagen; surface modification; Ti6Al4V

---

1. Introduction

Metallic load-bearing implants for bone contact rely on the formation of new bone tissue on the implant surface. Modifications of the surface which promote attachment, proliferation, and osteogenic differentiation of bone-forming cells are desirable. One strategy is the coating of the surfaces with fibrils of collagen, the main structural protein of mammalian tissue. Titanium and its alloys, including Ti6Al4V, are commonly used as implant materials for load-bearing applications.

Coatings of collagen types I, II, and III have been applied to improve adhesion and proliferation of bone-forming cells [1–3].

The collagen molecule is a long rigid complex structure consisting of three polypeptide chains which are connected to each other in the form of a triple helix configuration. Individual collagen units associate to fibrils under physiological conditions. This self-assembly of fibrils, also known as fibrillogenesis, can be induced under laboratory conditions by neutralizing an acidic solution of collagen molecules. This reduces electrostatic repulsions between the collagen molecules and enables fibrillogenesis to occur [4].

Collagen fibril coatings can be employed as artificial extracellular matrices into which other biologically active molecules can be incorporated. Various anionic polysaccharides, including glycosaminoglycans (GAG), are known to stimulate the attachment and proliferation of cells. Collagen fibril coatings containing GAG have been employed to improve cell adhesion and proliferation [5,6].

Recently, there has been growing interest in the use of marine polysaccharides. HE800 exopolysaccharides (EPS) is an unusual polysaccharide produced by the deep-sea hydrothermal bacterium *Vibrio diabolicus* [7]. This linear non-sulfated acidic polysaccharide is composed of a tetrasaccharide repeating unit containing N-acetyl-glucosamine (GlcNAc), two glucuronic acid (GlcA), and N-acetyl-galactosamine (GalNAc) residues [8]. HE800 EPS structure, which presents structural similarities to the GAG hyaluronic acid, confers to the EPS GAG-like properties. Native EPS of high-molecular weight (HMW) was shown to enhance in vivo bone regeneration [9] and stimulate collagen structuring by fibroblasts in reconstructed dermis [10]. GY785 EPS is a highly branched acidic heteropolysaccharide excreted by the deep-sea hydrothermal bacterium *Alteromonas infernus* [7]. This naturally slightly sulfated polysaccharide is composed of a nonasaccharide repeating unit with the main chain containing glucose (Glc), galacturonic acid (GalA), and galactose (Gal) residues. A short side chain constituted of two GlcA, Gal, and Glc is attached to a GalA residue of the main chain, bearing also a sulfate group [11]. Native HMW GY785 EPS and its low-molecular weight (LMW) chemically sulfated derivatives possess anti-coagulant [12] and anti-metastatic [13] properties, and favor chondrogenic differentiation of mesenchymal stem cells [14,15]. In summary, these EPS derivatives can inhibit some processes involved in tissue breakdown and inflammation, such as induction of matrix metalloproteases (MMP) by inflammatory cytokines (Interleukin-1β (IL-1β) and Tumor Necrosis Factor-alpha (TNF-α)) and complement cascade [10,12–15]. They can also promote in vitro cell proliferation and differentiation via major growth factors (Fibroblast Growth Factor (FGF)-2, Vascular Endothelial Growth Factor (VEGF), and Transforming Growth Factor (TGF)-β1) [11,13,14]. In similar way to heparin, EPS derivatives could also potentiate the osteogenic activities of Bone Morphogenetic Protein-2 (BMP-2) by regulating the binding to its receptors [16] or by exerting synergistic effects on osteoblasts combined with Wnt3 signaling protein involved in several development processes [17]. In contrast, they inhibit osteoclastogenesis and bone resorption. These derivatives play an important role in bone remodeling [18]. GAG-like properties of both EPS could therefore be exploited in elaboration of coatings enhancing the formation of new bone tissue on the implant surface.

In this study, Ti6Al4V samples were manufactured using an additive manufacturing method. Additive manufacturing allows the production of 3D structures with precise external dimensions and internal infrastructure, and can be used to fabricate a load-bearing implant with dimensions and architecture specifically tailored to the needs of an individual patient. The samples were subsequently coated with fibrils of collagen type I, both with and without derivatives of HE800 and GY785. The effect of the EPS derivatives on collagen fibril coating morphology and the attachment, morphology, and vitality of osteoblast-like MG63 cells was investigated.

2. Materials and Methods

2.1. HE800 and GY785 Exopolysaccharides (EPS) Production

Production and isolation of both EPS were previously described [7,19]. For HE800 and GY785 EPS production, respectively, *Vibrio diabolicus* and *Alteromonas infernus* were cultured in Zobell medium composed of 4 g/L of peptone, 1 g/L of yeast extract, and 33.3 g/L of aquarium salts at 25 °C and pH 7.4 in a fermenter containing 30 g/L of glucose, as a carbohydrate source. After 48 h of fermentation, the culture media were centrifuged (9000 g, 45 min), and the supernatants containing soluble EPS were ultrafiltrated on a 100 kDa cut-off membrane and freeze-dried.

2.2. Preparation of HE800 and GY785 EPS Derivatives

HE800 and GY785 derivatives were obtained by a free-radical depolymerization process using hydrogen peroxide, as previously described [20,21].

2.3. Characterization of EPS Derivatives

The properties of the EPS derivatives are shown in Table 1.

Table 1. Osidic composition (wt%), sulfur content, S (wt%), and weight-average molecular weight, Mw (g/moL), of HE800 and GY785 derivatives.

Exopolysaccharides (EPS) Derivative	Osidic Composition (wt%)						S (wt%)	Mw (g/moL)
	Gal	Glc	GalA	GlcA	GalNAc	GlcNAc		
HE800 derivative	0	0	0	19.8	10.6	10.8	0	280 000
GY785 derivative	19.2	16.8	6.9	9.3	0	0	3	240 000

2.3.1. Sugar Composition

Monosaccharide composition was determined according to the Kamerling et al. method [22], modified by Montreuil et al. [23]. Samples were hydrolyzed with 3 M MeOH/HCl for 4 h at 100 °C. Myo-inositol was used as an internal standard. The methyl glycosides obtained were then converted to trimethylsilyl derivatives with N,O-bis(trimethylsilyl)trifluoroacetamide and trimethylchlorosilane (BSTFA:TMCS) 99:1 (Merck). Gas Chromatography-Flame Ionisation Detector (GC-FID, Agilent Technologies 6890N, Santa Clara, CA, USA) was used to separate and quantify the per-*O*-trimethylsilyl methyl glycosides formed.

2.3.2. Molecular Weight

High-performance size-exclusion chromatography (HPSEC, Prominence, Shimadzu Co, Kyoto, Japan) coupled with multiangle light scattering (MALS, Dawn Heleos-II, Wyatt Technology, Santa Barbara, CA, USA) and differential refractive index (RI, Optilab, Wyatt technology) detectors was used to determine the weight-average molecular weight of the EPS derivatives. A refractive index increment dn/dc of 0.145 mL/g was applied to calculate the molecular weight.

2.3.3. Sulfate Content

Sulfate content in the samples was quantified by high-performance anion-exchange chromatography (HPAEC) using a Dionex DX-500 (Dionex, Sunnyvale, CA, USA), as previously described by Chopin et al. [20].

2.4. Atomic Force Microscopy (AFM): Sample Preparation and Imaging

HE800 and GY785 derivatives were firstly solubilized overnight at 1 mg/mL in water and then diluted at 5 µg/mL in water. Two microliters of each diluted solution were deposited onto a freshly cleaved mica surface. Samples were then immediately dried under ambient conditions before being imaged using a NanoWizard® atomic force microscope (AFM, JPK, Berlin, Germany) in intermittent contact mode at room temperature. In this imaging mode, rectangular cantilevers (Nanosensors NCL-W) with a spring constant of 40 N/m and a free resonance frequency of 165 kHz were used. The AFM tips, with a radius curvature of ~10 nm, were cleaned by UV-ozone treatment prior to AFM observation. JPK Data Processing software (JPK) was used for image processing and length measurements.

2.5. Production of Ti6Al4V Discs and Coating with Collagen Fibrils

The research used Ti6Al4V disks of 2 cm diameter and 2 mm thick produced using an additive manufacturing method (Electron Beam Melting, EBM) on an ARCAM EBM A2 (Arcam AB, Mölndal, Sweden) machine using the set of process parameters provided by the machine manufacturer, as described earlier [24].

Collagen fibril layers were formed by forming collagen hydrogels from acidic collagen solution, using the method of Karamachos et al. [25], on the surface of Ti6Al4V discs, as described in previous work [26]. The compositions of the hydrogels are shown in Table 2.

Table 2. The composition of the hydrogels produced. MEM–Minimum Essential Medium, ddH_2O–double-deionized water.

Components	Volume, µL
Collagen Type I (4 mg/mL)	280
10× MEM	40
HE800/GY785 derivative (5 mg/mL ddH_2O) or ddH_2O	80
1 M NaOH solution	~30

In brief, hydrogels were produced by mixing sterile solutions of collagen type I (BD Biosciences 354231, 4 mg/mL, San Jose, CA, USA), 10× Eagle's Minimum Essential Medium (MEM) (M0275, Sigma–Aldrich, Saint Louis, MO, USA), and EPS derivative solution (5 mg/mL) (or ddH_2O for control samples). Neutralization was performed by adding 2 µL increments of sterile-filtered 1 M sodium hydroxide solution until the color of the solution changed to purple (Figure S1). The purple solution was spread evenly on pre-autoclaved Ti6Al4V discs, and hydrogel formation took place at room temperature under sterile conditions for 2.5 h. Hydrogels were then removed from the surfaces of discs. Discs were rinsed three times in sterile double-deionized water (ddH_2O) and then dried under sterile conditions in a laminar flow hood, as described previously [27].

2.6. Scanning Electron Microscopy (SEM) of Ti6Al4V Discs Coated with Collagen Fibrils and EPS Derivatives

Scanning electron microscopy (SEM) was performed with a JEOL (JEOL Ltd., Tokyo, Japan) in secondary electron mode at an acceleration voltage of 5 keV. Prior to SEM analysis, Ti6Al4V discs were coated with a thin layer of gold. Samples were dried prior to gold coating under sterile conditions in a laminar flow hood (see Section 2.5).

2.7. Cell Biological Characterization of Ti6Al4V Discs Coated with Collagen Fibrils and EPS Derivatives

2.7.1. Cell Culture and Cell Seeding

The human osteosarcoma cell line MG-63 American type culture collection (ATCC) was routinely cultured in Dulbecco's Modified Eagle Medium (DMEM) supplemented with 10% foetal bovine serum (FBS) and 1% penicillin/streptomycin. Cells were incubated at 37 °C in a humidified 5% CO_2 environment until cultures reached 70–80% confluence. Following trypsinization, the trypan-blue exclusion assay was used to determine % viability prior to cell seeding. For all cell viability experiments, cells were resuspended in phenol-free media before a total of 3×10^4 cells were seeded on to Ti6Al4V alloy discs with or without collagen or collagen-EPS derivative coatings.

2.7.2. Cell Viability

To assess whether cells were able to attach and proliferate on Ti6Al4V alloy discs with and without collagen or collagen-EPS derivative coatings, the Presto-Blue cell viability assay was performed after cells were incubated on Ti6Al4V samples for 1, 4, and 7 days. Live/Dead imaging was performed after 7 days. Prior to both procedures, cell seeded Ti6Al4V alloy discs were washed with Dulbecco's phosphate buffered saline (DPBS) twice. The Presto-Blue cell viability reagent (ThermoFisher Scientific, Waltham, MA, USA) was diluted 1 in 10 in phenol-free DMEM before 1.5 ml was incubated with each sample for 4 h. The fluorescent signal of a 200 µL aliquot was read using an excitation of 560 nm and an emission of 590 nm. Diluted PrestoBlue reagent in the absence of cells or Ti6Al4V samples was used to measure background fluorescence which was subtracted from samples containing cells. In addition, PrestoBlue reagent was also incubated with Ti6Al4V alloy discs lacking cells to determine whether the samples interfered with the assay.

For Live/Dead imaging, cell-seeded Ti6Al4V alloy discs were incubated in DPBS containing 1 µg/mL Hoechst, 2 µM calcein-AM, and 4 µM ethidium homodimer-I for 30 min at room temperature. Individual fluorescent images were taken on an Axio Scope A1 LED microscope (Zeiss, Jena, Germany) and enhanced using ImageJ software.

3. Results

The results of SEM analysis are shown in Figure 1. The results demonstrated the formation of collagen fibril coatings on Ti6Al4V surfaces. Fibrils exhibited a banding morphology typical for collagen. It appeared that the presence of anionic EPS derivatives increased the diameter of collagen fibrils, but this cannot be concluded conclusively from the SEM data. On surfaces coated with fibrils formed in the presence of both EPS derivatives, short "threads" were observed associated with fibrils. Such threads were absent in samples coated with pure collagen fibrils.

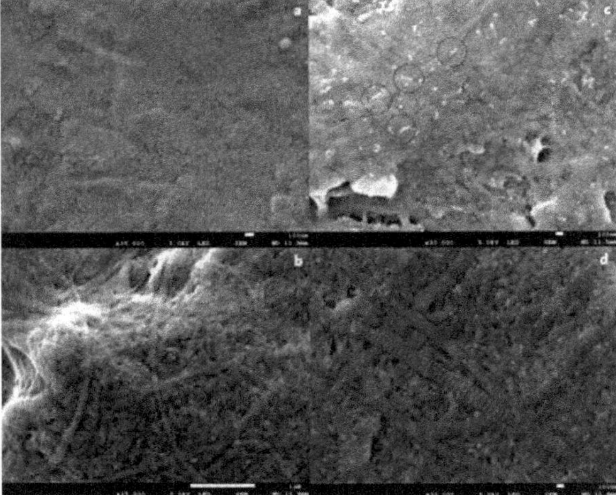

Figure 1. Scanning electron microscope (SEM) images of collagen fibril coatings on Ti6Al4V. (**a**) Bare Ti6Al4V; (**b**) collagen coating; (**c**) collagen + GY875 derivative coating; (**d**) collagen + HE800 derivative coating. Scale bars: (**a**) 100 nm; (**b**) 100 nm; (**c**) 1 µm; (**d**) 100 nm. Representative white "threads" on (**c**,**d**) have been indicated by the red circles.

The results of AFM analysis of dried, highly-diluted EPS derivative solutions are shown in Figure 2. The AFM images revealed that HE800 derivative formed threads of 702 ± 164 nm (N = 50) in length, which remained inter-connected due most likely to non-covalent interactions, such as water-mediated hydrogen bonds, van der Waals forces, and/or electrostatic interactions between anionic polysaccharide chains and residual ions remaining in the sample (Figure 2a). In contrast to HE800 derivative, GY785 derivative was present as short individual threads of 119 ± 31 nm in length (N = 80) (Figure 2b).

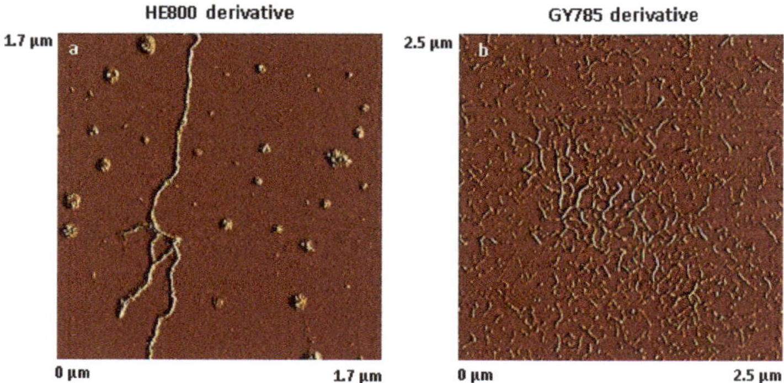

Figure 2. Atomic force microscope (AFM) images of dried, highly-diluted exopolysaccharides (EPS) solutions. (**a**) HE800 derivative (1.7 µm × 1.7 µm) and (**b**) GY785 derivative (2.5 µm × 2.5 µm).

The results of cell viability studies are shown in Figure 3. Values appeared to be similar after 1, 4, and 7 days of culture on all sample types, regardless of the surface on which they were cultured.

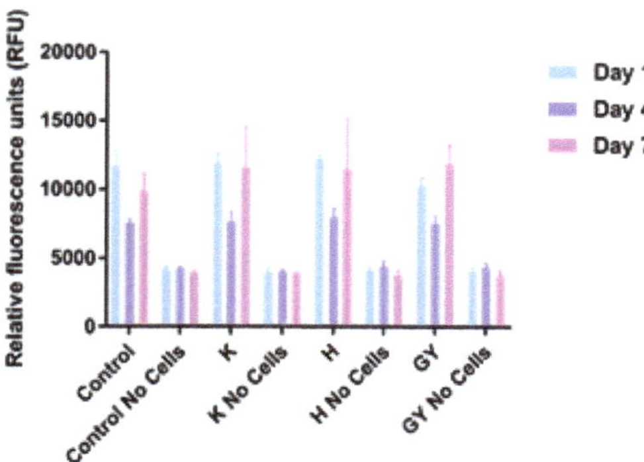

Figure 3. Proliferation assay of MG63 cells on Ti6Al4V samples after 1, 4, and 7 days. Error bars indicate standard deviation. Control: tissue culture polystyrene; H: collagen + HE800 derivative; K: bare Ti6Al4V; EY: collagen + GY875 derivative.

The results of Live/Dead staining and fluorescence microscopy are shown in Figure 4. Cells on all sample groups exhibited a spread morphology, which is characteristic for good adhesion. Nearly all cells appeared viable. Only a few dead, red-stained cells were observed.

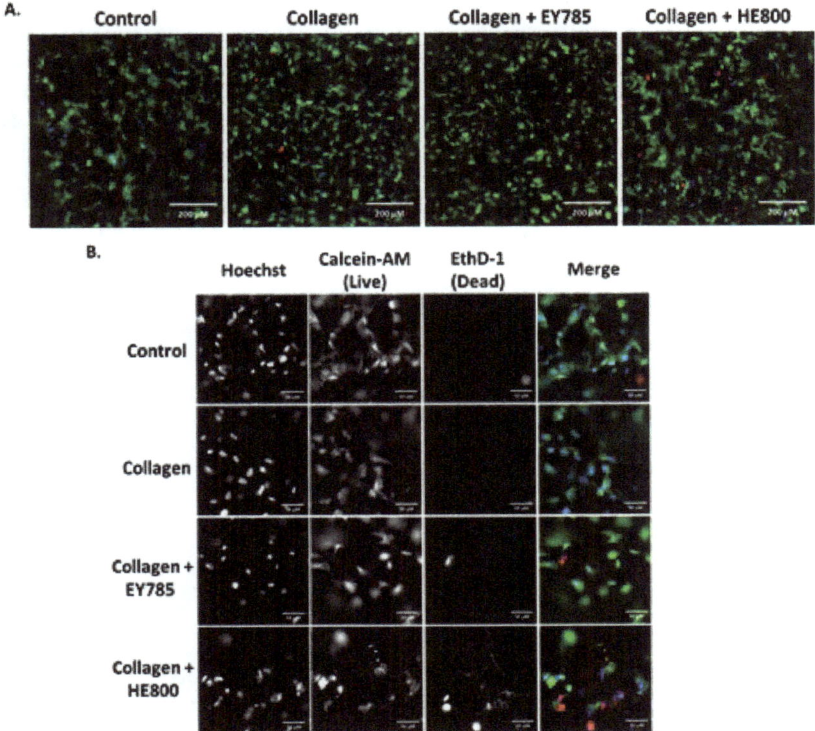

Figure 4. Fluorescence microscopy imaging of MG63 cells on Ti6Al4V samples after 7 days. Control: bare Ti6Al4V; blue: Hoechst staining; green: calcein-AM staining; red; Ethd-1 staining. (**A**) Images taken in widefield; (**B**) images taken at 5× magnification (with channels split). Scale bars: (**A**) 200 μm; (**B**) 50 μm.

4. Discussion

In this study, SEM analysis (Figure 1) demonstrated the formation of collagen fibril coatings on Ti6Al4V surfaces. It appeared that the presence of anionic EPS derivatives increased the thickness of collagen fibrils, but this cannot be concluded conclusively from the SEM data. No definite influence of EPS derivatives on banding morphology was observed.

It was previously shown that the addition of the native HMW HE800 EPS during collagenous matrix remodeling increased the formation of *D*-periodic striated collagen fibrils [10]. In other previous studies, formation of fibrillar collagen in the presence of anionic polysaccharides, such as alginate or GAG (e.g., heparin and hyaluronic acid), has been reported [28–30]. The kinetics of collagen self-assembly and the fibril thickness were highly affected by the nature of the polysaccharide due to electrostatic interactions between negatively charged polysaccharides and positively charged regions on collagen. Different authors have reported that polysaccharides influence the thickness of collagen fibrils. For example, the presence of alginate led to an increase in collagen fibers [30]. However, other authors have reported that other polysaccharides, such as GAG, decrease fibril thickness. Addition of the GAG chrondroitin sulfate reduced the thickness of fibrils of collagen types I and II in previous work [6]. Sulfated hyaluronic acid led to a decrease in collagen fibril diameter [28].

AFM analysis of dried, highly-diluted EPS derivative solutions (Figure 2) revealed the formation of "threads" which are in a similar size range to the threads observed on SEM images (Figure 1). On both SEM and AFM images revealed that HE800 derivative formed longer threads than GY785 derivative. The threads were present at different points on the fibrils. In other words, threads did not appear to associate preferentially with any region of the fibril. It is not clear, from the results of this study, whether formation of the threads takes place prior to or after EPS derivatives bind to the surface of fibrils. The nature of the bonding between EPS derivatives and fibrils in this study is not clear. Non-covalent interactions, such as water-mediated hydrogen bonds, van der Waals forces, or electrostatic interactions between anionic polysaccharide chains and positive charges on the surface of the fibrils, can be expected to play a role.

In this study, no obvious advantage of the collagen fibril coating nor the presence of EPS derivatives on cell viability and morphology was demonstrated (Figures 3 and 4). We decided to use the PrestoBlue cell viability reagent and Live/Dead staining, which have been used in hundreds of published studies to-date. The use of microscopy and Presto Blue has some limitations. We considered performing flow cytometry, but decided against it, because it would require cells to be removed from the EPS/collagen coated TiAl6V discs. Trypsinisation of cells where collagen is present could reduce the number of cells collected due to non-specific cleavage of proteins. In addition, increasing incubation times with trypsin may reduce cell viability.

It was previously shown that the addition of the native HMW HE800 EPS promoted human dermal fibroblast migration and proliferation [10]. Native HE800 and GY785 both promoted attachment of osteoblast and chondrocyte cell lines in previous work [31]. The derivatives used in this study are of smaller molecular weight. It has been reported that biological activity of polysaccharides, such as their effect on cell proliferation, can be affected by molecular weight [32].

Further work should focus on the use of primary cells, which may be more sensitive to differences in coating structure than cell lines. Osteogenic differentiation should also be investigated. From the point of view of physicochemical characterization of the coatings, it would be desirable to develop a technique to detect and quantify EPS chemically. There may be differences in coating thicknesses that might influence the results. Hence, it would be desirable to develop a technique to determine coating thickness.

5. Conclusions

Collagen fibril layers were formed successfully on Ti6Al4V discs produced by rapid prototyping. EPS derivatives were found associated with the fibrils on the coatings. Coatings did not markedly influence the attachment, morphology, and vitality of MG63 osteoblast-like cells cultured on the Ti6Al4V discs.

Supplementary Materials: The following are available online at http://www.mdpi.com/2072-666X/10/1/68/s1, Figure S1: Hydrogels formed by neutralization of acidic collagen solution containing exopolysaccharides (EPS). Left: hydrogel containing GY785; middle: hydrogel containing HE800; right: hydrogel without EPS.

Author Contributions: Conceptualization, T.E.L.D. and R.A.S.; methodology, T.E.L.D., K.N., and J.G.K.; formal analysis, K.N.; investigation, K.N., O.I.M., A.Z., S.C.-J., C.S., S.C., and A.K.; resources, A.Z., S.C.-J., C.S., S.C., and A.K.; writing—original draft preparation, T.E.L.D., O.I.M., K.N., S.C.-J., C.S., S.C., and A.Z.; writing—review and editing, T.E.L.D.; visualization, O.I.M. and S.C.; supervision, R.A.S., M.A.S., T.E.L.D., and J.G.K.; project administration, T.E.L.D., M.A.S., and R.A.S.; funding acquisition, T.E.L.D., O.I.M., M.A.S., and R.A.S.

Funding: This research was funded by the N8 Agrifood pump priming grant Food2Bone. Lancaster University Materials Science Institute and the Foreign and Commonwealth office (FCO), United Kingdom, are thanked for providing financial support (research stay of Oksana I. Mishukova at Lancaster University). The samples were prepared with financial support from Russian Science Foundation (project number 15-13-00043).

Acknowledgments: Sara Baldock, Lancaster University, is thanked for her assistance with SEM imaging.

Conflicts of Interest: The authors declare no conflict of interest. The funders had no role in the design of the study; in the collection, analyses, or interpretation of data; in the writing of the manuscript, or in the decision to publish the results.

References

1. Becker, D.; Geissler, U.; Hempel, U.; Bierbaum, S.; Scharnweber, D.; Worch, H.; Wenzel, K.W. Proliferation and differentiation of rat calvarial osteoblasts on type I collagen-coated titanium alloy. *J. Biomed. Mater. Res.* **2002**, *59*, 516–527. [CrossRef] [PubMed]
2. Bierbaum, S.; Hempel, U.; Geissler, U.; Hanke, T.; Scharnweber, D.; Wenzel, K.W.; Worch, H. Modification of Ti6Al4V surfaces using collagen i, iii, and fibronectin. Ii. Influence on osteoblast responses. *J. Biomed. Mater. Res. A* **2003**, *67*, 431–438. [CrossRef] [PubMed]
3. Geissler, U.; Hempel, U.; Wolf, C.; Scharnweber, D.; Worch, H.; Wenzel, K. Collagen type I-coating of Ti6Al4V promotes adhesion of osteoblasts. *J. Biomed. Mater. Res.* **2000**, *51*, 752–760. [CrossRef]
4. Gomez-Guillen, M.C.; Gimenez, B.; Lopez-Caballero, M.E.; Montero, M.P. Functional and bioactive properties of collagen and gelatin from alternative sources: A review. *J. Food Hydrocolloids* **2011**, *25*, 1813–1827. [CrossRef]
5. Bierbaum, S.; Douglas, T.; Hanke, T.; Scharnweber, D.; Tippelt, S.; Monsees, T.K.; Funk, R.H.; Worch, H. Collageneous matrix coatings on titanium implants modified with decorin and chondroitin sulfate: Characterization and influence on osteoblastic cells. *J. Biomed. Mater. Res. A* **2006**, *77*, 551–562. [CrossRef]
6. Douglas, T.; Heinemann, S.; Mietrach, C.; Hempel, U.; Bierbaum, S.; Scharnweber, D.; Worch, H. Interactions of collagen types I and II with chondroitin sulfates a-c and their effect on osteoblast adhesion. *Biomacromolecules* **2007**, *8*, 1085–1092. [CrossRef]
7. Raguénès, G.H.; Peres, A.; Ruimy, R.; Pignet, P.; Christen, R.; Loaëc, M.; Rougeaux, H.; Barbier, G.; Guezennec, J. *Alteromonas infernus* sp. Nov., a new polysaccharide-producing bacterium isolated from a deep-sea hydrothermal vent. *J. Appl. Microbiol.* **1997**, *82*, 422–430. [CrossRef]
8. Rougeaux, H.; Kervarec, N.; Pichon, R.; Guezennec, J. Structure of the exopolysaccharide of vibrio diabolicus isolated from a deep-sea hydrothermal vent. *Carbohydr. Res.* **1999**, *322*, 40–45. [CrossRef]
9. Zanchetta, P.; Lagarde, N.; Guezennec, J. A new bone-healing material: A hyaluronic acid-like bacterial exopolysaccharide. *Calcif. Tissue Int.* **2003**, *72*, 74–79. [CrossRef]
10. Senni, K.; Gueniche, F.; Changotade, S.; Septier, D.; Sinquin, C.; Ratiskol, J.; Lutomski, D.; Godeau, G.; Guezennec, J.; Colliec-Jouault, S. Unusual glycosaminoglycans from a deep sea hydrothermal bacterium improve fibrillar collagen structuring and fibroblast activities in engineered connective tissues. *Mar. Drugs* **2013**, *11*, 1351–1369. [CrossRef]
11. Roger, O.; Kervarec, N.; Ratiskol, J.; Colliec-Jouault, S.; Chevolot, L. Structural studies of the main exopolysaccharide produced by the deep-sea bacterium alteromonas infernus. *Carbohydr. Res.* **2004**, *339*, 2371–2380. [CrossRef] [PubMed]
12. Colliec Jouault, S.; Chevolot, L.; Helley, D.; Ratiskol, J.; Bros, A.; Sinquin, C.; Roger, O.; Fischer, A.M. Characterization, chemical modifications and in vitro anticoagulant properties of an exopolysaccharide produced by alteromonas infernus. *Biochim. Biophys. Acta* **2001**, *1528*, 141–151. [CrossRef]
13. Heymann, D.; Ruiz-Velasco, C.; Chesneau, J.; Ratiskol, J.; Sinquin, C.; Colliec-Jouault, S. Anti-metastatic properties of a marine bacterial exopolysaccharide-based derivative designed to mimic glycosaminoglycans. *Molecules* **2016**, *21*, 309. [CrossRef] [PubMed]
14. Merceron, C.; Portron, S.; Vignes-Colombeix, C.; Rederstorff, E.; Masson, M.; Lesoeur, J.; Sourice, S.; Sinquin, C.; Colliec-Jouault, S.; Weiss, P.; et al. Pharmacological modulation of human mesenchymal stem cell chondrogenesis by a chemically oversulfated polysaccharide of marine origin: Potential application to cartilage regenerative medicine. *Stem Cells* **2012**, *30*, 471–480. [CrossRef] [PubMed]
15. Rederstorff, E.; Rethore, G.; Weiss, P.; Sourice, S.; Beck-Cormier, S.; Mathieu, E.; Maillasson, M.; Jacques, Y.; Colliec-Jouault, S.; Fellah, B.H.; et al. Enriching a cellulose hydrogel with a biologically active marine exopolysaccharide for cell-based cartilage engineering. *J. Tissue Eng. Regen. Med.* **2017**, *11*, 1152–1164. [CrossRef] [PubMed]
16. Smith, R.A.A.; Murali, S.; Rai, B.; Lu, X.; Lim, Z.X.H.; Lee, J.J.L.; Nurcombe, V.; Cool, S.M. Minimum structural requirements for bmp-2-binding of heparin oligosaccharides. *Biomaterials* **2018**, *184*, 41–55. [CrossRef] [PubMed]
17. Ling, L.; Dombrowski, C.; Foong, K.M.; Haupt, L.M.; Stein, G.S.; Nurcombe, V.; van Wijnen, A.J.; Cool, S.M. Synergism between wnt3a and heparin enhances osteogenesis via a phosphoinositide 3-kinase/akt/runx2 pathway. *J. Biol. Chem.* **2010**, *285*, 26233–26244. [CrossRef]

18. Ruiz-Velasco, C.; Baud'huin, M.; Sinquin, C.; Maillasson, M.; Heyman, D.; Colliec-Jouault, S.; Padrines, M. Effects of a sulfated exopolysaccharide produced by alteromonas infernus on bone biology. *Glycobiology* **2011**, *21*, 781–795. [CrossRef]
19. Raguénès, G.; Christen, R.; Guezennec, J.; Pignet, P.; Barbier, G. *Vibrio diabolicus* sp. Nov., a new polysaccharide-secreting organism isolated from a deep-sea hydrothermal vent polychaete annelid, alvinella pompejana. *Int. J. Syst. Bacteriol.* **1997**, *47*, 989–995. [CrossRef]
20. Chopin, N.; Sinquin, C.; Ratiskol, J.; Zykwinska, A.; Weiss, P.; Cerantola, S.; Le Bideau, J.; Colliec-Jouault, S. A direct sulfation process of a marine polysaccharide in ionic liquid. *Biomed. Res. Int.* **2015**, *2015*, 508656. [CrossRef]
21. Senni, K.; Gueniche, F.; Yousfi, M.; Fioretti, F.; Godeau, G.; Colliec-Jouault, S. Sulfated Depolymerized Derivatives of Exopolysaccharides (eps) from Mesophilic Marine Bacteria, Method for Preparing Same, and Uses Thereof in Tissue Regeneration. U.S. Patent Application No. 11/629,579, 5 June 2008.
22. Kamerling, J.P.; Gerwig, G.J.; Vliegenthart, J.F.; Clamp, J.R. Characterization by gas-liquid chromatography-mass spectrometry and proton-magnetic-resonance spectroscopy of pertrimethylsilyl methyl glycosides obtained in the methanolysis of glycoproteins and glycopeptides. *Biochem. J.* **1975**, *151*, 491–495. [CrossRef] [PubMed]
23. Montreuil, J.; Bouquelet, S.; Debray, H.; Fournet, B.; Spik, G.; Strecker, G. A pratical approach. In *Glycoptoteins in Carbohydrate Analysis*; Chaplin, M.F., Kennedy, J.F., Eds.; IRL Press: Oxford, UK, 1986; pp. 143–204.
24. Popov, V.; Muller-Kamskii, G.; Kovalevsky, A.; Dzhenzhera, G.; Strokin, E.; Kolomiets, A.; Ramon, J. Design and 3d-printing of titanium bone implants: Brief review of approach and clinical cases. *J. Biomed. Eng. Lett.* **2018**, *8*, 337–344. [CrossRef] [PubMed]
25. Karamichos, D.; Brown, R.A.; Mudera, V. Complex dependence of substrate stiffness and serum concentration on cell-force generation. *J. Biomed. Mater. Res. A* **2006**, *78*, 407–415. [CrossRef] [PubMed]
26. Vandrovcova, M.; Douglas, T.E.L.; Heinemann, S.; Scharnweber, D.; Dubruel, P.; Bacakova, L. Collagen-lactoferrin fibrillar coatings enhance osteoblast proliferation and differentiation. *J. Biomed. Mater. Res. Part A* **2015**, *103*, 525–533. [CrossRef] [PubMed]
27. Douglas, T.E.L.; Hempel, U.; Zydek, J.; Vladescu, A.; Pietryga, K.; Kaeswurm, J.A.H.; Buchweitz, M.; Surmenev, R.A.; Surmeneva, M.A.; Cotrut, C.M.; et al. Pectin coatings on titanium alloy scaffolds produced by additive manufacturing: Promotion of human bone marrow stromal cell proliferation. *Mater. Lett.* **2018**, *227*, 225–228. [CrossRef]
28. Rother, S.; Salbach-Hirsch, J.; Moeller, S.; Seemann, T.; Schnabelrauch, M.; Hofbauer, L.C.; Hintze, V.; Scharnweber, D. Bioinspired collagen/glycosaminoglycan-based cellular microenvironments for tuning osteoclastogenesis. *ACS Appl. Mater. Interfaces* **2015**, *7*, 23787–23797. [CrossRef]
29. Salchert, K.; Oswald, J.; Streller, U.; Grimmer, M.; Herold, N.; Werner, C. Fibrillar collagen assembled in the presence of glycosaminoglycans to constitute bioartificial stem cell niches in vitro. *J. Mater. Sci. Mater. Med.* **2005**, *16*, 581–585. [CrossRef]
30. Tsai, S.W.; Liu, R.L.; Hsu, F.Y.; Chen, C.C. A study of the influence of polysaccharides on collagen self-assembly: Nanostructure and kinetics. *Biopolymers* **2006**, *83*, 381–388. [CrossRef]
31. Rederstorff, E.; Weiss, P.; Sourice, S.; Pilet, P.; Xie, F.; Sinquin, C.; Colliec-Jouault, S.; Guicheux, J.; Laib, S. An in vitro study of two gag-like marine polysaccharides incorporated into injectable hydrogels for bone and cartilage tissue engineering. *Acta Biomater.* **2011**, *7*, 2119–2130. [CrossRef]
32. Zhao, N.; Wang, X.; Qin, L.; Guo, Z.; Li, D. Effect of molecular weight and concentration of hyaluronan on cell proliferation and osteogenic differentiation in vitro. *Biochem. Biophys. Res. Commun.* **2015**, *465*, 569–574. [CrossRef]

Article

Internal Structure of Matrix-Type Multilayer Capsules Templated on Porous Vaterite $CaCO_3$ Crystals as Probed by Staining with a Fluorescence Dye

Lucas Jeannot [1], Michael Bell [2], Ryan Ashwell [2], Dmitry Volodkin [2,3] and Anna S. Vikulina [2,4,*]

1 Robert Schuman University Institute of Technology (IUT Robert Schuman), University of Strasbourg, 72 Route Du Rhin, 67411 Illkirch CEDEX, France; lucas.jeannot68@gmail.com

2 School of Science and Technology, Nottingham Trent University, Clifton Lane, Nottingham NG11 8NS, UK; michael.bell2013@my.ntu.ac.uk (M.B.); ryan.ashwell2014@my.ntu.ac.uk (R.A.); dmitry.volodkin@ntu.ac.uk (D.V.)

3 Department of Chemistry, Lomonosov Moscow State University, Leninskiye Gory 1-3, 119991 Moscow, Russia

4 Department Cellular Biotechnology & Biochips, Branch Bioanalytics and Bioprocesses (Fraunhofer IZI-BB), Fraunhofer Institute for Cell Therapy and Immunology, Am Mühlenberg 13, 14476 Potsdam-Golm, Germany

* Correspondence: anna.vikulina@izi-bb.fraunhofer.de; Tel.: +49-331-5818-7122

Received: 28 September 2018; Accepted: 23 October 2018; Published: 25 October 2018

Abstract: Multilayer capsules templated on decomposable vaterite $CaCO_3$ crystals are widely used as vehicles for drug delivery. The capsule represents typically not a hollow but matrix-like structure due to polymer diffusion into the porous crystals during multilayer deposition. The capsule formation mechanism is not well-studied but its understanding is crucial to tune capsule structure for a proper drug release performance. This study proposes new approach to noninvasively probe and adjust internal capsule structure. Polymer capsules made of poly(styrene-sulfonate) (PSS) and poly(diallyldimethylammonium chloride) (PDAD) have been stained with fluorescence dye rhodamine 6G. Physical-chemical aspects of intermolecular interactions required to validate the approach and adjust capsule structure are addressed. The capsules consist of a defined shell (typically 0.5–2 μm) and an internal matrix of PSS-PDAD complex (typically 10–40% of a total capsule volume). An increase of ionic strength and polymer deposition time leads to the thickening of the capsule shell and formation of a denser internal matrix, respectively. This is explained by effects of a polymer conformation and limitations in polymer diffusion through the crystal pores. We believe that the design of the capsules with desired internal structure will allow achieving effective encapsulation and controlled/programmed release of bioactives for advanced drug delivery applications.

Keywords: layer-by-layer; self-assembly; mesoporous; calcium carbonate; fluorescence

1. Introduction

The layer-by-layer (LbL) assembly of oppositely charges polyelectrolytes is a simple but powerful method allowing the design of multilayer polymer architectures [1–4]. Typically, this method assumes the alternating polymer deposition on core materials which can be either flat surfaces [5,6] or 3D structures [2–4]. The LbL coating of sacrificial 3D templates (cores) such as polystyrene [7,8], melamine formaldehyde [9,10], manganese carbonate [10], and calcium carbonate particles [1,11,12], includes further removal of the colloidal cores and subsequent formation of the multilayer capsules. Such capsules have a polymer shell and inner cavity that may be loaded with various kinds of therapeutic molecules including hydrophilic and hydrophobic low-molecular weight drugs, (bio)polymers, proteins and enzymes, hormones, and DNA [13].

The major advantages and the key properties of the vaterite calcium carbonate crystals are (i) highly developed mesoporous internal structure that offers a large surface for encapsulation of molecules of interest; (ii) decomposition of these cores at mild conditions using slightly acidic solvents or chelating agents (e.g., EDTA and citric acid), and (iii) simple, reproducible, and inexpensive method of crystal synthesis in lab [13]. The capsules templated on the $CaCO_3$ cores have been successfully applied for various biomedical applications including intracellular and extracellular drug delivery [14,15] and medical diagnostics [16]. A number of recent works on $CaCO_3$ templated capsules is devoted to the investigation of drug release performance of the capsules [17–21]. Different release behavior for the capsules of identical composition but templated on the cores of different nature ($MnCO_3$ and melamine formaldehyde) revealed that the capsules prepared using different core materials can be either hollow or filled with an internal polymer matrix present into the capsule lumen [10]. This depends on the porosity of the cores and ability of polymers to diffuse into the core.

Hollow capsules consist of semipermeable shells and empty internal cavity that can be filled with encapsulated molecules. The release from such structures is usually limited by the diffusion across the capsule shell. Instead, matrix-type capsules are filled with a polymer network that can host and restrain encapsulated molecules and therefore to govern the release of the molecules [22,23].

Modern methods used for the investigation of the internal structure of the capsules are usually based on imaging technologies and include characterization of capsule mechanical properties using atomic force microscopy (AFM) [24]; capsule structure using scanning electron microscopy (SEM) including cryo-SEM [25] and environmental SEM [26] to analyze fully hydrated and chemically unmodified state of the capsules. However, these methods can be destructive or require drying of the capsules. Some additional information can also be obtained by probing the capsule permeability using fluorescent recovery after photobleaching [27].

Herein, we investigate the internal structure of $CaCO_3$-templated multilayer capsules composed from model synthetic well-studied polymers poly(styrene-sulfonate) and poly(diallyldimethylammonium chloride) or PSS/PDAD for short. Deposition of PSS/PDAD multilayers onto the cores of different nature, e.g., polystyrene [28], melamine formaldehyde [29], calcium phosphate [30], and vaterite particles [31] is widely used for the fabrication of multilayer PSS/PDAD capsules. Both PDAD and PSS polymers are synthetic; although they may not be easily biodegradable in blood compositions, these polymers possess negligible cell toxicity when used in low concentrations suitable for a wide range of biomedical applications. For instance, quantum dots protected by a layer of PDAD have been shown to have low toxicity and are used for the cell analysis detection and imaging [32]. Gold nanorods coated with either PSS/PDAD or single PDAD layer have also negligible effects on cell functions and viability [33]. It also has been reported that PSS/PDAD nanocapsules do not change the cell culture metabolic conditions as was probed using breast cancer cells [34].

We present a new approach to identify the capsule structure that is based on the postloading of preformed capsules with the fluorescent dye, rhodamine 6G (R6G). We focus on analysis of interpolymer PSS-PDAD interaction and the interactions of the polymers with the dye. This is investigated via fluorescence characteristics of the dye in the presence of polymers and their complexes. A way to tune the capsule internal structure by variation of the capsule preparation conditions such as ionic strength and polymer deposition time is considered. We hope that in future this study will allow preprogramming the release profile for drug delivery and other biological applications based on the utilization of the $CaCO_3$-templated capsules.

2. Materials and Methods

Calcium chloride dehydrate ($CaCl_2{\cdot}2H_2O$), sodium carbonate (Na_2CO_3), sodium chloride (NaCl), poly(styrene-sulfonate) (PSS, average MW 70 kDa), poly(diallyldimethylammonium chloride) (PDAD, molecular weight 200–350 kDa), rhodamine 6G (R6G), and ethylenediaminetetraacetic acid sodium salt (EDTA) were purchased from Sigma-Aldrich (Seelze, Germany). All chemicals were used without further purification. TRIS-buffered saline (TBS, 10X), pH 7.4 (J60764), contained 250 mM TRIS, 27 mM potassium chloride, and 1.37 M sodium chloride, was from Alfa Aesar (Heysham, UK). Stock TRIS buffer solution was diluted 10 times for the experiments. All solutions were prepared using Millipore water having a resistivity higher than 18.2 MΩ·cm.

2.1. Fabrication of $CaCO_3$ Vaterite Crystals

$CaCO_3$ crystals were synthesized as described with the slight modifications [35]. A 0.33 M solution of Na_2CO_3 in water was added to the equal volume of 0.33 M $CaCl_2$ in water and agitated at 650 rpm for 30 s. After mixing the solution was left to crystallize for 10 min. For the washing, the suspension of the crystals was centrifuged at 1000× g for 3 min and supernatant was removed. $CaCO_3$ was then washed by resuspension in water followed by re-centrifuging and supernatant extraction. The crystals were dried in the oven preheated at 70 °C for 1–2 h.

2.2. LbL-Based PSS/PDAD Capsule Formation

Dry $CaCO_3$ crystals (10 mg) were suspended in 0.5 mL of NaCl of differing concentration (0.05 M, 0.3 M, and 0.1 M). Once suspended in solution, 1 mL of 2 mg·mL^{-1} PSS (dissolved in the NaCl solution with respective concentration 0.05 M, 0.3 M, or 0.1 M) was added to the suspension of calcium carbonate cores. The cores were incubated and shaken in this mixture for 3 min, 10 min or 20 min following by centrifugation at 1000× g for 3 min. The supernatant was then removed and the particles were washed twice with 1.5 mL of NaCl solution with respective concentration, re-suspended and centrifuged under the same conditions. For addition of the second polymer layer, PDAD, the same process as for the PSS layer was sequentially repeated. The capsules with n = 1 to 6 number of layers have been fabricated. Crystals coated with polyelectrolyte layers have been analyzed at the same day as multilayers have been prepared.

$CaCO_3$ cores has been removed by dissolution in 0.2 M EDTA (with the pH adjusted to 7.4) prior to the loading with R6G and further analysis.

2.3. Postloading of PSS/PDAD Capsules

Suspension of the capsules containing approximately 10^3–10^4 capsules (estimated based on the average capsule size and assuming the 100% yield for both core fabrication and capsule formation) was incubated with R6G (final concentration 0.1–8 µM) for 30 min and the imaging of the capsules has been performed directly in the presence of R6G in the supernatant.

2.4. Fluorescence Microscopy

Analysis of the microparticles prepared in this study was carried out using fluorescence microscopy (EVOS FL, Thermo Fisher Scientific, Waltham, MA, USA). The imaging was performed by keeping imaging parameters (acquisition time, laser power, magnification) constant. The excitation wavelength used was 530 nm.

2.5. R6G Binding to PSS, PDAD and Their Complex in the Solution

For the first set of experiments, aqueous solution of R6G was rapidly added to water or PSS or PDAD dissolved in water. Final concentration of R6G varied in the range of 0.2 to 2 mM while polymer concentration was fixed at 0.10 mg·mL^{-1} for PSS and 0.08 mg·mL^{-1} for PDAD. For the second set of experiments, 0.5 mM·R6G was rapidly added to pre-formed PSS/PDAD complex (mass ratio 1:1, PSS concentration 0.04–0.2 mg·mL^{-1} or different mass ratios for 0.1 mg·mL^{-1} PSS). After intensive shaking for 1 min, all samples have been filtrated using Amicon Ultra-0.5 with ultracel-3 Membrane (Merck Millipore, Darmstadt, Germany) with a threshold of 3 kDa by centrifugation at 15,000× g

for 20 min. The supernatants were collected and transferred to 25 mM TRIS buffer solution pH 7.4 containing 137 mM NaCl for the measurements. Absorbance spectra have been recorded from 2 μL drops of non-diluted samples using NanoDrop One Microvolume UV–Vis Spectrophotometer (Thermo Fisher Scientific). Measurements were performed in triplicates.

2.6. Characterization of the Crystals and Microcapsules

Analysis of the morphology of vaterite crystals microcapsules prepared in this study was carried out using scanning electron microscopy (SEM, Zeiss DSM 40, Goettingen, Germany). $CaCO_3$ crystals and $(PSS/PDAD)_2PSS$ capsules were dried and analyzed by light optical microscopy (EVOS FL, Thermo Fisher Scientific) on the same day.

3. Results and Discussion

3.1. $CaCO_3$ Templates: Internal Structure

Vaterite microcrystals were obtained by conventional method of mixing equimolar solutions of $CaCl_2$ and Na_2CO_3. According to SEM images (Figure 1a), dried crystals had spherical shape with a diameter of 8.6 ± 3.5 μm (n = 50). The pore size in the microspheres prepared by similar procedure has previously been reported [36] and was found to be in the range of 5 to 40 nm. The crystals have highly developed internal structure having total surface area of about 10 $m^2 \cdot g^{-1}$ [37]. Herein, porous internal structure of vaterite crystals is evidenced by the SEM imaging of the broken crystals (Figure 1b). The channel-like structure of interconnected pores inside $CaCO_3$ crystals allows them to host an enormous amount of encapsulates or, in the same way, to fill the crystals with a polymeric matrix [31,38–40].

Figure 1. SEM images of $CaCO_3$ vaterite crystals (**a**) demonstrating internal structure of the broken crystal (**b**).

3.2. Formation of PSS/PDAD Capsules

The well-investigated polyelectrolyte pair of PSS and PDAD (Figure 2) has been used to prepare multilayer capsules. For all the experiments, PSS has been used as a first layer. Sequential polymer deposition has been followed by the dissolution of $CaCO_3$ core by the addition of 0.2 M EDTA. Figure 3 shows the light transmittance images of $CaCO_3$ crystals coated with $(PSS/PDAD)_2/PSS$ multilayers during core dissolution in real time. The crystals of a larger size have been used for this experiment for the purpose of better visualization.

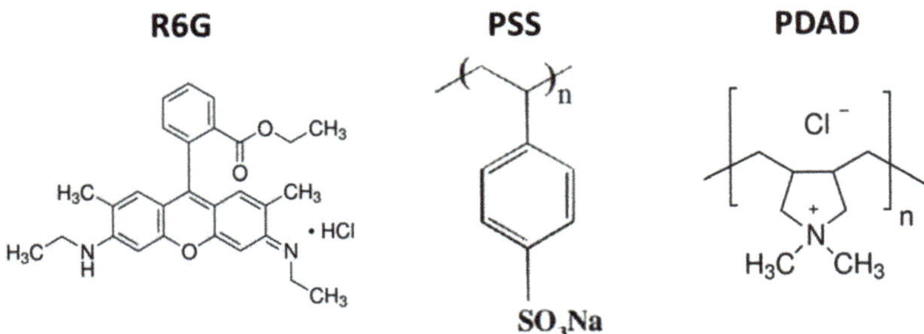

Figure 2. Chemical formulae of R6G and polyelectrolytes (poly(styrene-sulfonate) (PSS), poly(diallyldimethylammonium chloride) (PDAD)) used in this study.

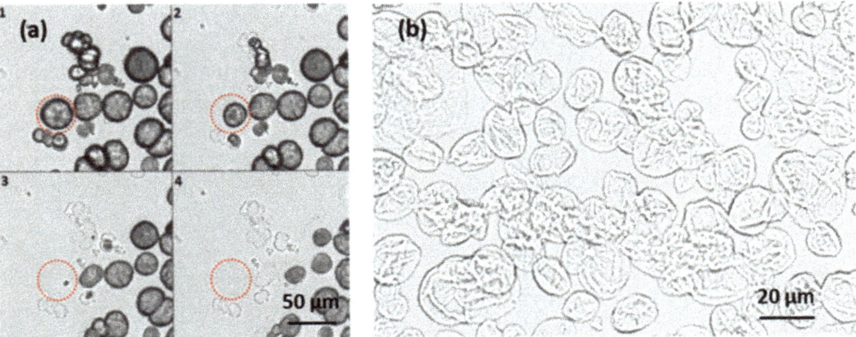

Figure 3. (**a**) Optical images of capsule formation via addition of 0.2M EDTA solution to $CaCO_3$ crystals coated with $(PSS/PDAD)_2/PSS$ multilayers. The images follow a chronological order from 1 to 4 with 30 s interval and (**b**) optical image of dried $(PSS/PDAD)_2/PSS$ capsules.

First, we focused on capsule stability and an internal structure. The investigation of the stability of the crystals coated with different number of PSS/PDAD layers allowed us to reveal the optimal conditions for the formation of the microcapsules (Figure 4). It is known that storage of the vaterite crystals in water for long time (overnight or more) results in recrystallization of vaterite to more stable calcite polymorph [41]. Calcite crystals have typical cubic shape that allows to easy distinguish them from spherical vaterite crystals. Uncoated vaterite crystals undergo complete recrystallization to calcite while stored overnight (Figure 4b,c). We found that the degree of recrystallization of vaterite crystals significantly dropped down for three or more deposited polymer layers (Figure 4a). This can be explained by stabilization of the crystals coated with multilayers, similar effect was observed for capsules made of PSS and poly(allylamine hydrochloride (PAH) [42].

Figure 3b shows more detailed morphology of $(PSS/PDAD)_2/PSS$ capsules. Apparently, the capsules have a relatively smooth morphology since the capsules appear rather uniform and flat upon drying. Despite visual flattening, the capsules are usually not hollow inside but have a polymer complex in the internal lumen due to permeation of polymers through crystal pores during the LbL coating procedure [1,2]. The complex can affect capsule properties.

An increase of the number of layers (from three to six) led to the strengthening of the capsule shell that resulted in less prominent capsule shrinkage during core dissolution (Figure 4d). The shrinkage takes place due to annealing the polymer structure into the formed capsules. This annealing is driven by closure of some voids between polymers in order to create more ionic pairs in the polymer complex. As an example, images of $CaCO_3$ crystals coated with three polymer layers before and after addition of 0.2 M EDTA are shown in Figure 4e,f. The more layers deposited, the more pronounced is the shrinkage (Figure 4d); most probably this can be explained by the following. There is more filling the pores with polymer complex for higher numbers of deposited layers. The polymer complex restricts physically the shrinkage of the capsules. The effect of the shrinkage can be completely eliminated for the capsules formed by more than five layers, however, in this case the dissolution of the core required significantly longer time (Figure 4d). Altogether, these results suggest that the optimal PSS/PDAD capsules templated on the $CaCO_3$ cores are assembled from four or five layers.

Interestingly, $CaCO_3$-templated PSS/PDAD capsules appear to possess better mechanical stability compared to those templated on other cores, e.g., melamine formaldehyde [38]. It is also of note that PSS/PDAD capsules prepared in this study were not prone to the swelling as it was previously demonstrated for PSS/PDAD melamine formaldehyde-templated capsules of a similar size [39]. This can be explained by low osmotic pressure generated inside the capsules during the core dissolution step because of a quick release of ions of the dissolved $CaCO_3$ core.

The results described above allow us to assume that the formed polymeric matrix inside $CaCO_3$ template should remain its structure after the elimination of the core. However, the structure may be affected by shrinkage which cannot be avoided, however, a small amount of shrinkage for capsules with four and five layers can be accepted and an influence of the layer number can thus be studied. Staining of the capsules with the dye having high affinity to one of the polymers used, i.e., PSS, could help identify the distribution of polymers within the capsule interior. This approach will be further used. It is noninvasive and is based on binding of a fluorescent probe R6G (Figure 2) to free permanent charges on the PSS backbone. Such an approach first requires a quantitative analysis of PSS interaction with the fluorescent dye. These issues are further considered in the next two sections.

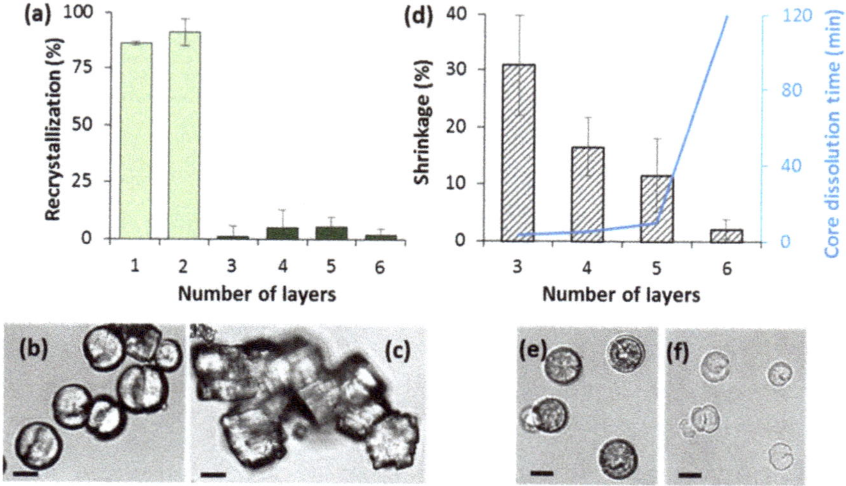

Figure 4. (**a**) Degree of the recrystallization as a percentage of calcite crystals for $CaCO_3$ vaterite crystals coated with different number of PSS/PDAD layers after storage in TRIS buffer overnight. Error bars are standard deviations (SD) for $n = 3$ samples. (**b**,**c**): Optical images of freshly prepared uncoated $CaCO_3$ crystals (b) and these crystals stored in TRIS buffer overnight (**c**). (**d**) Time required for the dissolution of $CaCO_3$ core by the addition of 0.2 M EDTA and degree of the shrinkage of PSS/PDAD capsules composed of different number of layers. Error bars are SD (calculated for at least 10 capsules). (**e**,**f**): $CaCO_3$ crystals coated with (PSS/PDAD)/PSS layers before (**e**) and after 5 min of incubation with 0.2 M EDTA (**f**). Scale bars are 10 µm.

3.3. Fluorescence of R6G in the Presence of PSS and PDAD

One of the most common and well-studied fluorescent dyes, R6G (Figure 2), has been chosen here as a marker to understand the polymer distribution inside the polymer capsules. R6G is known to have high affinity to PSS due to hydrophobic interactions as well as ionic contacts between R6G and PSS. Besides this, the R6G molecule has a small size (MW 442) and a high diffusion coefficient of ~4.3×10^{-10} $m^2 \cdot s^{-1}$ [43] that eliminates diffusional limitations and allows the post-loading of capsules with this dye. This made R6G an ideal candidate for the task above.

Prior to the investigation of the interaction of R6G with the polymer complex, the molecular complexes of R6G with both capsule components, PSS and PDAD, have been formed and studied.

Figure 5a shows the spectra of free R6G and its complexes with PDAD and PSS: yellow, red, and gray lines, respectively. Concentrations of 0.10 $mg \cdot mL^{-1}$ for PSS and 0.08 $mg \cdot mL^{-1}$ for PDAD correspond to 0.5 mM of polymer monomer units (MW of the monomers is 206 for sodium salt of PSS and 162 for PDAD). This allowed the formation of equimolar complex of 0.5 mM R6G with both polymers. The measurements were performed in TRIS buffer solution pH 7.4 containing 137 mM NaCl, the same medium as used for capsule fabrication. The wavelengths for maximum adsorption of a monomer and a dimer were found to be 533 nm and 500 nm, respectively. This is insignificantly lower than those values reported for R6G dissolved in water (526 nm for a monomer and 498 nm for a dimer) [44] that may be explained by the use of TRIS-buffer in our study. The addition of PDAD does not lead to any significant changes in R6G spectrum, while the strong attraction of R6G to PSS backbone resulted in the shift of the whole absorbance spectra towards longer wavelengths and the reversal of intensity for the monomer and dimer adsorption maxima. Further elimination of R6G bound to PSS via ultracentrifugation allows retrieving the shape of initial R6G spectrum (Figure 5a, black line).

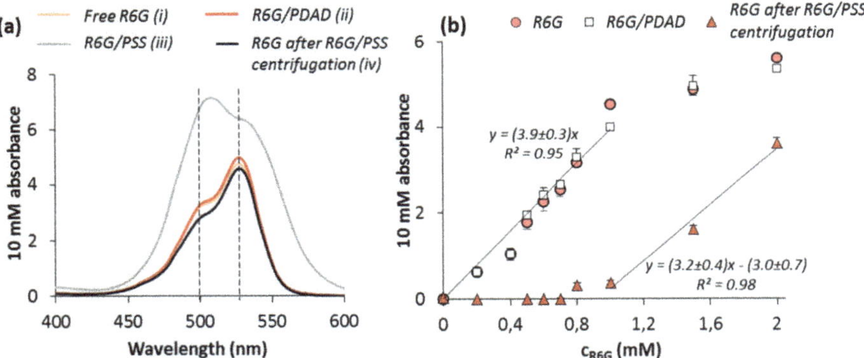

Figure 5. (**a**) Absorbance spectra of R6G (**i**) in absence of polymers; (**ii**) in presence of PDAD; (**iii**) in presence of PSS; and (**iv**) after incubation with PSS that was further removed by ultracentrifugation. (**b**) Concentration dependence of 10 mM absorbance of freeR6G at 533 nm, R6G in presence of PDAD and after incubation with PSS that was further removed by ultracentrifugation. Twenty-five millimolar TRIS buffer solution pH 7.4 containing 137 mM NaCl was used as a solvent.

The linear dependence of R6G absorbance at its maximum on the dye concentration (Figure 5b) was found for the range up to approximately 1 mM for both, free R6G and R6G in the presence of PDAD. Further increase of R6G concentration is most likely associated with R6G self-quenching and consequent deviation from the linear law [45]. Keeping in mind the shift of the maxima in the spectra of R6G bound to PSS, the PSS-R6G complex was separated from free R6G by ultracentrifugation prior to the measurements. In contrast to PDAD, the addition of PSS resulted in binding of all R6G molecules for the concentration up to 1 mM. For the concentration of R6G higher than 1 mM, linear dependence with the same slope as for free R6G has been constructed with linear coefficients 3.2 ± 0.4 for R6G/PSS and 3.9 ± 0.3 for R6G (Figure 5b). This can be explained by the saturation of all PSS binding cites by R6G at polymer:R6G molar ratio of approximately 1:2. Taking into account the reversal of the monomer and dimer maxima observed for the R6G-PSS complex (Figure 5a), it can be assumed that the interaction of R6G with PSS leads to its dimerization. This is in agreement with previously reported findings for R6G [46,47] and other dyes of a similar structure [48].

3.4. Interaction of R6G with PSS-PDAD Complex and Multilayers

One can assume that R6G and PDAD may compete for binding to the PSS molecule and therefore the loading of PSS/PDAD multilayers with R6G may cause the weakening of interpolymer interaction and affect the structure of multilayer capsules. In order to probe and compare the force of R6G-PSS and PDAD-PSS interaction, firstly, R6G was added to pre-formed polymer complex of a varied concentration (Figure 6a). The complex was formed at PSS:PDAD mass ratio of 1:1. Apparently, the linear decrease of free R6G concentration in solution (R^2 = 0.974) in the contact with polymer complex can be explained by quantitative binding of R6G to free binding sites of PSS. Importantly, it appears that R6G does not destroy or interpose the pre-formed PSS/PDAD complex which is also evidenced by the linearity of the observed concentration curve (Figure 6a). For the second set of experiments, the addition of R6G to PSS/PDAD complex, formed for different polymer ratios (Figure 6a), also revealed linear dependence of the amount of bound R6G from the free binding sites of PSS that reaches the saturation when all PSS binding sites are occupied by PDAD.

This obviously indicates the anchorage of R6G to free PSS binding sites. Importantly, it seems that the equilibrium in in the interaction between R6G, PSS, and PDAD is shifted towards the formation of the polyelectrolyte complex PSS-PDAD. Therefore, the addition of R6G to the pre-formed PDAD-PSS complex does not lead to disintegration of the latest. This gives us the possibility to use R6G as a marker for staining of PSS inside the PSS/PDAD multilayers.

In contrast to polymer complex, polyelectrolyte multilayers templated on vaterite cores consist of unknown amount of polymers. Because of this, the concentration of R6G to be used for the staining of the capsules required adjustment. Figure 6a shows the dependence of fluorescence signal accumulated inside the capsules on the concentration of R6G solution in the contact with them. Similarly to R6G interaction with polymer complex in solution, the increase of R6G concentration leads to the increase of fluorescent signal accumulated in the capsules (Figure 6b). This is also accompanied by the increase of background fluorescence. Rapid accumulation of fluorescence in the capsules is followed by the plateau that corresponds to the region of saturation of the capsules with R6G. Herein that is also of worth to note that self-quenching of R6G fluorescence is not likely at this concentration range (Figure 5b). Based on these results, an R6G concentration of 8 μM has been chosen for further experiments.

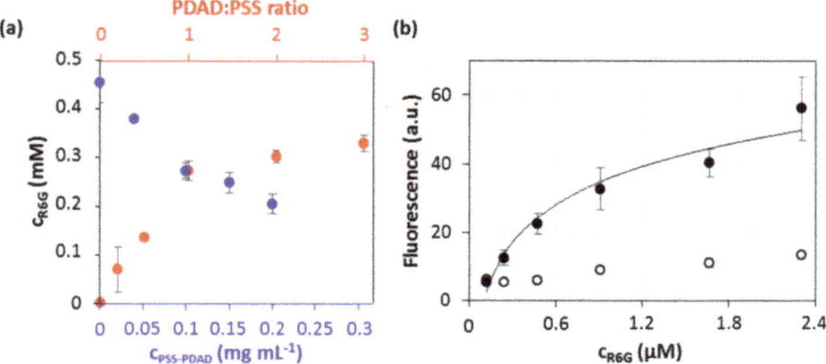

Figure 6. (**a**) Concentration of free R6G in the solution after incubation with the PSS-PDAD complex (the complex was further removed by ultracentrifugation) as a function of concentration of PSS-PDAD complex (PSS:PDAD mass ratio 1:1)—blue axis—and as a function of PDAD:PSS mass ratio (PSS concentration of 0.1 mg·mL^{-1})—red axis. (**b**) Cumulative fluorescence of the capsules (black circles) and background fluorescence (empty circles) as a function of initial concentration of R6G added to the suspension of $(PSS/PDAD)_2/PSS$ capsules. SD are given for n = 4.

3.5. Imaging of the Internal Capsule Structure

Using the approach developed above, pre-formed PSS/PDAD multilayer capsules were postloaded with R6G and their internal structure has been investigated via fluorescence imaging as described in Figure 7. Capsules were characterized by (i) shell thickness calculated as a width at half-peak height and (ii) capsule filled volume (red filled area). To estimate the latest, two fluorescence peaks of the capsule shell were fitted with two Gaussian functions (gray line) and their total area was subtracted from the total area under the profile and correlated with the maximum capsule volume (yellow filled area).

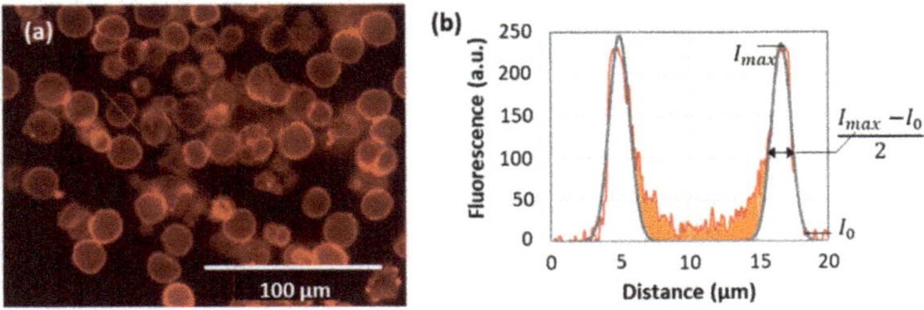

Figure 7. (**a**) Fluorescence images of $(PSS/PDAD)_2/PSS$ capsules postloaded with R6G. (**b**) Typical mathematical treatment of the fluorescence profile depicted as a white line in (a) and taken across the capsule.

The effect of the last layer in polymer deposition sequence has further been studied to understand the capsule formation mechanism. $(PSS/PDAD)_2$ and $(PSS/PDAD)_2/PSS$ capsules have been assembled on 15 ± 5 µm ($n = 40$) crystals and had the same size of 14 ± 2 µm ($n = 40$). The deposition of PSS as the last layer resulted in the saturation of the whole capsule with PSS that is evidenced by more than two times higher cumulative fluorescence of $(PSS/PDAD)_2/PSS$ capsules while compared with $(PSS/PDAD)_2$ (Figure 8a,b). Interestingly, this did not affect the overall distribution of PSS: the filled volume was found to be $(47 \pm 1)\%$ for $(PSS/PDAD)_2/PSS$ capsules and $(46 \pm 5)\%$ for $(PSS/PDAD)_2$ capsules prepared under the same conditions (Figure 8c). The only explanation can be that both polymers can permeate inside the pores of the crystals and form a polymer matrix. At the same time, up until five deposited layers, multiple polymer deposition of polymer molecules is accompanied by their spontaneous redistribution between the capsule interior matrix and the capsule shell. More layers deposited may create diffusion limitations that will reduce an increment in the densification of the internal matrix and make the shell thicker.

Figure 9 shows the schematic structure of the capsules based on the results of the analysis of the staining profiles. PSS/PDAD multilayers form the shell of the capsule and the internal matrix inside the capsule. For the low filling of the capsule interior with polymeric matrix, the capsules are expected to behave as a semipermeable barrier. In this case, the capsule can be assumed as a hollow sphere and the shell of the capsule is supposed to play a role of a membrane that regulates transport and release of encapsulated drugs. The synthesis and release kinetics studies for this type of capsules have been reported, for instance, for chitosan/alginate $CaCO_3$-templated capsules [49] or PAH/PSS capsules templated on nanoporous anodic alumina [50]. On the other hand, the filling of the capsules with the polymer leads to the formulation of filled (matrix-type) capsules. Matrix capsules are typically prone to a different release mechanism that is mostly determined by the composition and internal molecular structure of the matrix [22]. Some examples include poly(L-glutamic acid)/chitosan microcapsules templated on melamine formaldehyde [51], chitosan/alginate capsules built up on liposomes [52], or carrageenan/chitosan capsules deposited onto oil nanoemulsion droplets [53].

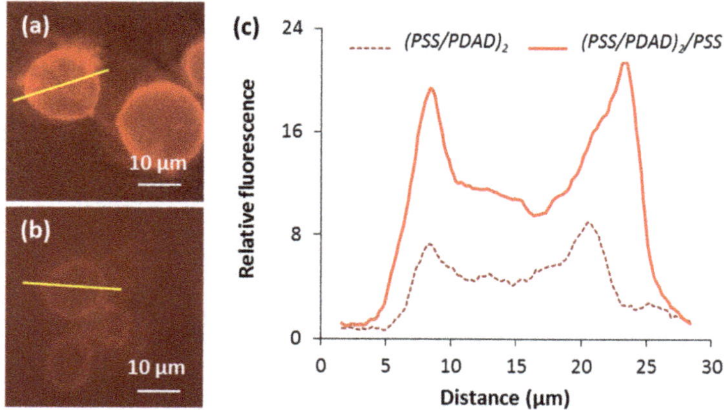

Figure 8. Fluorescence images of (**a**) $(PSS/PDAD)_2/PSS$ and (**b**) $(PSS/PDAD)_2$ capsules stained with R6G under the same conditions. (**c**) Fluorescence profiles across the capsules show the distribution of R6G (correspond to yellow lines in (**a**,**b**)).

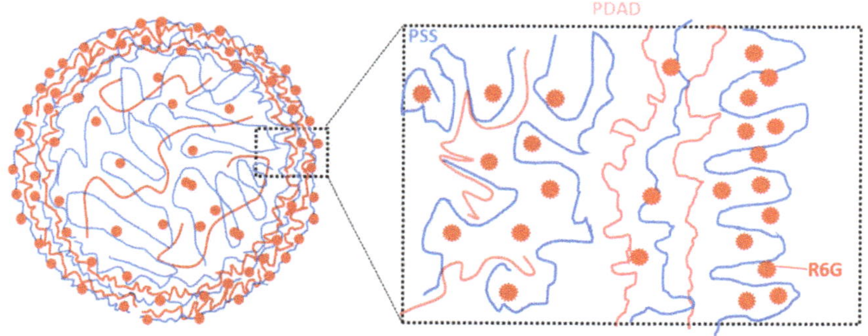

Figure 9. The scheme of the structure of PSS/PDAD multilayer capsule stained with R6G. PSS and PDAD molecules form internal matrix and the shell of the capsule. R6G binds to backbones of the PSS molecules.

3.6. How to Adjust Capsule Internal Structure?

In order to better understand the mechanism of the formation of a polymer matrix in the capsule and to evaluate factors that can affect the internal structure of the capsules, the effect of the preparation conditions on the distribution of PSS inside the capsules was investigated.

First, the effect of ionic strength has been studied (Figure 10a,b). The increase of NaCl concentration in the medium during capsule formation resulted in the thickening of the capsule shell (Figure 10a) whereas the filled volume remained the same (Figure 10b). Previously it has been shown that shell thickening leads to the deceleration of the drug release from core-shell polyelectrolyte capsules [54,55].

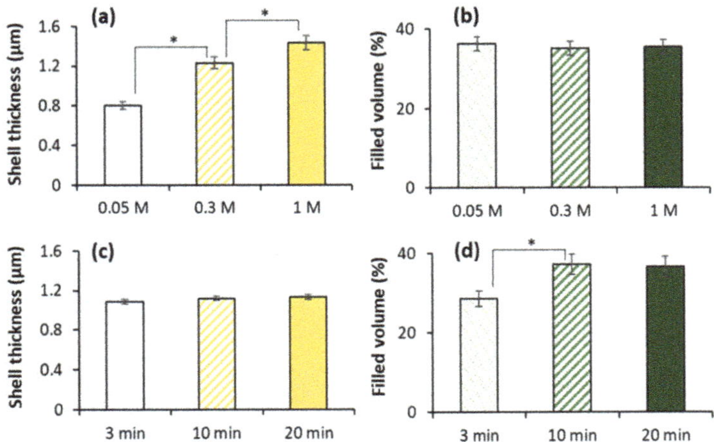

Figure 10. The influence of the ionic strength in the incubation media (**a**,**b**) and deposition time for each polymer layer (**c**,**d**) on $(PSS/PDAD)_2/PSS$ capsule shell thickness and filled volume. Error bars are SD for at least $n = 3$. * indicates statistical difference of the values ($p < 0.95$).

On the contrary, deposition time for each polymer step influences the amount of PSS loaded inside vaterite templates (Figure 10d), while the thickness of the capsule wall is not affected (Figure 10c). It seems that the increase of the deposition time allows PSS of the first deposition layer to diffuse deeper inside the crystals and to form the polymer matrix. At the same time, the formation of PSS/PDAD multilayers requires less time that the diffusion of PSS inside the core, therefore polyelectrolyte complex is formed on the crystal surface at shorter times and the shell thickness is not affected by variation of the deposition time. Formulation of a densely packed polyelectrolyte networks results in reduction in the cumulative release rate due to strong electrostatic interactions as it has been shown for liposome-templated capsules [53].

The combination of two approaches to vary capsule shell thickness and polymer filling ratio presented in Figure 10 will further allow to control the structure of PSS/PDAD capsules. This is an important step in the understanding of the mechanism of capsule formation and will help to tune release profiles from drug-loaded capsules that is crucial for biomedical and clinical uses.

4. Conclusions

This study demonstrated development of novel approaches for the investigation of internal structure of polyelectrolyte capsules based on nondestructive postloading with fluorescent probes. The fabrication of PSS/PDAD capsules composed from different number of layers was described in terms of capsule mechanical stability assessed by the degree of recrystallization of vaterite $CaCO_3$ and the changes of the capsule volume (the effect of shrinkage) during core elimination. The structure of PSS/PDAD capsules templated onto vaterite $CaCO_3$ crystals has been investigated. It was shown that polymers partially penetrate in the internal pores of vaterite cores and are not liberated after core removal forming internal matrix of the capsule.

The internal structure of PSS/PDAD capsules can easily be controlled by varying the conditions of polymer deposition. The increase of ionic strength of surrounding medium during LbL deposition results in the thickening of the capsule shell while the structure of internal PSS matrix remains unaffected. On the other hand, the increase of the time of polymer deposition allows polymer molecules to better fill in the internal pores of $CaCO_3$ core and to form more polymer matrix in the capsule after the core removal. This is not accompanied by any changes in the thickness of the capsule wall. Taking together, these two approaches can be successfully applied to further tune the structure of the capsules.

The results of this study may open new perspectives to control and pre-program release rate and profile that is essential for further utilization of multilayer capsules as drug delivery carriers. In addition, this approach can be transferred to planar multilayers in order to better understand their growth mechanism and transport through such highly charged systems [56]. This staining approach can be used to adjust properties of the multilayers to be employed as mimics of the extracellular matrix and as carriers for hosting bioactives and delivery of bioactives to cells as actively studied in our laboratory [57–59].

Author Contributions: A.S.V. and D.V. conceived and designed the experiments; L.J., M.B., R.A., and A.S.V. participated in the experimental design and performed the experiments; all authors contributed to data analysis; D.V. contributed to funding acquisition, materials, and analysis tools; A.S.V. coordinated the work and wrote the first paper draft. A.S.V. and D.V. were the supervisors. L.J and D.V. contributed to the writing of the final manuscript.

Acknowledgments: The work was performed within the framework of the M.V. Lomonosov Moscow State University state task, part 2 (government grant AAAA-A16-116052010081-5). This work was supported in part by M.V., Lomonosov Moscow State University Program of Development. L.J. acknowledges Erasmus+ EU program (project "Hollow and matrix-type capsules made of polymer multilayers for drug delivery"). A.V. thanks the support from QR fund (NTU). The authors thank Kathryn Kroon for help with SEM imaging.

Conflicts of Interest: The authors declare no conflicts of interest.

References

1. Volodkin, D.V.; Larionova, N.I.; Sukhorukov, G.B. Protein encapsulation via porous $CaCO_3$ microparticles templating. *Biomacromolecules* **2004**, *5*, 1962–1972. [CrossRef] [PubMed]
2. Volodkin, D.V.; Petrov, A.I.; Prevot, M.; Sukhorukov, G.B. Matrix Polyelectrolyte Microcapsules: New System for Macromolecule Encapsulation. *Langmuir* **2004**, *20*, 3398–3406. [CrossRef] [PubMed]
3. Johnston, A.P.; Cortez, C.; Angelatos, A.S.; Caruso, F. Layer-by-layer engineered capsules and their applications. *Curr. Opin. Colloid Interface Sci.* **2006**, *11*, 203–209. [CrossRef]
4. Yu, W.; Chen, Y.; Mao, Z. Hollow Polyelectrolyte Microcapsules as Advanced Drug Delivery Carriers. *J. Nanosci. Nanotechnol.* **2016**, *16*, 5435–5446. [CrossRef] [PubMed]
5. Volodkin, D.; von Klitzing, R.; Moehwald, H. Polyelectrolyte Multilayers: Towards Single Cell Studies. *Polymers* **2014**, *6*, 1502–1527. [CrossRef]
6. Izumrudov, V.A.; Mussabayeva, B.K.; Murzagulova, K.B. Polyelectrolyte multilayers: Preparation and applications. *Russ. Chem. Rev.* **2018**, *87*, 192–200. [CrossRef]
7. Caruso, R.A.; Susha, A.; Caruso, F. Multilayered Titania, Silica, and Laponite Nanoparticle Coatings on Polystyrene Colloidal Templates and Resulting Inorganic Hollow Spheres. *Chem. Mater.* **2001**, *13*, 400–409. [CrossRef]
8. Katagiri, K.; Shishijima, Y.; Koumoto, K.; Inumaru, K. Preparation of pH-Responsive Hollow Capsules via Layer-by-Layer Assembly of Exfoliated Layered Double Hydroxide Nanosheets and Polyelectrolytes. *J. Nanosci. Nanotechnol.* **2018**, *18*, 110–115. [CrossRef] [PubMed]
9. Skirtach, A.G.; de Geest, B.G.; Mamedov, A.; Antipov, A.A.; Kotov, N.A.; Sukhorukov, G.B. Ultrasound stimulated release and catalysis using polyelectrolyte multilayer capsules. *J. Mater. Chem.* **2007**, *17*, 1050–1054. [CrossRef]
10. Sukhorukov, G.B.; Shchukin, D.G.; Dong, W.-F.; Möhwald, H.; Lulevich, V.V.; Vinogradova, O.I. Comparative Analysis of Hollow and Filled Polyelectrolyte Microcapsules Templated on Melamine Formaldehyde and Carbonate Cores. *Macromol. Chem. Phys.* **2004**, *205*, 530–535. [CrossRef]
11. Bukreeva, T.V.; Marchenko, I.V.; Parakhonskiy, B.V.; Grigor'ev, Y.V. Formation of silver nanoparticles on shells of polyelectrolyte capsules using silver-mirror reaction. *Colloid J* **2009**, *71*, 596–602. [CrossRef]
12. Feoktistova, N.; Rose, J.; Prokopović, V.Z.; Vikulina, A.S.; Skirtach, A.; Volodkin, D. Controlling the Vaterite $CaCO_3$ Crystal Pores. Design of Tailor-Made Polymer Based Microcapsules by Hard Templating. *Langmuir* **2016**, *32*, 4229–4238. [CrossRef] [PubMed]
13. Volodkin, D. $CaCO_3$ templated micro-beads and -capsules for bioapplications. *Adv. Colloid Interface Sci.* **2014**, *207*, 306–324. [CrossRef] [PubMed]

14. Muñoz Javier, A.; Kreft, O.; Semmling, M.; Kempter, S.; Skirtach, A.G.; Bruns, O.T.; del Pino, P.; Bedard, M.F.; Rädler, J.; Käs, J.; et al. Uptake of Colloidal Polyelectrolyte-Coated Particles and Polyelectrolyte Multilayer Capsules by Living Cells. *Adv. Mater.* **2008**, *20*, 4281–4287. [CrossRef]
15. Anandhakumar, S.; Nagaraja, V.; Raichur, A.M. Reversible polyelectrolyte capsules as carriers for protein delivery. *Colloids Surf. B* **2010**, *78*, 266–274. [CrossRef] [PubMed]
16. Trichet, V.; Layrolle, P.; Escriou, V. Lipid nanoparticles for siRNA delivery in lungs. *Nanomedicine* **2012**, *7*, 181–183. [CrossRef] [PubMed]
17. Kurapati, R.; Raichur, A.M. Composite cyclodextrin–calcium carbonate porous microparticles and modified multilayer capsules: Novel carriers for encapsulation of hydrophobic drugs. *J. Mater. Chem. B* **2013**, *1*, 3175. [CrossRef]
18. Radhakrishnan, K.; Raichur, A.M. Biologically triggered exploding protein based microcapsules for drug delivery. *Chem. Commun.* **2012**, *48*, 2307–2309. [CrossRef] [PubMed]
19. Marchenko, I.; Yashchenok, A.; Borodina, T.; Bukreeva, T.; Konrad, M.; Möhwald, H.; Skirtach, A. Controlled enzyme-catalyzed degradation of polymeric capsules templated on $CaCO_3$: influence of the number of LbL layers, conditions of degradation, and disassembly of multicompartments. *J. Controll. Release* **2012**, *162*, 599–605. [CrossRef] [PubMed]
20. Tiwari, S.; Mishra, B. Multilayered membrane-controlled microcapsules for controlled delivery of isoniazid. *Daru J. Fac. Pharm. Tehran Univ. Med. Sci.* **2011**, *19*, 41–46.
21. Wang, C.; Ye, S.; Sun, Q.; He, C.; Ye, W.; Liu, X.; Tong, Z. Microcapsules for controlled release fabricated via layer-by-layer self-assembly of polyelectrolytes. *J. Exp. Nanosci.* **2008**, *3*, 133–145. [CrossRef]
22. Siegel, R.A.; Rathbone, M.J. Overview of Controlled Release Mechanisms. In *Fundamentals and Applications of Controlled Release Drug Delivery*; Siepmann, J., Siegel, R.A., Rathbone, M.J., Eds.; Springer US: Boston, MA, USA, 2012; pp. 19–43.
23. Dash, S.; Murthy, P.N.; Nath, L.; Chowdhury, P. Kinetic modeling on drug release from controlled drug delivery systems. *Acta Pol. Pharm.* **2010**, *67*, 217–223. [PubMed]
24. Kozlovskaya, V.; Kharlampieva, E.; Drachuk, I.; Cheng, D.; Tsukruk, V.V. Responsive microcapsule reactors based on hydrogen-bonded tannic acid layer-by-layer assemblies. *Soft Matter* **2010**, *6*, 3596. [CrossRef]
25. Behra, M.; Schmidt, S.; Hartmann, J.; Volodkin, D.V.; Hartmann, L. Synthesis of porous PEG microgels using $CaCO_3$ microspheres as hard templates. *Macromol. Rapid Commun.* **2012**, *33*, 1049–1054. [CrossRef] [PubMed]
26. Krajčovič, T.; Bučko, M.; Vikartovská, A.; Lacík, I.; Uhelská, L.; Chorvát, D.; Neděla, V.; Tihlaříková, E.; Gericke, M.; Heinze, T.; et al. Polyelectrolyte Complex Beads by Novel Two-Step Process for Improved Performance of Viable Whole-Cell Baeyer-Villiger Monoxygenase by Immobilization. *Catalysts* **2017**, *7*, 353. [CrossRef]
27. Uhlig, K.; Madaboosi, N.; Schmidt, S.; Jäger, M.S.; Rose, J.; Duschl, C.; Volodkin, D.V. 3d localization and diffusion of proteins in polyelectrolyte multilayers. *Soft Matter* **2012**, *8*, 11786. [CrossRef]
28. Marchenko, I.V.; Parakhonsky, G.V.; Bukreeva, T.V.; Plotnikov, G.S.; Baranov, A.N.; Saletsky, A.M. Embedding of fluorescent dyes into polyelectrolyte capsules for remote destruction of the capsule shell by laser irradiation. In Proceedings of the Saratov Fall Meeting 2009, Saratov, Russia, 21 September 2009; p. 75470I. [CrossRef]
29. Gao, C.; Leporatti, S.; Donath, E.; Möhwald, H. Surface Texture of Poly(styrenesulfonate sodium salt) and Poly(diallyldimethylammonium chloride) Micron-Sized Multilayer Capsules: A Scanning Force and Confocal Microscopy Study. *J. Phys. Chem. B* **2000**, *104*, 7144–7149. [CrossRef]
30. Elizarova, I.S.; Luckham, P.F. Fabrication of polyelectrolyte multilayered nano-capsules using a continuous layer-by-layer approach. *J. Colloid Interface Sci.* **2016**, *470*, 92–99. [CrossRef] [PubMed]
31. Han, Y.; Bu, J.; Zhang, Y.; Tong, W.; Gao, C. Encapsulation of photosensitizer into multilayer microcapsules by combination of spontaneous deposition and heat-induced shrinkage for photodynamic therapy. *Macromol. Biosci.* **2012**, *12*, 1436–1442. [CrossRef] [PubMed]
32. Hu, H.-Y.; Dou, X.-R.; Jiang, Z.-L.; Tang, J.-H.; Xie, L.; Xie, H.-P. Cytotoxicity and cellular imaging of quantum dots protected by polyelectrolyte. *J. Pharm. Anal.* **2012**, *2*, 293–297. [CrossRef] [PubMed]
33. Hauck, T.S.; Ghazani, A.A.; Chan, W.C.W. Assessing the effect of surface chemistry on gold nanorod uptake, toxicity, and gene expression in mammalian cells. *Small* **2008**, *4*, 153–159. [CrossRef] [PubMed]

34. Bazylińska, U.; Pietkiewicz, J.; Saczko, J.; Nattich-Rak, M.; Rossowska, J.; Garbiec, A.; Wilk, K.A. Nanoemulsion-templated multilayer nanocapsules for cyanine-type photosensitizer delivery to human breast carcinoma cells. *Eur. J. Pharm. Sci.* **2012**, *47*, 406–420. [CrossRef] [PubMed]
35. Balabushevich, N.G.; Sholina, E.A.; Mikhalchik, E.V.; Filatova, L.Y.; Vikulina, A.S.; Volodkin, D. Self-Assembled Mucin-Containing Microcarriers via Hard Templating on $CaCO_3$ Crystals. *Micromachines* **2018**, *9*, 307. [CrossRef]
36. Vikulina, A.S.; Feoktistova, N.A.; Balabushevich, N.G.; Skirtach, A.G.; Volodkin, D. The mechanism of catalase loading into porous vaterite $CaCO_3$ crystals by co-synthesis. *Phys. Chem. Chem. Phys.* **2018**, *20*, 8822–8831. [CrossRef] [PubMed]
37. Parakhonskiy, B.V.; Yashchenok, A.M.; Donatan, S.; Volodkin, D.V.; Tessarolo, F.; Antolini, R.; Möhwald, H.; Skirtach, A.G. Macromolecule loading into spherical, elliptical, star-like and cubic calcium carbonate carriers. *Chemphyschem* **2014**, *15*, 2817–2822. [CrossRef] [PubMed]
38. Gao, C.; Leporatti, S.; Moya, S.; Donath, E.; Möhwald, H. Stability and Mechanical Properties of Polyelectrolyte Capsules Obtained by Stepwise Assembly of Poly(styrenesulfonate sodium salt) and Poly(diallyldimethyl ammonium) Chloride onto Melamine Resin Particles. *Langmuir* **2001**, *17*, 3491–3495. [CrossRef]
39. Gao, C.; Leporatti, S.; Moya, S.; Donath, E.; Möhwald, H. Swelling and shrinking of polyelectrolyte microcapsules in response to changes in temperature and ionic strength. *Chemistry* **2003**, *9*, 915–920. [CrossRef] [PubMed]
40. Anandhakumar, S.; Debapriya, M.; Nagaraja, V.; Raichur, A.M. Polyelectrolyte microcapsules for sustained delivery of water-soluble drugs. *Mater. Sci. Eng. C* **2011**, *31*, 342–349. [CrossRef]
41. Bots, P.; Benning, L.G.; Rodriguez-Blanco, J.-D.; Roncal-Herrero, T.; Shaw, S. Mechanistic Insights into the Crystallization of Amorphous Calcium Carbonate (ACC). *Cryst. Growth Des.* **2012**, *12*, 3806–3814. [CrossRef]
42. Sergeeva, A.; Sergeev, R.; Lengert, E.; Zakharevich, A.; Parakhonskiy, B.; Gorin, D.; Sergeev, S.; Volodkin, D. Composite Magnetite and Protein Containing $CaCO_3$ Crystals. External Manipulation and Vaterite $\rightarrow$ Calcite Recrystallization-Mediated Release Performance. *ACS Appl. Mater. Interfaces* **2015**, *7*, 21315–21325. [CrossRef] [PubMed]
43. Gendron, P.-O.; Avaltroni, F.; Wilkinson, K.J. Diffusion coefficients of several rhodamine derivatives as determined by pulsed field gradient-nuclear magnetic resonance and fluorescence correlation spectroscopy. *J. Fluoresc.* **2008**, *18*, 1093–1101. [CrossRef] [PubMed]
44. Terdale, S.; Tantray, A. Spectroscopic study of the dimerization of rhodamine 6G in water and different organic solvents. *J. Mol. Liq.* **2017**, *225*, 662–671. [CrossRef]
45. Arbeloa, F.; Ojeda, P.; Arbeloa, I. Flourescence self-quenching of the molecular forms of Rhodamine B in aqueous and ethanolic solutions. *J. Lumin.* **1989**, *44*, 105–112. [CrossRef]
46. Moreno-Villoslada, I.; Fuenzalida, J.P.; Tripailaf, G.; Araya-Hermosilla, R.; Pizarro, G.D.C.; Marambio, O.G.; Nishide, H. Comparative study of the self-aggregation of rhodamine 6G in the presence of poly(sodium 4-styrenesulfonate), poly(*N*-phenylmaleimide-co-acrylic acid), poly(styrene-alt-maleic acid), and poly(sodium acrylate). *J. Phys. Chem. B* **2010**, *114*, 11983–11992. [CrossRef] [PubMed]
47. Peyratout, C.; Donath, E.; Daehne, L. Electrostatic interactions of cationic dyes with negatively charged polyelectrolytes in aqueous solution. *J. Photochem. Photobiol. A* **2001**, *142*, 51–57. [CrossRef]
48. Ben Mahmoud, S.; Hichem, A.; Essafi, W. Spectrophotometric study of the interaction of methylene blue with poly(styrene-co-sodium styrene sulfonate). *Mediterr. J. Chem* **2016**, *5*, 493–506. [CrossRef]
49. Shen, H.; Li, F.; Wang, D.; Yang, Z.; Yao, C.; Ye, Y.; Wang, X. Chitosan-alginate BSA-gel-capsules for local chemotherapy against drug-resistant breast cancer. *Drug Des. Dev. Ther.* **2018**, *12*, 921–934. [CrossRef] [PubMed]
50. Wu, Q.-X.; Lin, D.-Q.; Yao, S.-J. Design of chitosan and its water soluble derivatives-based drug carriers with polyelectrolyte complexes. *Mar. drugs* **2014**, *12*, 6236–6253. [CrossRef] [PubMed]
51. Yan, S.; Zhu, J.; Wang, Z.; Yin, J.; Zheng, Y.; Chen, X. Layer-by-layer assembly of poly(L-glutamic acid)/chitosan microcapsules for high loading and sustained release of 5-fluorouracil. *Eur. J. Pharm. Biopharm.* **2011**, *78*, 336–345. [CrossRef] [PubMed]
52. Cuomo, F.; Ceglie, A.; Piludu, M.; Miguel, M.G.; Lindman, B.; Lopez, F. Loading and protection of hydrophilic molecules into liposome-templated polyelectrolyte nanocapsules. *Langmuir* **2014**, *30*, 7993–7999. [CrossRef] [PubMed]

53. Rochín-Wong, S.; Rosas-Durazo, A.; Zavala-Rivera, P.; Maldonado, A.; Martínez-Barbosa, M.; Vélaz, I.; Tánori, J. Drug Release Properties of Diflunisal from Layer-By-Layer Self-Assembled κ-Carrageenan/Chitosan Nanocapsules: Effect of Deposited Layers. *Polymers* **2018**, *10*, 760. [CrossRef]
54. Cuomo, F.; Lopez, F.; Piludu, M.; Miguel, M.G.; Lindman, B.; Ceglie, A. Release of small hydrophilic molecules from polyelectrolyte capsules: effect of the wall thickness. *J. Colloid Interface Sci.* **2015**, *447*, 211–216. [CrossRef] [PubMed]
55. Yoshida, K.; Ono, T.; Kashiwagi, Y.; Takahashi, S.; Sato, K.; Anzai, J.-I. pH-Dependent Release of Insulin from Layer-by-Layer-Deposited Polyelectrolyte Microcapsules. *Polymers* **2015**, *7*, 1269–1278. [CrossRef]
56. Vikulina, A.S.; Aleed, S.T.; Paulraj, T.; Vladimirov, Y.A.; Duschl, C.; von Klitzing, R.; Volodkin, D. Temperature-induced molecular transport through polymer multilayers coated with PNIPAM microgels. *Phys. Chem. Chem. Phys.* **2015**, *17*, 12771–12777. [CrossRef] [PubMed]
57. Prokopović, V.Z.; Vikulina, A.S.; Sustr, D.; Duschl, C.; Volodkin, D. Biodegradation-Resistant Multilayers Coated with Gold Nanoparticles. Toward a Tailor-made Artificial Extracellular Matrix. *ACS Appl. Mater. Interfaces* **2016**, *8*, 24345–24349. [CrossRef] [PubMed]
58. Prokopovic, V.Z.; Vikulina, A.S.; Sustr, D.; Shchukina, E.M.; Shchukin, D.G.; Volodkin, D.V. Binding Mechanism of the Model Charged Dye Carboxyfluorescein to Hyaluronan/Polylysine Multilayers. *ACS Appl. Mater. Interfaces* **2017**, *9*, 38908–38918. [CrossRef] [PubMed]
59. Madaboosi, N.; Uhlig, K.; Schmidt, S.; Vikulina, A.S.; Möhwald, H.; Duschl, C.; Volodkin, D. A "Cell-Friendly" Window for the Interaction of Cells with Hyaluronic Acid/Poly-L-Lysine Multilayers. *Macromol. Biosci.* **2018**, *18*, 1700319. [CrossRef] [PubMed]

Article

Self-Assembled Mucin-Containing Microcarriers via Hard Templating on $CaCO_3$ Crystals

Nadezhda G. Balabushevich [1], Ekaterina A. Sholina [1], Elena V. Mikhalchik [2], Lyubov Y. Filatova [1], Anna S. Vikulina [3] and Dmitry Volodkin [1,3,*]

1 Department of Chemistry, Lomonosov Moscow State University, Leninskiye Gory 1-3, 119991 Moscow, Russia; nbalab2008@gmail.com (N.G.B.); sholina-katya@mail.ru (E.A.S.); luboff.filatova@gmail.com (L.Y.F.)
2 Federal Research and Clinical Centre of Physical-Chemical Medicine, Malaya Pirogovskaya, 1A, 119992 Moscow, Russia; lemik2007@yandex.ru
3 Nottingham Trent University, School of Science and Technology, Clifton Lane, Nottingham NG11 8NS, UK; anna.vikulina@ntu.ac.uk
* Correspondence: dmitry.volodkin@ntu.ac.uk; Tel.: +44-115-848-3140

Received: 9 May 2018; Accepted: 11 June 2018; Published: 19 June 2018

Abstract: Porous vaterite crystals of $CaCO_3$ are extensively used for the fabrication of self-assembled polymer-based microparticles (capsules, beads, etc.) utilized for drug delivery and controlled release. The nature of the polymer used plays a crucial role and discovery of new perspective biopolymers is essential to assemble microparticles with desired characteristics, such as biocompatibility, drug loading efficiency/capacity, release rate, and stability. Glycoprotein mucin is tested here as a good candidate to assemble the microparticles because of high charge due to sialic acids, mucoadhesive properties, and a tendency to self-assemble, forming gels. Mucin loading into the crystals via co-synthesis is twice as effective as via adsorption into preformed crystals. Desialylated mucin has weaker binding to the crystals most probably due to electrostatic interactions between sialic acids and calcium ions on the crystal surface. Improved loading of low-molecular-weight inhibitor aprotinin into the mucin-containing crystals is demonstrated. Multilayer capsules $(mucin/protamine)_3$ have been made by the layer-by-layer self-assembly. Interestingly, the deposition of single mucin layers $(mucin/water)_3$ has also been proven, however, the capsules were unstable, most probably due to additional (to hydrogen bonding) electrostatic interactions in the case of the two polymers used. Finally, approaches to load biologically-active compounds (BACs) into the mucin-containing microparticles are discussed.

Keywords: $CaCO_3$; mucin; adsorption; co-synthesis; layer-by-layer; protamine; aprotinin

---

1. Introduction

The layer-by-layer (LbL) adsorption of oppositely-charged polymers (polyelectrolytes) onto matrices of various nature are actively used for the immobilization of biologically-active compounds (BACs) [1–4]. Bio-friendly loading of BACs into the vaterite $CaCO_3$ crystals has been shown to be effective for encapsulation of fragile BACs into polymer-based microparticles assembled onto these sacrificial crystals [5–7]. The crystals coated by polymers can be eliminated in the presence of chelating agents, such as EDTA (ethylenediaminetetraacetic acid) or citric acid, or at pH below neutral.

The following synthetic biopolymers have been utilized for the preparation of biologically/medically-relevant microparticles: poly-L-lysine, poly-L-arginine, poly-L-glutamic acid, and poly-L-aspartic acid [8]. Natural biopolymers and their derivatives have also been used: sodium alginate, chitosan, pectin, gelatine, carrageenan, hyaluronic acid, chondroitin sulfate, dextran, and cellulose [8]. In general, natural polypeptides and polysaccharides are weak polyelectrolytes and can adopt multiple conformations in response to changes in the solution pH and temperature. The folding and unfolding of their chains is driven by a balance of a number of the internal interactions (hydrogen bonds, hydrophobic, electrostatic, etc.), providing fascinating properties for microparticles made of self-assembled natural biopolymers.

Nowadays one of the main research directions in microencapsulation through self-assembled nano- and micro-particles is devoted to looking for new biopolymers providing the microparticles with desired properties, such as biocompatibility, biodegradability, and mucoadhesiveness [9]. One of the major challenges is to efficiently load and adjust a sustained release of BACs that are typically weakly bound to polymer-based microparticles.

In the past decade an interest into mucins, mucoadhesive glycoproteins, has significantly increased [10–15]. Mucins are the main components of mucous membranes in the gastrointestinal tract, as well as nasal and oral mucous membranes. The mucins possess a function of a barrier for pathogens, as well as drugs.

Mucins are large, extracellular glycoproteins with molecular weights ranging from 0.5 to 20 MDa. Membrane-bound and secreted mucins are both highly glycosylated [10]. They consist of 80% carbohydrates: N-acetylgalactosamine, N-acetylglucosamine, fucose, galactose, and sialic acid (N-acetylneuraminic acid), and traces of mannose and sulfate. The oligosaccharide chains are attached to the hydroxyl side chains of serine and threonines in the protein core. They consist of 5–15 monomers, exhibiting moderate branching. The protein core itself is arranged into distinct regions. The central glycosylated region is comprised of multiple tandem repeats rich in serine, threonine, and proline (STP repeats). At the amino and carboxy terminals, as well as between STP repeats, there are regions of the second type with a small number of O-glycosylation and a few N-glycosylation sites. The content of cysteine in these regions is more than 10% and it participates in disulfide bond formation with subsequent dimerization and polymerization of the dimers to form multimers (Figure 1). Due to this structure a number of various interactions are present in mucins including electrostatics, hydrophobic interactions, and hydrogen bonding; these interactions largely define the properties of mucins [16].

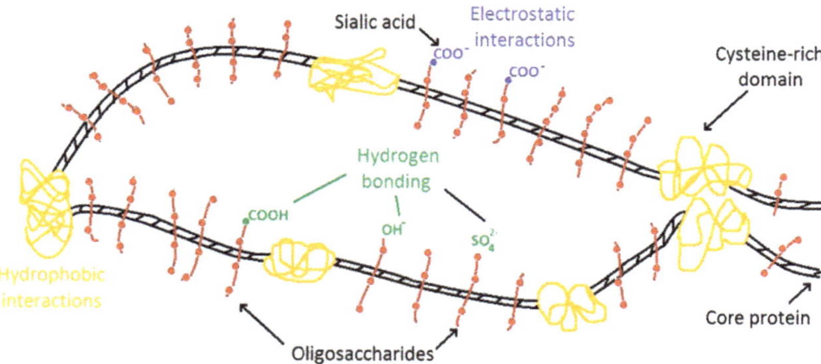

Figure 1. Schematic structure of mucin glycoproteins and their potentially mucoadhesive elements. More details can be found in [16].

Conformation of mucin is affected by pH and ionic strength [17]. Formation of self-assembled aggregates and gels is typical for mucins and is driven by the formation of S–S bonds and the interpenetration of the end glucose chains of mucin molecules [18] These processes, however, strongly depend on the source of mucin, purity, pH, and ionic strength [11,12,19].

Adsorption of mucin onto various solid surfaces has been reported, for instance onto hydrophilic and hydrophobic silica [20]. Elipsometry measurements have shown that adsorption is more pronounced at hydrophobic surfaces giving 1.2 and 3.8 mg of mucin per m^2 of the hydrophilic and hydrophobic silica, respectively. Mucin can be strongly bound to 282 nm-sized nanoparticles of polystyrene resulting in 2.2 mg of mucin per one m^2 of the nanoparticles [21]. After binding the nanoparticles were coated with 4–6 nm layer of mucin and possessed hydrophilic properties.

The negative charge of mucin at neutral and alkaline pH is due to the presence of sialic acids (pKa = 2.6) located on the ends of the chains of a polysaccharide backbone [11]. Mucins from various sources have been employed to assemble multilayers using the following polycations: polyallylamine hydrochloride, poly-L-lysine, polyethylenimine, methylcellulose, chitosan, and lactoperoxidases (Mw 78 kDa, pI 8.3) [20–26]. The thickness of the formed multilayers depends on pH and ionic strength.

Conformation of mucin molecules depends of the concentration of calcium ions [27–30]. Ca^{2+} ions have an effect on mucin aggregation, viscosity, and permeation of a mucous membrane for BACs and nanoparticles [28]. It is also known that Ca^{2+} has a significant influence on the formation of gallstones [27,31]. At the same time, to the best of our knowledge, there are no reports devoted to the interaction between mucin and the vaterite $CaCO_3$ crystals.

This work aims at the development of new approaches for BAC microencapsulation based on the vaterite $CaCO_3$ crystals and mucin. Here the mucin from the porcine stomach has been used for the naturally-derived mucin widely employed for research [11,25]. The porcine gastric mucin has been shown to be structurally related to human gastric mucin [32] and is, therefore, a decent substitute of human gastric mucin because of its high availability and reduced number of ethical issues required for the research. Moreover, mucins derived from a stomach have a tendency for gelation in acidic medium which may be important in order to form polymer-based microparticles when the $CaCO_3$ crystals are eliminated in acidic medium. In order to test the applicability of mucin for the microencapsulation we focus on the following important aspects of the encapsulation process: (i) development of simple and robust analytical approaches (with no use of radiolabels and fluorescent probes) to determine native mucin in the presence of crystal and BACs; (ii) assessment of mucin loading into the crystals and multilayer capsules prepared based on mucin; and (iii) analysis of aprotinin (model BAC poorly loaded into the crystals [33,34]) encapsulation into the crystals in the presence of mucin.

2. Materials and Methods

Anhydrous calcium chloride, ≥93.0% (C1016), anhydrous sodium carbonate Na_2CO_3, ≥99.0% (S7795), commercial mucin from porcine stomach, Type III, (m1778), bound sialic acid 0.5–1.5%, protamine from salmon, fluorescein isothiocyanate isomer 1 (FITC), 2,4,6-trinitrobenzenesulfonic acid (TNBS), *N*-acetylneuraminic acid from *Escherichia coli*, ethylenediaminetetraacetic acid (EDTA), aprotinin, *N*-benzoyl-L-arginine ethyl ester (BAEE), gel filtration molecular weight markers kit MW 12–200 kDa (Sigma-Aldrich, St. Louis, MO, USA); trypsin from bovine pancreas (Fluka, Dresden, Germany); 5,5′-dithiobis(2-nitrobenzoic acid) (DTNB, Serva, Heidelberg, Germany); cysteine, >99.5% (BioUltra Sigma-Aldrich, St. Louis, MO, USA); Sephadex G-200 (Pharmacia, Stockholm, Sweden) were used. All chemicals for buffers were laboratory grade and purchased from Sigma–Aldrich (St. Louis, MO, USA). Before use, mucin solutions were sonicated for 30 min using the ultrasonic bath (Elmasonic S15H, Singen, Germany).

2.1. Analytical Determination of Mucin

The concentration of the glycoprotein mucin in solution was determined spectrophotometrically at the wavelengths of 214 nm and 260 nm, as well as by Schiff's method via measurements of adsorption at 555 nm [35]. Analytical size exclusion chromatography in the Biofox 17 SEC 8 × 300 mm column (Bio-Works, Uppsala, Sweden) has been used utilizing the Smartline chromatographic system (Knauer, Berlin, Germany) in a solution of 0.15 M NaCl (Table S1). Preliminarily, the column was calibrated using solutions of purified mucin with different concentrations (0.01–0.1 mg mL^{-1}) and proteins with different molecular weights (Figure S1, Table S1). A total of 0.2 mL of the mucin solution with a concentrations of 0.01–1.0 mg mL^{-1} were used for the chromatography analysis at the elution rate of 0.5 mL min^{-1}. Absorbance of the eluted solutions was measured using the UV detector at wavelengths of 214 nm and 260 nm. After that the maximum absorbance of the high-molecular weight fraction was measured (the time of elution was 9.3–9.7 min).

2.2. Purification of Mucin via Chromatography

A total of 15–45 mL of the mucin solution (1–5 mg mL^{-1}) was subjected to the column filled with Sephadex G-200 (dimension 2.5 × 35 cm) using the chromatographic system Bio-Logic LP (Bio-Rad, Hercules, CA, USA) in the solution of ammonia (pH of 9.0). The elution rate was 0.5 mL min^{-1}, collection time for one fraction of eluted solution 12 min. Absorbance was determined in the obtained fractions at wavelengths of 214, 260, and 480 nm. The fractions containing mucin, as identified by absorbance and specific determination by the Schiff's method (wavelength 555 nm), were combined and freeze-dried.

2.3. Synthesis of Mucin-FITC

Fifty milligrams of commercially available mucin was dissolved in 10 mL of 0.5 M carbonate buffer with pH 9.0. One millilitre of FITC solution (1 mg mL^{-1}) in dimethylformamide was added drop by drop and the whole mixture was incubated for 24 h at 4 °C. The solution was chromatographed at the column filled with Sephadex G-200, as described above (Figure S2). The molar modification degree of the free amino groups in mucin-FITC was determined by titration with TNBS [36] and was found to be 17 ± 2% relative to purified mucin.

2.4. Synthesis of Desialylated Mucin

Fifteen milligrams of commercially available mucin was dissolved in 50 mL of 0.01 M HCl and incubated for 3 h at 80 °C. The solution was chromatographed at the column filled with Sephadex G-200, as described above (Figure S3). The content of sialic acids, determined by Hess's method [37] using the calibration curve for N-acetylneuraminic acid, was found to be 2.30 ± 0.10, 1.80 ± 0.10, and 0.41 ± 0.05% for commercial, purified, and desialylated mucin, respectively.

2.5. Mucin Loading into the $CaCO_3$ Crystals by Adsorption

The $CaCO_3$ crystals were formed according to the standard procedure [5] by mixing of equimolar solutions of $CaCl_2$ and Na_2CO_3. The formed crystals were washed twice with a pure water and dried at 70 °C. Thirty milligrams of the dry $CaCO_3$ crystals were mixed with 1.5 mL of 1 mg mL^{-1} mucin solution. The suspension was agitated on a shaker for 30 min. The precipitate was separated by centrifugation (2 min, 1000× *g*) followed by removal of the supernatant and washed twice with 1.5 mL water. The content of mucin was analysed in the supernatant and the washing solutions.

Efficiency of mucin incorporation was calculated using the following equation:

$$\eta = \frac{(c_0 - c_e)}{c_0}$$

where η is efficiency of protein incorporation, and c_0 and c_e are the initial and equilibrium protein concentrations, respectively (mg mL^{-1}).

The amount of the loaded protein at equilibrium was calculated using the following equation:

$$q_e = \frac{(c_0 - c_e)\cdot V}{m}$$

where q_e is the adsorption capacity (mg g^{-1}), c_0 and c_e are the initial and equilibrium protein concentrations, respectively (mg mL^{-1}); V is the volume (mL) of the protein solution; and m is the mass (g) of $CaCO_3$.

2.6. Mucin and Aprotinin Loading into the $CaCO_3$ Crystals by Co-Synthesis

Three millilitres of the solution containing 1.67 mg mL^{-1} mucin and/or 0.167 mg mL^{-1} aprotinin were added to 1 mL of the solution 1 M $CaCl_2$. The solution was stirred for 5 min (100 rpm), 1 mL of the solution of 1M Na_2CO_3 was added and further stirred for an additional 45 s. Then the suspension was incubated for 15 min. The precipitate was separated by centrifugation (2 min, 1000× g) and washed twice with 5 mL of water. If necessary, the particles were dried. The mass of the precipitate was determined, and the mass of $CaCO_3$ was calculated assuming the complete process of crystallization of the insoluble crystals. The concentrations of mucin were determined in the supernatant and washing solutions and, if necessary, the activity of aprotinin was also determined. The concentration of active aprotinin in the solutions was determined by inhibition of trypsin using BAEE substrate as described in [34,38].

2.7. Preparation of Polyelectrolyte Microcapsules

The LbL deposition of either mucin (control experiment) or mucin and protamine (a pair of polymers used) has been performed on the synthetized $CaCO_3$ crystals containing glycoprotein mucin preloaded by co-synthesis. In either case, the polymer concentration was 0.5 mg mL^{-1} and the crystals concentration was 20 mg mL^{-1}. The coated crystals were washed twice with water and their ζ-potential was measured after each adsorption step. The $CaCO_3$ matrix in the prepared particles coated with three mucin-protamine bilayers was dissolved by dropwise addition of an equimolar amount of 0.2 M EDTA (to solubilize all the crystals), and then the formed polyelectrolyte capsules were washed three times with water. The content of mucin in the microcapsules was determined by the analysis of its content in solutions obtained during the capsule preparation procedure.

2.8. Characterisation of the Crystals and Microcapsules

Analysis of microparticles prepared in this study was carried out using optical microscopy (Carl Zeiss, Jena, Germany), scanning electron microscopy (SEM, Zeiss DSM 40, Jena, Germany), and fluorescence microscopy (EVOS FL, Thermo Fisher Scientific, Waltham, MA, USA). Determination of the hydrodynamic diameter of commercial mucin (0.1 mg mL^{-1}) was carried out using DLS (Malvern Zetasizer Nano ZS, Malvern, UK) and nanoparticle tracking analysis (NTA, Malvern NanoSight NS500, Malvern, UK). A suspension of microparticles (0.5 mg mL^{-1}), and solutions of 0.1 mg mL^{-1} mucin or protamine were used for the analysis of the ζ-potential (ZP) using DLS.

3. Results and Discussion

3.1. Analysis of Mucin Purity via Permeation Gel Chromatography

The first step in this study was to develop an effective approach to determine mucin concentration in alkaline media (pH 7–10) and in the presence of BAC such as proteins and peptides. This is necessary because the $CaCO_3$ crystals dispersed in water provide the alkaline pH due to hydrolysis of the $CaCO_3$.

The classical way to determine mucin is based on Schiff's method [35]. OH groups of mucin are oxidised by periodic acid to aldehydes (Figure S4). Our results indicate that pH of the analysed solution has a significant effect on the purple-violet colour of solutions (A_{555}) obtained using this staining method (Figure S5). Thus, this method has been found as not useful for the determination of mucin concentration at alkaline pH.

A number of reports have identified some protein-based low-molecular-weight impurities in commercial samples of mucin [11,13,15,39]. Absorbance spectra of mucin used in this study show the presence of two well-defined peaks at 214 and 216 nm as shown in the Figure 2, line 1.

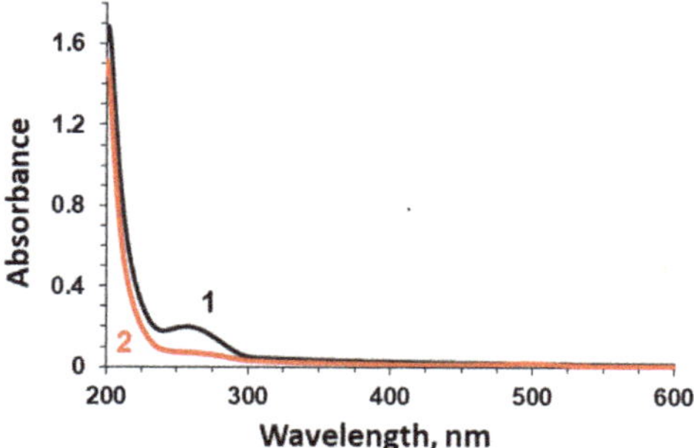

Figure 2. Absorption spectra of commercial (black line 1) and purified (red line 2) 0.1 mg mL^{-1} mucin.

Analytical size exclusion chromatography has been employed for further analysis of commercial mucin revealing the presence of two fractions: low- and high-molecular-weight fractions (Figure 3, line 1). Only the high-molecular-weight fraction had absorption at both 214 and 260 nm and gave a specific staining by Schiff's method. Another fraction (6–30 kDa) significantly contributed to the absorption at 260 nm is most probably the low-molecular-weight impurity of a protein nature.

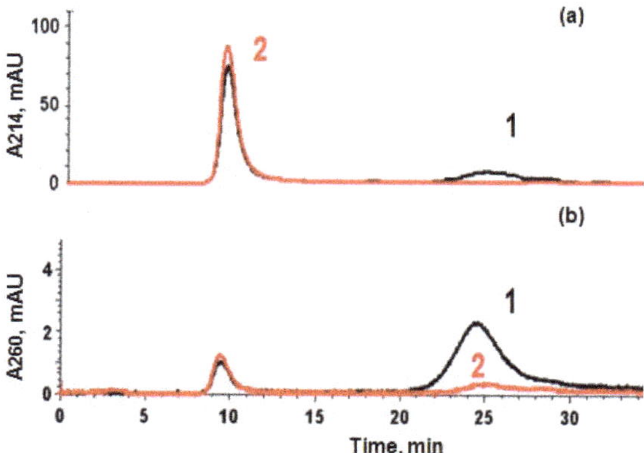

Figure 3. Elution profiles registered at 214 (**a**) and 260 nm (**b**) by analytical exclusion chromatography of commercial (black line 1) and purified (red line 2) mucins on a Biofox 17 SEC column (8 × 300 mm).

To purify the commercial mucin, gel-permeation chromatography on Sephadex G-200 has been used (Figure 4). The high-molecular-weight fraction has been separated and lyophilized. This fraction has been further used in this study and is called purified mucin. Absorbance spectra (Figure 1) and chromatography profiles (Figure 3) of commercial and purified mucin confirm no protein-based impurities in the purified mucin. In addition, the purified mucin has been titrated by TNBS [36] showing a significant reduction of primary amino groups in purified mucin (38 per a mucin molecule) compared to the commercial one (164 per mucin molecule).

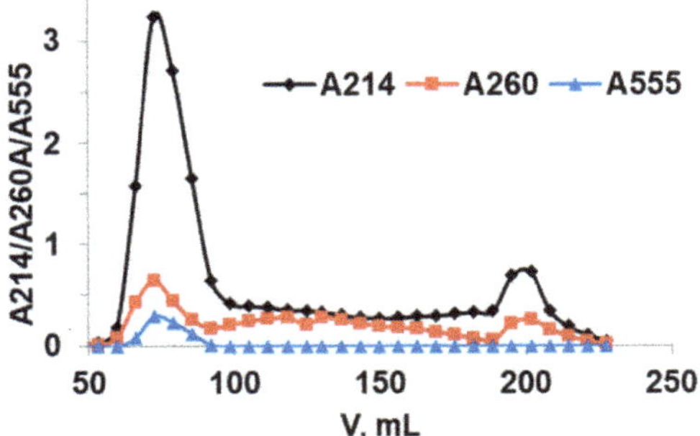

Figure 4. Elution profile obtained by gel filtration on Sephadex G-200 of commercial mucin.

Further, the purified mucin has been used for the construction of calibration lines in order to determine glycoproteins in this study. Based on the results of gel-permeation chromatography, the content of glycoprotein in the commercial mucin has been found to be 87 $\pm$ 3% that is in line with results obtained using the Schiff's method (content is 84 $\pm$ 2%).

3.2. Loading of Mucin into $CaCO_3$ Crystals (Adsorption and Co-Synthesis)

Loading of mucin into the crystals has been achieved via two methods, namely adsorption (mucin is adsorbed into performed crystals) or co-synthesis (mucin is trapped in the crystals during crystal synthesis). Figure 5 shows the schematics of the approaches used. Of note, the conditions of mucin loading have been kept identical for both loading methods, such as the same time of exposure of mucin to the crystals (15 min), the same final crystal and mucin concentrations (20 and 1 mg mL^{-1}, respectively), etc. This gives an option to compare the methods.

Optical and SEM microscopies did not reveal significant differences in morphology and size (3–5 μm) of the crystals before and after mucin loading (Figure 6). This may be related to the rather low concentration of mucin used. We believe that analysis of the surface of the crystals is sufficient to conclude about the crystal internal structure change. This is because we have found (by atomic force microscopy (AFM) measurements and other relevant techniques) that there is direct correlation between the size of nanocrystallines on the surface of the crystals and those in the internal volume [40]. It would be of interest to analyse the morphology of the crystals using AFM in the future in order to get more insights into the internal structure of the crystals. In this work we use optical microscopy and SEM images for identification of the polymorph form of the crystals since the vaterite form has a typical spherical shape, compared to cubic calcite. The same crystals, as obtained in this study, have been previously analysed by X-ray diffraction analysis (XRD) revealing the vaterite form [40]. In the future we plan to utilize XRD as a powerful tool to analyse the structure of the crystals loaded with substantial amounts of mucin under various conditions. This will open a way to probe the effect of the preparation conditions on the crystal structure.

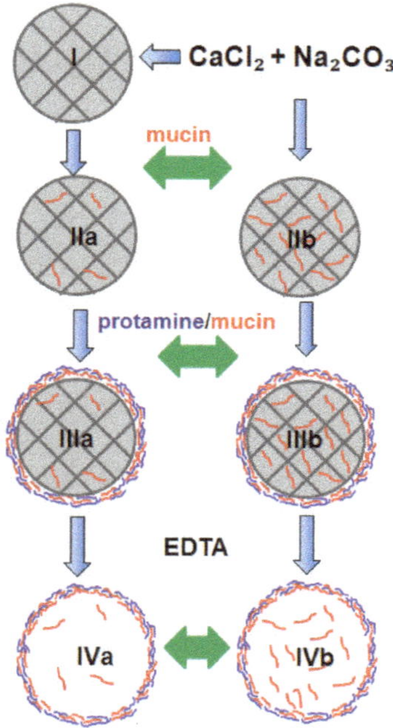

Figure 5. Schematics of formulation of microparticles (crystals and polymer capsules) using mucin: I–$CaCO_3$ crystals; II–the crystals with mucin loaded by adsorption (**a**) or co-synthesis (**b**); III–crystals coated with three layer of oppositely charged protamine and mucin; IV–polymer-based microcapsules obtained after dissolution of the coated crystals using EDTA. Green arrows indicate a step of introduction of a BAC to be further encapsulated into the microparticles.

Figure 6. SEM (**a**,**b**) and optical (**c**) microscopy images of $CaCO_3$ crystals with adsorbed (**a**) and co-synthetized mucin (**b**,**c**).

Loading efficiency for mucin loading into the crystals and the amount of mucin released from the crystals during washing with water have been worked out by analysis of supernatants of solutions obtained during the mucin loading procedure (Table 1). Loading by co-synthesis resulted in almost twice higher content of mucin in the crystals. Moreover, mucin loaded by co-synthesis has, to a higher extent, been retained in the crystals during crystal washing with water. These results corroborate well with findings reported for proteins [33] showing the same trend. Based on the surface area of $CaCO_3$ crystals obtained using the same procedure as in this study (8.8 $m^2\ g^{-1}$ [41]) the mucin adsorption per a unit surface area of the crystals can be calculated. This value has been found to be 1.25 mg m^{-2}, which is similar to mucin adsorption onto SiO_2, which gives 1.2 mg m^{-2} [21]. Better understanding of the effect of the hydrophobicity of the surface of the crystals and limitations for mucin diffusion through the pores of the crystals onto the loading amount can be realised in our future work.

Table 1. Characteristics of the incorporation of commercial mucin (1 mg mL^{-1}) into the $CaCO_3$ crystals (20 mg mL^{-1}).

Loading Method	Efficiency of Mucin Loading, %		Release after Washing, % of Loaded	ZP, mV
	Spectrophotometry	Analytical Chromatography		
Adsorption	12 ± 2	10 ± 1	11 ± 1	$-(15 \pm 3)$
Co-synthesis	22 ± 3	18 ± 2	5 ± 1	$-(11 \pm 2)$

Further, we have analysed an average hydrodynamics diameter of mucin in water solution (1 mg mL^{-1}). DLS has revealed two populations of molecules of sizes 40 and 250 nm, respectively. This is in a line with results obtained for commercial mucin from other sources [19]. The fraction with larger size was diminished as a result of ultrasound treatment, dilution of mucin solution, increase of pH (Figure S6). Both fractions are characterized by a negative value of the zeta potential (−15 mV). We believe that the fraction with the larger size (250 nm) corresponds to the aggregated mucin and the smaller fraction (40 nm) belongs most probably to single mucin molecules.

Literature reports indicate that intermolecular interaction between mucin molecules can be enhanced in the presence of Ca^{2+} [28] and hydrodynamic radius of mucin in the solution of $CaCl_2$ is reduced compared to that in water [29]. We hypothesize that the smaller fraction of mucin molecules (single mucin molecules) can diffuse through pores of the crystals (pore size in the range 5–40 nm [42] and the larger fraction would be located presumably on the crystal surface. This may be valid for both methods of mucin loading into the crystal, i.e., adsorption and co-synthesis. In order to prove this assumption, we have analysed the mucin distribution in the crystals using fluorescence microscopy and mucin-FITC (Figure 7). The results demonstrate that mucin is predominately located on edges of the crystals and partially penetrates inside the crystal pores, which supports the assumption above. In the previous work we have found that 70 kDa poly(sodium 4-styrenesulfonate) and poly(allylamine hydrochloride) do not permeate well through pores of the crystals made at 22 °C (same protocol as for the crystals prepared in this study) [40]. These polymers are supposed to have smaller size than 40 nm (the small fraction of mucin identified as mentioned above). However, multiple adsorption of these synthetic polymers by the LbL manner and their high charge may limit their diffusion into the crystals.

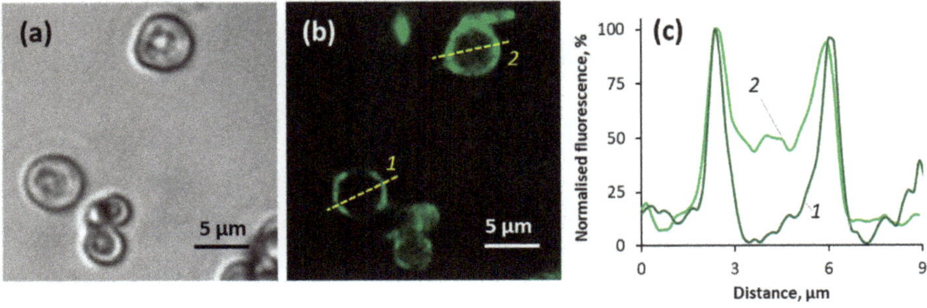

Figure 7. Transmittance (**a**) and fluorescence (**b**) microscopy images of vaterite crystals (4 mg mL^{-1} $CaCO_3$) after incubation with mucin-FITC (1 mg mL^{-1}) for 15 min; and (**c**)-fluorescence profiles taken along the yellow lines 1 and 2 across the particles in the image (**b**).

Loading of desialylated mucin into the $CaCO_3$ crystals has been investigated in order to probe an effect of sialic acids. The loading efficiency of the desialylated mucin was found to be 33 and 36% lower compared to that for the loading of commercial mucin using adsorption and co-synthesis, respectively. At the same time, the loss of mucin during washing in water increased by 10–15% for either loading approaches. Thus, we can conclude that the presence of sialic acids in mucin most probably improves the loading and retention of mucin in crystals, which can be explained by the interaction of the acids with Ca^{2+} provided from the crystals. Binding of desialylated mucin with the crystals can be driven by the interaction of Ca^{2+} with the protein-based part of mucin [29].

3.3. Encapsulation of Aprotinin into Mucin-Containing $CaCO_3$ Crystals

Low-molecular-weight protein aprotinin (MW 6.5 kDa, pI 10.5) is an inhibitor of proteolytic enzymes and is actively used as a medicine [43]. It has recently been shown [33,34] that positively-charged aprotinin does not lose its biological activity in the presence of the $CaCO_3$ crystals. However, its loading into the crystals by adsorption or co-synthesis is rather low compared to other proteins, such as insulin and catalase; moreover, the retention of aprotinin in the crystals is not high in washing steps with water.

We have further tested whether aprotinin loading into the crystals can be improved via co-loading of mucin into the crystals. Co-loading of mucin resulted in an increase of aprotinin content by a factor of three, giving a high content of aprotinin in the crystals of 1.5 ± 0.2 mg g^{-1} after two washing steps with water. We believe that the formation of inter-polyelectrolyte complex between mucin and aprotinin is a reason of better retention of aprotinin in the formed hybrid crystals.

3.4. Mucin-Containing Polymer Multilayer Capsules

In this part of the work we have considered an option to utilize mucin for the formulation of multilayer capsules made of mucin and protamine as oppositely-charged bio-polymers. For this, mucin-containing crystals were coated by mucin and protamine layers in the LbL manner. The peptide protamine (5 kDa, pI 10.5) [44] has been previously used as a polycation [34,45] and, similar to aprotinin, it did not show high affinity to the $CaCO_3$ crystals [33]. The zeta potential of protamine in water has been found to be $+(7 \pm 3)$ mV. During the LbL polymer deposition onto the crystal, the Zeta potential of the coated crystals has been reversed from negative values (mucin deposition) to positive ones (protamine deposition), as shown in Figure 8b. This proves the formation of the mucin-protamine complex upon the coating procedure. Protamine has absorption maxima at 214 nm and this is why the inclusion of mucin into the crystals has been determined using gel-permeation chromatography. Protamine adsorption resulted in the removal of a part of previously-deposited mucin (Figure 8a). However, the trend of the bio-polymer deposition demonstrated an increase of a total amount of the adsorbed polymers as a whole. The deposition process is most probably driven by the formation of both electrostatic interactions between the polymers and hydrogen bonding as well. The similar deposition behaviour has been reported for bovine submaxillary mucin in combination with polyallylamine hydrochloride [23] or lactoperoxidase [20].

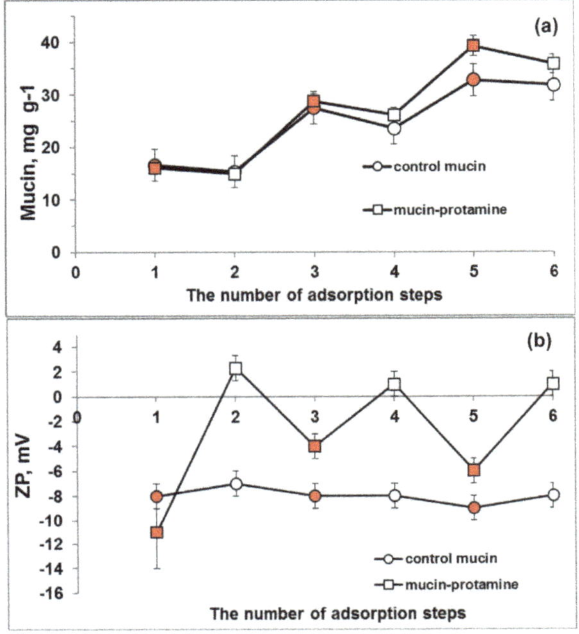

Figure 8. The content of mucin (**a**) and the zeta potential (**b**) of mucin-containing crystals (n = 1) as a function of a number of deposition steps of either mucin or water (control) or mucin and protamine pairs. The adsorption of mucin is indicated with a red colour.

Deposition of one polymer, namely mucin, has been used as a control experiment (Figure 8). In this case, the crystals have been incubated stepwise in mucin solution and in water. The deposited sequence can be shown as $CaCO_3$(mucin)$_3$ since the incubation in water is typically used as an intermediate step between the deposition of polymers to wash out weakly-adsorbed polymer molecules. The zeta potential of the coated particles has not been changed with an increase of the number of deposited layers of mucin (Figure 8b). At the same time, the total amount of adsorbed mucin has been increased with each deposition layer (Figure 8a). This is most probably due to adsorption of more than one layer of mucin. The Ellman method [46] did not reveal any free SH-groups in the commercial mucin sample, meaning that disulphide bonds cannot be responsible for intermolecular interactions between the mucin molecules. It is of note that an amount of adsorbed mucin in the controlled experiment was slightly lower than that for the mucin-protamine coating (Figure 8a). This stimulated us to hypothesize that the mucin-protamine interactions are driven by both electrostatics and hydrogen bonding (Figure 1) and the sequential deposition of only mucin (control experiment) takes place due to only hydrogen bonding allowing the deposition of multiple layers of only mucin. In future research, in order to probe the hydrogen bonding, we plan to prepare the crystals with higher mucin content and utilize Fourier-transform infrared (FTIR) spectroscopy as a convenient way of analysing mucin [47]. In this work, the rather low loading concentration of mucin limits us for FTIR study.

The coated crystals $CaCO_3$(mucin-protamine)$_3$ and $CaCO_3$(mucin)$_3$ crystals-were stable upon storage in water at 4 °C for a month without any sign of recrystallization (Figure 9 and Figure S7). The presence of 0.2M EDTA resulted in dissolution of the carbonate crystals followed by formation of stable (mucin-protamine)$_3$ capsules (Figure 9b). At the same time, stable capsules solely made of mucin have not been formed. This confirms that electrostatic interactions play a crucial role in the formation of stable multilayers and may be the contribution of electrostatics is stronger than that of hydrogen bonds, stressing the dominating role of electrostatics.

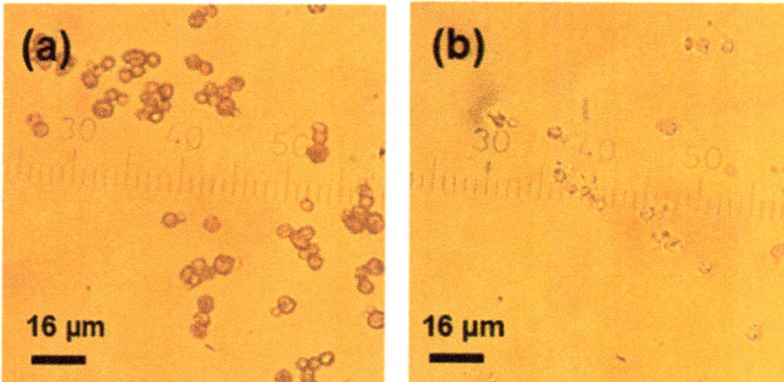

Figure 9. Optical microscopy images of polymer-coated crystals $CaCO_3$(mucin-protamine)$_3$ (**a**) and multilayer capsules (mucin-ptotamine)$_3$ formed after addition of EDTA to the coated crystals (**b**). The $CaCO_3$ crystals contain mucin loaded by co-synthesis.

3.5. BAC Loading into Mucin-Containing Microparticles

Based on the results obtained using proteins (protamine, aprotinin), one can expect that the mucin-containing particles may be effective for loading of BAC, in particular small-molecular-weight and positively charged BACs that are difficult to load in substantial amounts. A rationality for the loading of such BACs should be driven by the nature and a charge of the components (Figure 2, indicated by green arrows). One can expect three main methods to load BACs: (i) via adsorption or co-synthesis into the crystals together with mucin; (ii) as polycations during sequential deposition of multilayers; and (iii) post-loading into preformed mucin-containing polymer capsules. The presence of mucin in the $CaCO_3$ crystals or multilayer capsules makes the crystals and capsules extremely attractive for mucoadhesive delivery of BACs through a mucous membrane [15,48–50]. In regards to this, the LbL assembly will provide a strong tool for introduction of various components into the assembled structures in order to tune a function of the finally-assembled structure. Novel approaches to probe diffusion into the formed multilayers are essential to understand and control a structure of the assemblies and release characteristics of BACs from the assembled structures [51,52].

We believe that the approaches demonstrated here for assembly of mucin-containing polymer-based microparticles can be further used for the design of composite multifunctional micro- and nano-particles with required applications. For instance, introduction of the protein-repelling agent polyethylene glycol into the particles by the approaches developed earlier [53,54] may reduce protein-binding to the particles if undesired. Loading of a thermo-sensitive polymer, such as poly(N-isopropilacrylamide), can give an option to assemble particles able to change their size and hydrophobicity upon the temperature increase at physiologically-relevant conditions [55]. Utilization of mucin-containing $CaCO_3$ crystals for fabrication of porous self-assembled alginate hydrogels [56–58] may be beneficial to design the hydrogels for the engineering of tissue having contact with mucosa. New perspectives are open to assemble mucin-containing particles of various shapes via hard templating onto protein aggregates as demonstrated earlier [59–62]. This approach may further be considered in our upcoming study.

4. Conclusions

This study demonstrated the development of novel approaches for encapsulation of mucin into self-assembled microstructures, i.e., polymer-based microparticles templated onto vaterite $CaCO_3$ crystals. The following analytical methods for determination of mucin in alkaline solutions and in the presence of the crystals and proteins have been adopted: spectrophotometrical analysis of adsorption at 214 nm and analytical permeation chromatography using a Biofox 17 SEC column. The analytical methods used demonstrated the presence of protein-based low-molecular-weight fraction in commercial mucin samples that had been removed to obtain purified mucin for further use.

Loading of mucin into the crystals by co-synthesis has been shown to be more effective than the loading by adsorption (the loading capacity 11 and 6 mg of mucin per gram of the crystals, respectively). Mucin can aggregate in water solution giving aggregates with a hydrodynamic diameter of a few hundreds of nm, the diameter of single mucin molecules (in equilibrium with the aggregates) is around 40 nm. As proven by optical fluorescence imaging and DLS mucin molecules can adsorb onto and diffuse into the porous crystals that have pore size of the same dimensions (pores are in the range 20–60 nm). Most probably the bulky and highly hydrated mucins are able to squeeze out through the pores. Interestingly, desialylated mucin demonstrated weaker binding to the crystals that can be explained by an enhancement of the binding via the interaction of calcium ions with sialic acids present in the mucin backbone. Moreover, the presence of the sialic acids improved retention of mucin in the crystals upon water washings.

Loading of mucin into performed $CaCO_3$ crystals can be achieved via LbL deposition of a single mucin or mucin-protamine pair, as proven by monitoring of the zeta potential and the amount of deposited compounds. Stable multilayer capsules can, however, be formed after crystal elimination by EDTA only for mucin-protamine pairs. This can be explained by additional electrostatic interactions for the pair compared to only hydrogen bonding in the case of mucin deposition as a single component.

Successful inclusion of BACs into mucin-containing microparticles is demonstrated for aprotinin as an important protease inhibitor. Finally, approaches for BAC loading are discussed in regards to the encapsulation strategy proposed based on the $CaCO_3$ crystal. The results of this study may open new perspectives to utilize mucin as an important mucoadhesive polymer for effective encapsulation of various compounds, including BACs.

Supplementary Materials: The following are available online at http://www.mdpi.com/2072-666X/9/6/307/s1, Figure S1: Calibration curve used for the determination of the concentration of mucin by analytical exclusion chromatography using Biofox 17 SEC in 0.15 M NaCl solution by measurement of absorbance of eluted samples at 214 nm with a release time of 9.3–9.7 min; Figure S2: Gel-permeation chromatography of mucin-FITC using Sephadex G-200; Figure S3: Gel-permeation chromatography of desialylated mucin using Sephadex G-200; Figure S4: Scheme of quantitative determination of mucin by the Schiff method; Figure S5. Influence of the tested media on the determination of mucin by the Schiff method; Figure S6. Typical hydrodynamic diameter distribution for commercial mucin (1 mg mL^{-1}, H_2O, 250 °C); Figure S7. Optical microscopy images of vaterite crystals before (a) and (b) after coating with (mucin)3; Table S1. Calibration curves used in the work.

Author Contributions: Conceived and designed the experiments: N.G.B., E.A.S., E.V.M., L.Y.F., A.S.V., and D.V.; performed the experiments: N.G.B., E.A.S., and A.S.V.; analyzed the data: N.G.B., E.A.S., E.V.M., L.Y.F., A.S.V., and D.V.; contributed reagents/materials/analysis tools: N.G.B., E.A.S., E.V.M., L.Y.F., and A.S.V.; wrote the paper: N.G.B., and D.V. All the authors discussed the results and commented on the manuscript.

Acknowledgments: The work was performed within the framework of the M.V. Lomonosov Moscow State University state task, part 2 (government grant AAAA-A16-116052010081-5). This work was supported in part by M.V. Lomonosov Moscow State University Program of Development. The authors thank S.A. Gusev for SEM imaging. A.S.V and D.V. also would like to acknowledge the QR Fund (2017/2018) from Nottingham Trent University.

Conflicts of Interest: The authors declare no conflict of interest.

References

1. Delcea, M.; Mohwald, H.; Skirtach, A.G. Stimuli-responsive LbL capsules and nanoshells for drug delivery. *Adv. Drug Deliv. Rev.* **2011**, *63*, 730–747. [CrossRef] [PubMed]
2. Volodkin, D.; Skirtach, A.; Möhwald, H. *LbL Films as Reservoirs for Bioactive Molecules Bioactive Surfaces;* Börner, H.G., Lutz, J.-F., Eds.; Springer: Berlin/Heidelberg, Germany, 2011; Volume 240, pp. 135–161.
3. Ariga, K.; Lvov, Y.M.; Kawakami, K.; Ji, Q.M.; Hill, J.P. Layer-by-Layer Self-assembled Shells for Drug Delivery. *Adv. Drug Deliv. Rev.* **2011**, *63*, 762–771. [CrossRef] [PubMed]

4. Balabushevich, N.G.; Izumrudov, V.A.; Larionova, N.I. Protein microparticles with controlled stability prepared via layer-by-layer adsorption of biopolyelectrolytes. *Polym. Sci. Ser. A* **2012**, *54*, 540–551. [CrossRef]
5. Volodkin, D.V.; Larionova, N.I.; Sukhorukov, G.B. Protein Encapsulation via Porous $CaCO_3$ Microparticles Templating. *Biomacromolecules* **2004**, *5*, 1962–1972. [CrossRef] [PubMed]
6. Volodkin, D.V.; Petrov, A.I.; Prevot, M.; Sukhorukov, G.B. Matrix Polyelectrolyte Microcapsules: New System for Macromolecule Encapsulation. *Langmuir* **2004**, *20*, 3398–3406. [CrossRef] [PubMed]
7. Volodkin, D. $CaCO_3$ templated micro-beads and-capsules for bioapplications. *Adv. Colloid. Interface Sci.* **2014**, *207*, 306–324. [CrossRef] [PubMed]
8. Díez-Pascual, A.M.; Shuttleworth, P.S. Layer-by-Layer Assembly of Biopolyelectrolytes onto Thermo/pH-Responsive Micro/Nano-Gels. *Materials* **2014**, *7*, 7472–7512. [CrossRef] [PubMed]
9. Ariga, K.; McShane, M.; Lvov, Y.M.; Ji, Q.M.; Hill, J.P. Layer-by-layer assembly for drug delivery and related applications. *Expert Opin. Drug Deliv.* **2011**, *8*, 633–644. [CrossRef] [PubMed]
10. Dekker, J.; Rossen, J.; Buller, H.; Einerhand, A. The MUC family: An obituary. *Trends Biochem. Sci.* **2002**, *27*, 126–131. [CrossRef]
11. Lee, S.; Muller, M.; Rezwan, K.; Spencer, N.D. Porcine gastric mucin (PGM) at the water/poly(dimethylsiloxane) (PDMA) interface: Influence of pH and ionic strength on its conformation, adsorption and aqueous lubrication properties. *Langmuir* **2005**, *21*, 8344–8353. [CrossRef] [PubMed]
12. Bansil, R.; Turner, B.S. Mucin structure, aggregation, physiological functions and biomedical applications. *Curr. Opin. Colloid Interface Sci.* **2006**, *11*, 164–170. [CrossRef]
13. Sandberg, T.; Blom, H.; Caldwell, K.D. Potential use of mucins as biomaterial coatings. I. Fractionation, characterization, and model adsorption of bovine, porcine, and human mucins. *J. Biomed. Mater. Res. A* **2009**, *91*, 762–772. [CrossRef] [PubMed]
14. Leal, J.; Smyth, H.D.C.; Ghosh, D. Physicochemical properties of mucus and their impact on ransmucosal drug delivery. *Int. J. Pharm.* **2017**, *532*, 555–572. [CrossRef] [PubMed]
15. Bansil, R.; Turner, B.S. The biology of mucus: Composition, synthesis and organization. *Adv. Drug Deliv. Rev.* **2018**, *124*, 3–15. [CrossRef] [PubMed]
16. Yang, X.; Forier, K.; Steukers, L.; Vlierberghe, S.; Dubruel, P.; Braeckmans, K.; Glorieu, S.; Nauwynck, H.J. Immobilization of Pseudorabies Virus in Porcine Tracheal Respiratory Mucus Revealed by Single Particle Tracking. *PLoS ONE* **2012**, *7*, e51054. [CrossRef] [PubMed]
17. Cao, X.; Bansil, R.; Bhaskar, K.; Turner, B.; LaMont, J.; Niu, N.; Afdhal, N.H. pHdependent conformational change of gastric mucin leads to sol–gel transition. *Biophys. J.* **1999**, *76*, 1250–1258. [CrossRef]
18. Taylor, C.; Allen, A.; Dettmar, P.; Pearson, J. The gel matrix of gastric mucus is maintained by a complex interplay of transient and nontransient associations. *Biomacromolecules* **2003**, *4*, 922–927. [CrossRef] [PubMed]
19. Nikogeorgos, N.; Madsen, J.B.; Lee, S. Influence of impurities and contact scale on the lubricating propertiesof bovine submaxillary mucin (BSM) films on a hydrophobic surface. *Colloids Surf. B Biointerfaces* **2014**, *122*, 760–766. [CrossRef] [PubMed]
20. Svensson, O.; Arnebrant, T. Mucin layers and multilayers—Physicochemical properties and applications. *Curr. Opin. Colloid Interface Sci.* **2010**, *15*, 395–405. [CrossRef]
21. Shi, L.; Caldwell, K.D. Mucin Adsorption to Hydrophobic Surfaces. *J. Colloid Interface Sci.* **2000**, *224*, 372–381. [CrossRef] [PubMed]
22. Svensson, O.; Lindh, L.; Cardenas, M.; Arnebrant, T. Layer-by-layer assembly of mucin and itosan—Influence of surface properties, concentration and type of mucin. *J. Colloid Interface Sci.* **2006**, *299*, 608–616. [CrossRef] [PubMed]
23. Ahn, J.; Crouzier, T.; Ribbeck, K.; Rubher, M.F.; Cohen, R.E. Turning the properties of mucin via layer-by-layer assembly. *Biomacromolecules* **2015**, *16*, 228–235. [CrossRef] [PubMed]
24. Lindh, L.; Svendsen, I.E.; Svensson, O.; Cárdenas, M.; Arnebrant, T. The salivary mucin MUC5B and lactoperoxidase can be used for layer-by-layer film formation. *J. Colloid Interface Sci.* **2007**, *310*, 74–82. [CrossRef] [PubMed]
25. Nikogeorgos, N.; Patil, N.J.; Zappone, B.; Lee, S. Interaction of porcine gastric mucin with various polycations and its influence on the boundary lubrication properties. *Polymer* **2016**, *100*, 158–168. [CrossRef]
26. Nowald, C.; Penk, A.; Chiu, H.-Y.; Bein, T.; Huster, D.; Lieleg, O. A Selective ucin/Methylcellulose Hybrid Gel with Tailored Mechanical Properties. *Macromol. Biosci.* **2016**, *16*, 567–579. [CrossRef] [PubMed]

27. Berg, A.A.; Buul, J.D.; Tytgat, G.N.J.; Groen, A.K.J.; Ostrow, D. Mucins and calcium phosphate precipitates additively stimulate cholesterol crystallization. *J. Lipid Res.* **1998**, *39*, 1744–1751. [PubMed]
28. Raynal, B.D.; Hardingham, T.E.; Sheehan, J.K.; Thornton, D.J. Calcium-dependent protein interactions in MUC5B provide reversible cross-links in salivary mucus. *J. Biol. Chem.* **2003**, *278*, 28703–28710. [CrossRef] [PubMed]
29. Su, Y.; Xu, Y.; Yang, L.; Wenga, S.; Soloway, R.D.; Wang, D.; Wua, J. Spectroscopic studies of the effect of the metal ions on the structure of mucin. *J. Mol. Struct.* **2009**, *920*, 8–13. [CrossRef]
30. Amborta, D.; Johanssona, M.E.V.; Gustafssona, J.K.; Nilssonb, H.E.; Ermunda, A.; Johanssona, B.R.; Koeckb, P.J.B.; Hebertb, H.; Hanssona, G.C. Calcium and pH-dependent packing and release of the gel-forming MUC2 mucin. *PNAS* **2012**, *109*, 5645–5650. [CrossRef] [PubMed]
31. Yamasaki, T.; Chijiiwa, K.; Endo, M. Isolation of mucin from human hepatic bile and its induced effects on precipitation of cholesterol and calcium carbonate in vivo. *Deg. Dis. Sci.* **1993**, *38*, 909–915. [CrossRef]
32. Turner, B.S.; Bhaskar, K.R.; Hadzopoulou-Cladaras, M.; LaMont, J.T. Cysteine-rich regions of pig gastric mucin contain von Willebrand factor and cystine knot domains at the carboxyl terminal. *Biochim. Biophys. Acta-Gene Struct. Expr.* **1999**, *1447*, 77–92. [CrossRef]
33. Balabushevich, N.G.; Guerenu, A.V.; Feoktistova, N.A.; Volodkin, D. Protein loading into porous $CaCO_3$ microspheres: Adsorption equilibrium and bioactivity retention. *Phys. Chem. Chem. Phys.* **2015**, *17*, 2523–2530. [CrossRef] [PubMed]
34. Balabushevich, N.G.; Lopez de Guerenu, A.V.; Feoktistova, N.A.; Volodkin, D. Protein-Containing Multilayer Capsules by Templating on Mesoporous $CaCO_3$ Particles: POST- and PRE-Loading Approaches. *Macromol. Biosci.* **2016**, *16*, 95–105. [CrossRef] [PubMed]
35. Mantle, M.; Allen, A. A colorimetric assay for glycoproteins based on the periodic acid/Schiff stain. *Biochem. Soc. Trans.* **1978**, *6*, 607–609. [CrossRef] [PubMed]
36. Fields, R. The rapid determination of amino groups with TNBS. *Methods Enzymol.* **1972**, *25*, 464–468. [PubMed]
37. Hess, E.L.; Coburn, A.F.; Bates, R.C.; Murphy, P. A New Method for Measuring Sialic Acid Levels in Serum and Its Application to Rheumatic Fever. *J. Clin. Investig.* **1957**, *36*, 449–455. [CrossRef] [PubMed]
38. Balabushevitch, N.G.; Kildeyeva, N.R.; Moroz, N.A.; Trusova, S.P.; Virnik, A.D.; Khromov, G.L.; Larionova, N.I. Regulating aspects of biosoluble and insoluble film release systems containing protein proteinase inhibitor. *Appl. Biochem. Biotechnol.* **1996**, *61*, 129–138. [CrossRef]
39. Shomig, V.J.; Kasdorf, B.T.; Scholz, K.; Bidmon, K.; Lieleg, O.; Berensmeier, S. An optimazid purification process for porcine gastric mucin with preservation of its natural functional properties. *RSC Adv.* **2016**, *6*, 44932–44943. [CrossRef]
40. Feoktistova, N.; Rose, J.; Prokopovic, V.Z.; Vikulina, A.S.; Skirtach, A.; Volodkin, D. Controlling the vaterite $CaCO_3$ crystal pores. Design of tailor-made polymer based microcapsules by hard templating. *Langmuir* **2016**, *32*, 4229–4238. [CrossRef] [PubMed]
41. Paulraj, T.; Feoktistova, N.; Velk, N.; Uhlig, K.; Duschl, C.; Volodkin, D. Microporous Polymeric 3D Scaffolds Templated by the Layer-by-Layer Self-Assembly. *Macromol. Rapid Commun.* **2014**, *35*, 1408–1413. [CrossRef] [PubMed]
42. Vikulina, A.S.; Feoktistova, N.A.; Balabushevich, N.G.; Skirtach, A.G.; Volodkin, D.V. The mechanism of catalase loading into porous vaterite $CaCO_3$ crystals by co-synthesis. *Phys. Chem. Chem. Phys.* **2018**, *20*, 8822–8831. [CrossRef] [PubMed]
43. Fritz, H.; Wunderer, G. Biochemistry and Applications of Aprotinin, the Kallikrein Inhibitor from Bovine Organs. *Arzneim. Forsch./Drug Res.* **1983**, *33*, 479–494.
44. Balhorn, R. The protamine family of sperm nuclear proteins. *Genome Biol.* **2007**, *8*, 227–234. [CrossRef] [PubMed]
45. Balabushevich, N.G.; Zimina, E.P.; Larioniva, N.I. Encapsulation of catalase in polyelectrolyte microspheres composed of melamine formaldehyde, dextran sulfate, and protamine. *Biochem. Mosc.* **2004**, *69*, 763–769. [CrossRef]
46. Ellman, G.L. Tissue sulfhydryl groups. *Arch. Biochem. Biophys.* **1959**, *82*, 70–77. [CrossRef]
47. Boateng, J.S.; Pawar, H.V.; Tetteh, J. Evaluation of *in vitro* wound adhesion characteristics of composite film and wafer based dressings using texture analysis and FTIR spectroscopy: A chemometrics factor analysis approach. *RSC Adv.* **2015**, *5*, 107064–107075. [CrossRef]

48. Sandberg, T.; Karlsson, O.M.; Carlsson, J.; Feiler, A.; Caldwell, K.D. Potential use of mucins as biomaterial coatings. II. Mucin coatings affect the conformation and neutrophil-activating properties of adsorbed host proteins—Toward a mucosal mimic. *J. Biomed. Mater. Res. A* **2009**, *91*, 773–785. [CrossRef] [PubMed]
49. Lechanteur, A.; Neves, J.; Sarmento, B. The role of mucus in cell-based models used to screen mucosal drug delivery. *Adv. Drug Del. Rev.* **2018**, *124*, 50–63. [CrossRef] [PubMed]
50. Builders, P.F.; Kunle, O.O.; Okpaku, L.C.; Builders, M.I.; Attama, A.A.; Adikwu, M.U. Preparation and evaluation of mucinated sodium alginate microparticles for oral delivery of insulin. *Eur. J. Pharm. Biopharm.* **2008**, *70*, 777–783. [CrossRef] [PubMed]
51. Uhlig, K.; Madaboosi, N.; Schmidt, S.; Jager, M.S.; Rose, J.; Duschl, C.; Volodkin, D.V. 3d localization and diffusion of proteins in polyelectrolyte multilayers. *Soft Matter* **2012**, *8*, 11786–11789. [CrossRef]
52. Sustr, D.; Hlavacek, A.; Duschl, C.; Volodkin, D. Multi-Fractional Analysis of Molecular Diffusion in Polymer Multilayers by FRAP. A New Simulation-Based Approach. *J. Phys. Chem B* **2018**, *122*, 1323–1333. [CrossRef] [PubMed]
53. Behra, M.; Schmidt, S.; Hartmann, J.; Volodkin, D.V.; Hartmann, L. Synthesis of Porous PEG Microgels Using $CaCO_3$ Microspheres as Hard Templates. *Macromol. Rapid Commun.* **2012**, *33*, 1049–1054. [CrossRef] [PubMed]
54. Behra, M.; Azzouz, N.; Schmidt, S.; Volodkin, D.V.; Mosca, S.; Chanana, M.; Seeberger, P.H.; Hartmann, L. Magnetic porous sugar-functionalized PEG microgels for efficient isolation and removal of bacteria from solution. *Biomacromolecules* **2013**, *14*, 1927–1935. [CrossRef] [PubMed]
55. Feoktistova, N.; Stoychev, G.; Ionov, L.; Volodkin, D. Porous thermo-responsive pNIPAM microgels. *Eur. Polym. J.* **2015**, *68*, 650–656. [CrossRef]
56. Sergeeva, A.; Feoktistova, N.; Prokopovic, V.; Gorin, D.; Volodkin, D. Design of porous alginate hydrogels by sacrificial $CaCO_3$ templates: Pore formation mechanism. *Adv. Mater. Interfaces* **2016**, *2*, 1500386. [CrossRef]
57. Sergeeva, A.; Sergeev, R.; Lengert, E.; Zakharevich, A.; Parakhonskiy, B.; Gorin, D.; Sergeev, S.; Volodkin, D. Composite magnetite and protein containing $CaCO_3$ crystals. External manipulation and vaterite calcite recrystallization-mediated release performance. *ACS Appl. Mater. Interfaces* **2015**, *7*, 21315–21325. [CrossRef] [PubMed]
58. Sergeeva, A.S.; Gorin, D.A.; Volodkin, D.V. In-situ assembly of Ca-alginate gels with controlled pore loading/release capability. *Langmuir* **2015**, *13*, 10813–10821. [CrossRef] [PubMed]
59. Volodkin, D.V.; Balabushevitch, N.G.; Sukhorukov, G.B.; Larionova, N.I. Model systems for controlled protein release: PH-sensitive polyelectrolyte microparticles. *S.T.P. Pharma Sci.* **2003**, *13*, 163–170.
60. Balabushevich, N.G.; Pechenkin, M.A.; Zorov, I.N.; Shibanova, E.D.; Larionova, N.I. Mucoadhesive Polyelectrolyte Microparticles Containing Recombinant Human Insulin and Its Analogs Aspart and Lispro. *Biochem. Mosc.* **2011**, *76*, 327–331. [CrossRef]
61. Balabushevich, N.G.; Pechenkin, M.A.; Shibanova, E.D.; Volodkin, D.V.; Mikhalchik, E.V. Multifunctional Polyelectrolyte Microparticles for Oral Insulin Delivery. *Macromol. Biosci.* **2013**, *13*, 1379–1388. [CrossRef] [PubMed]
62. Parakhonskiy, B.V.; Yashchenok, A.M.; Donatan, S.; Volodkin, D.V.; Tessarolo, F.; Antolini, R.; Möhwald, H.; Skirtach, A.G. Macromolecule Loading into Spherical, Elliptical, Star-Like and Cubic Calcium Carbonate Carriers. *Chem. Phys. Chem.* **2014**, *15*, 2817–2822. [CrossRef] [PubMed]

Article

A Facile Interfacial Self-Assembly of Crystalline Colloidal Monolayers by Tension Gradient

Dong Feng, Ding Weng and Jiadao Wang *

State Key Laboratory of Tribology, Tsinghua University, Beijing 100084, China; fd14@mails.tsinghua.edu.cn (D.F.); dingweng@mail.tsinghua.edu.cn (D.W.)

* Correspondence: jdwang@mail.tsinghua.edu.cn; Tel.: +86-010-6279-6458

Received: 11 May 2018; Accepted: 10 June 2018; Published: 13 June 2018

Abstract: Many self-assembly approaches of colloidal monolayers have flourished but with some shortages, such as complexity, time-consumption, parameter sensitivity, and high-cost. This paper presents a facile, rapid, well-controlled, and low-cost method to prepare monolayers by directly adding silica particle suspensions containing water and ethanol to different liquids. A detailed analysis of the self-assembly process was conducted. The particles dove into water firstly, then moved up under the effect of the buoyancy and the tension gradient. The tension gradient induced the Marangoni convection and the relative motion between the water and the particles. At last, the particles were adsorbed at the air-water interface to minimize the free energy. The quality of the monolayers depended on the addition of sodium dodecyl sulfonate or ethanol in the water subphase. An interfacial polymerization of ethyl 2-cyanoacrylate was used to determine the contact angles of the particles at different subphase surfaces. The value of the detachment energy was positively associated with the contact angle and the surface tension. When the detachment energy decreased to a certain value, some particles detached from the surface, leading to the formation of a quasi-double layer. We also observed that the content of ethanol in suspensions influenced the arrangement of particles.

Keywords: self-assembly; air-liquid interface; tension gradient; Marangoni convection; nanoparticle; monolayer

---

1. Introduction

Two-dimensional ordered arrangements of colloidal particles, commonly referred to as crystalline colloidal monolayers, have a wide range of applications in surface-enhanced Raman scattering (SERS) [1], patterned surface fabrication [2], photonic crystals [3], wetting property modification [4] and so on. Recently, numerous methods have been developed to prepare crystalline monolayers on various substrates. The main approaches are based on (1) the wettability of colloidal suspensions on the substrates or (2) the wettability of colloidal particles at the air-liquid interfaces. In the first case, the methods include spin coating [5], dip coating [6], and convective coating [7]. The suspensions form wetting films on the substrates and the order of the monolayers depends on the evaporation process of the suspensions. Thus, these methods are sensitive to small variations of ambient humidity or temperature, and careful adjustments of the experimental parameters have to be carried out [8]. Most predominantly, it is difficult to rearrange particles once they contact the substrates, thus resulting in defects. As for the second case, particles with some extent hydrophobicity float at the air-liquid interface, which provides the required particle mobility for defect-free packing. These processes generally consist of floating particles at the interfaces, crystallizing the particle films with a mobile barrier (Langmuir–Blodgett technique) [9] or a surfactant [10], and transferring the films onto substrates. Many methods are used to float the particles at the interfaces. One method consists of dispersing the particles in an organic liquid and slowly pipetting them onto the water surface. A monolayer formed at the water surface after evaporation of the solvent, which was time consuming [11]. Retsch et al. [12] deposited sparsely distributed particles on a parent substrate, followed by an immersion step, during which the particles detached from the substrate and floated on water. This process was a little complex and hard to control. Other methods require the formation of the meniscuses to enable the particles spread on the water surfaces. Zhang and co-workers [13] reported on the fabrication of 2D arrays of colloidal particles by a needle tip flow method. The position of the needle tip was controlled carefully to enable the spreading of the particles along the meniscus formed between the tip and the water surface. Another method was adding water to a level where a meniscus was formed around the periphery of a glass slide in a Petri dish, followed by dropping colloidal suspensions on the glass slide. Once the suspensions contacted water, the particles spread at the water surface [14]. In these methods, the precise control was needed to form the meniscus. Also, the supply of the particles should be below a certain value to prevent breaking of the water surface [15] and the efficiency was limited.

In this paper, we developed a facile, fast, and cost-effective method for the fabrication of large-area crystalline colloidal monolayers. SiO_2 particles dispersed in a mixture of water and ethanol were directly added to the liquid. The monolayers could still form even though the suspensions dove into the liquid with high speed and big impact. They were subsequently transferred to a target substrate. Particle image velocimetry (PIV) was used to observe the moving-up of particles after the suspensions were added to the water, the mechanism of which was attributed to the buoyancy and the tension gradient. Scanning electron microscope (SEM) images of transferred monolayers indicated that the concentration variation of sodium dodecyl sulfonate (SDS) or ethanol in the water subphase influenced the order of the monolayers. The contact angles of SiO_2 at different subphase surfaces were determined by a trapping technique. Then the influence of the wettability of particles on the formation of the monolayer was discussed. In previous studies, ethanol in the suspension was commonly thought to play a role as a spreading agent [16,17]. We found that the arrangement of particles was affected by the content of ethanol in the suspension.

2. Materials and Methods

2.1. Materials

The hydrophobic silica particle suspensions (2.5% *w/v*) with diameter of 900 nm were purchased from Tianjin Baseline Chromtech Research Center (Tianjin, China). The functional group on the surfaces of the particles was vinyl. The solvents were the mixtures of water and ethanol (1:1, *v/v*). Silicon (100) wafers (Zhejiang Lijing Silicon Material Co. Ltd., Quzhou, China) were cut into 1.5×1.5 cm^2 and cleaned ultrasonically in acetone, ethanol, and deionized water for 20 min each in series. Acetone and ethanol were obtained from Beijing Chemical Works (Beijing, China). SDS was from J&K Scientific Ltd. (Beijing, China). The instant adhesive MC100 was produced by the 3M Company (Bracknell, UK). All other chemicals were of analytical grade and were used as purchased.

2.2. Monolayer Preparation

Figure 1a–e shows the process flow for the self-assembly of the monolayer at the air-liquid interface and the subsequent transfer to a silicon wafer. The hydrophobic silica suspensions were directly added to the liquid (Figure 1a). The suspensions dove into the liquid and then the particles moved up to the air-liquid interface (Figure 1b). Patches of monolayers formed and floated at the interface (Figure 1c). With the further additions of the suspensions, the patches continued to grow in size until the whole interface was covered. The monolayer could be seen by the bright, colorful Bragg reflexes (>63 cm^2, Figure 1f). A silicon wafer was immersed into the subphase and elevated under a shallow angle to transfer the silica monolayer (Figure 1d). Drying was performed under a dip angle of approximately 45° (Figure 1e). As a result, a hexagonally close-packed silica particle monolayer was obtained on the substrate. Even though the suspension was added at a height of 1.46 m above the water surface and parts of the suspension dove to a depth nearly 3 cm from the surface, the particles could still float up to form a monolayer (Video S1). Thus, the method was not sensitive to the dropping height. In the follow-up experiments, the dropping height was about 1 cm.

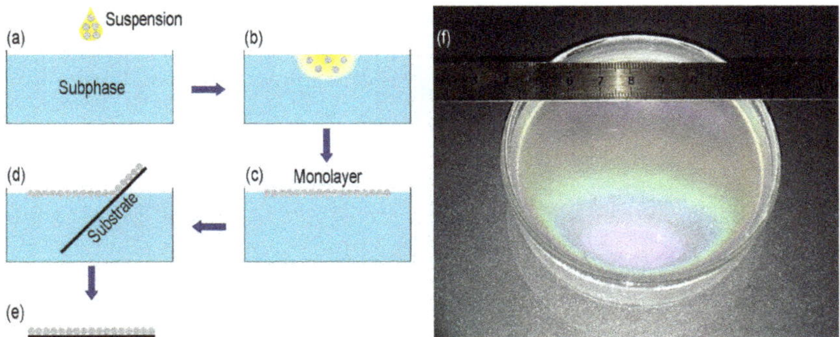

Figure 1. Schematic illustration of the monolayer fabrication process. (**a**) Addition of the hydrophobic silica suspension to the liquid; (**b**) Moving up of the particles to the air-liquid interface; (**c**) Self-assembly of a close-packed silica particle monolayer; (**d**) Pick-up of the monolayer with a silicon wafer; (**e**) Drying of the monolayer; (**f**) Photograph of the monolayer obtained from 900 nm particles.

2.3. Determination of the Contact Angle of Nanoparticles at the Interface

Inspired by the trapping technique based on the anionic polymerization reaction of butylcyanoacrylate [18], we used the instant adhesive MC100 (3M Company, Bracknell, UK) to visualize the interfacial position of silica particles. The main constituent of the adhesive was ethyl 2-cyanoacrylate (ECA) and the content of ECA was 90–95%. As shown in Figure 2a, a few drops of the instant adhesive were applied thinly and evenly to the bottom of a Petri dish. The Petri dish was upended on a weighing bottle containing a silica monolayer at the air-liquid interface and a closed space was formed. The instant adhesive was heated to 60 °C by a heating plate. The ECA monomers diffused through the vapor phase and contacted the liquid. The anions formed due to the reaction between the hydroxide ions (OH^-) and the monomers, which initiated polymerization reaction (Figure 2b) [19]. As more monomers were supplied, a layer of poly(ethyl 2-cyanoacrylate) (PECA) appeared. The monomers could diffuse into the layer, leading to the growth of polymer into the subphase (Figure 2c). A thin polymer membrane formed around particles if there was a liquid film at the surfaces of particles (Figure 2d). Figure 2e shows the reaction scheme of the polymerization. After five minutes of reaction, the monolayer with PECA was transferred to a silicon wafer. In side-view images, the interfacial position of the particles could be visualized by SEM. This approach ensured that the monolayer and the interface were not perturbed. To illustrate that the polymerization process had no impact on the wettability, we performed contact angle measurement of a water drop (30 μL) on a silicon wafer. The contact angle was 61.1° (Figure 3c). Then the polymerization process occurred at the surface of the drop. Compared with Figure 3a, Figure 3b shows that a polymer membrane formed on the drop surface. The contact angle was 61.3° (Figure 3d). There was no significant change in the contact angle. This illustrated that the time required to form the membrane was so short that the wetting state of particles at the interface would not be influenced. The top of the drop became flat because the surface tension of water disappeared after the drop surface was covered by the membrane. In order to study whether the thickness of the polymer layer affected the contact angle of nanoparticles at the interface, we prolonged the polymerization reaction time to 20 min. The monolayer was obtained at the deionized water surface. Figure 4b shows that the membrane was much thicker than that when the time was 5 min in Figure 4a. The contact angle of the colloids did not change and the lateral arrangement of the colloids was not disturbed by the growing polymer film.

Overall, the polymerization process was thought to possess little impact on the interfacial position of the silica particles.

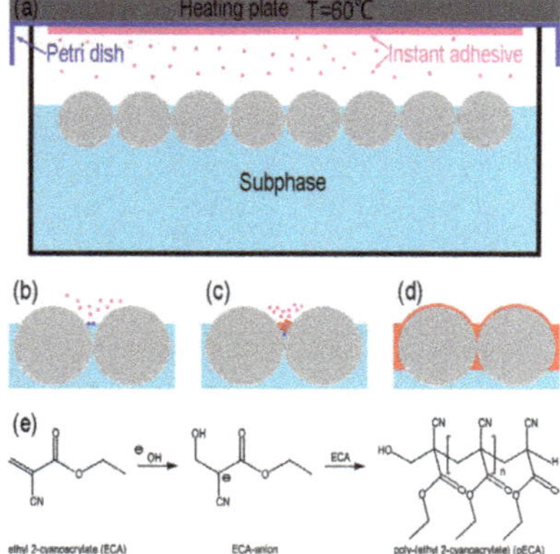

Figure 2. Interfacial polymerization of ethyl 2-cyanoacrylate (ECA) used to determine the contact angle of silica particles at the interface. (**a**) Scheme of the experimental setup; (**b–d**) Schematic of the reaction: The ECA monomers (pink) diffuse via the vapor phase to the air-liquid interface. The ECA anions (blue) are formed upon contact with the liquid. The anionic polymerization occurs and the polymer (red) is generated. The polymerization proceeds with the addition of monomers and the polymer membrane eventually covers the surfaces of the liquid and particles; (**e**) The polymerization reaction of ECA initiated by hydroxide ions in the liquid.

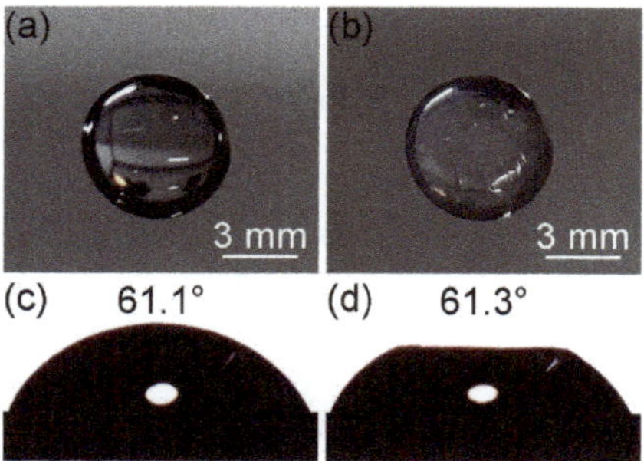

Figure 3. The appearance of a water drop on a silicon wafer before (**a**) and after (**b**) the polymer membrane forms on the drop surface; (**c**,**d**) are contact angle measurements of the water drop in (**a**,**b**), respectively.

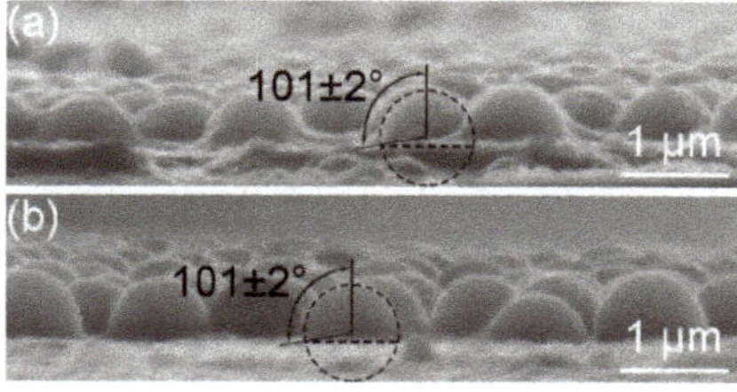

Figure 4. Side-view scanning electron microscope (SEM) images of the monolayers with the poly(ethyl 2-cyanoacrylate) (PECA), which are obtained at the deionized water surfaces. Different polymerization reaction times are used. (**a**) 5 min and (**b**) 20 min.

2.4. Characterization

SEM (FEI Quanta 200 FEG, FEI Company, Hillsboro, OR, USA) was used to image the morphologies of the colloidal films. The motion of particles during the film formation process was recorded by PIV (MicroVec, Inc., Beijing, China) from a side view. The PIV setup included a 532 nm diode pump solid state laser (DPSSL) with a maximum power of 5 W and a 12–14 bit camera with a maximum capture rate of 258 fps, fitted with a 100 mm f/2.8 Tokina lens. When PIV was used to capture the motion of particles from above, the laser was replaced by a high brightness cold light source (XD-300, Nanjing Yanan Special Light Factory, Nanjing, China). The surface tension data were obtained using the hanging plate method with a Dataphysics OCA-20 (DataPhysics Instruments GmbH, Filderstadt, Germany) analyzer. Water contact angles were measured on a Dataphysics OCA 25 (DataPhysics Instruments GmbH, Filderstadt, Germany) instrument at room temperature.

3. Results and Discussion

3.1. Self-Assembly Process of the Colloidal Monolayer at the Air-Water Interface

To clarify the formation mechanism of the monolayer, the motion of particles was recorded by PIV from a side view. The suspension drop dove into the water after it was added above the air-water interface. Then, the particles moved up to the air-water interface (Figure 5a). Another common method to produce monolayers was the usage of a partially immersed glass slide to add colloidal suspensions to the air–water interface, which was thought to be an effective approach to let the particles flow gently at the interface [20,21]. We used the above method to observe the motion of the particles after the suspension was injected onto the glass slide. It was found that some particles still dove into the water and then moved up to the interface (Figure 5b), which illustrated that the use of the glass slide would not have a significant impact on the quality of the monolayer. In the pink dashed line frame, no particles existed and the velocity distribution appeared because the glass slide reflected light. The reason why the particles could move up to the surface is attributed to three aspects (Figure 5c). Firstly, the suspension is under the effect of buoyancy due to the lower density of ethanol. Thus, the suspension does not sink into water. Secondly, a radiated concentration distribution of ethanol from high to low at the interface is formed when ethanol starts to dissolve into water, which is centered on the position where the suspension is dripped (the supplying point). Therefore, a radiated surface tension gradient from low to high is generated because the surface tension of ethanol is much lower than that of water. Variations in surface tension result in the Marangoni convection [22,23], which drives the particles to move up. Thirdly, the concentration of ethanol is higher at the position closer to the interface because of the convective transport and the dissolution process. The tension becomes lower and lower from the interior of subphase to the surface. The particles move up relative to water due to the stresses resulting from the variations of tension at the particle surfaces [24]. Without the relative motion, it is hard for the particles to move across and be absorbed at the air-water interface. In Figure 6, the motion of particles at the interface was recorded by PIV from a top view. The numerical values at the top right corners of the images represent the time of occurrence and the initial time was the moment when the drop of the suspension contacted the air-water interface. Silica particles moved radially under the effect of Marangoni convection and aggregated into a ring-like structure whose center was the supplying point (Figure 6a). As more particles moved to the interface, the ring became wide (Figure 6b). When the dissolution of ethanol reached the steady state, the surface tension gradient disappeared and the ring-like monolayer shrank to the supplying point (Figure 6c). In the end, the ring transformed into a piece of monolayer with a few small fragments around (Figure 6d). The time required for the whole process was about 0.92 s.

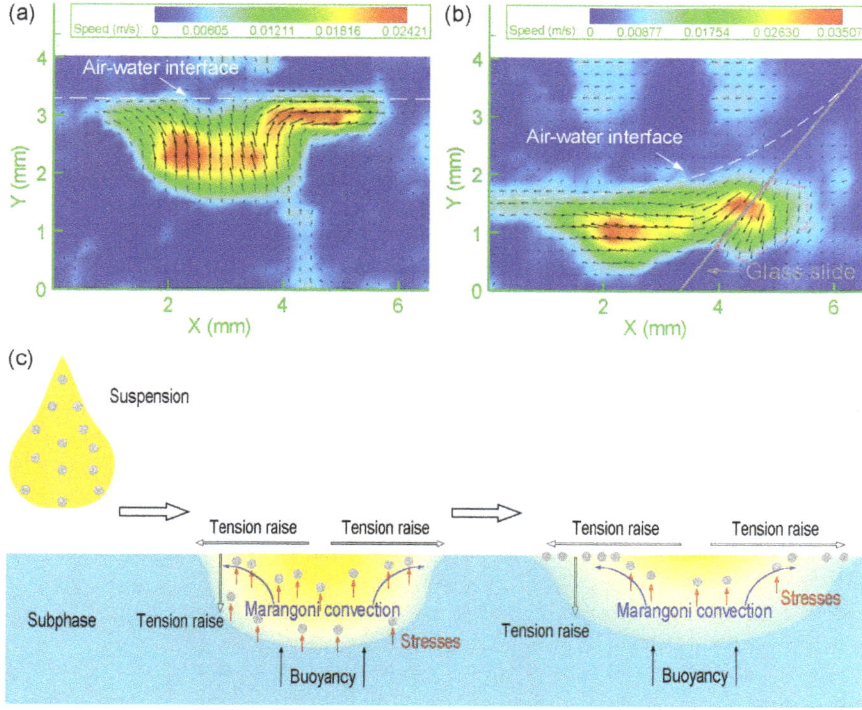

Figure 5. Transient motion of particles after the suspension is added to water (**a**) directly and (**b**) via a partially immersed glass slide at a tilt angle of approximately 45° with respect to the water surface. The direction of the arrows indicates the motion direction of the particles. The color scale bar shows the magnitude of the velocity (m/s); (**c**) Schematic illustration of the mechanism for particles moving up to the air-subphase interface.

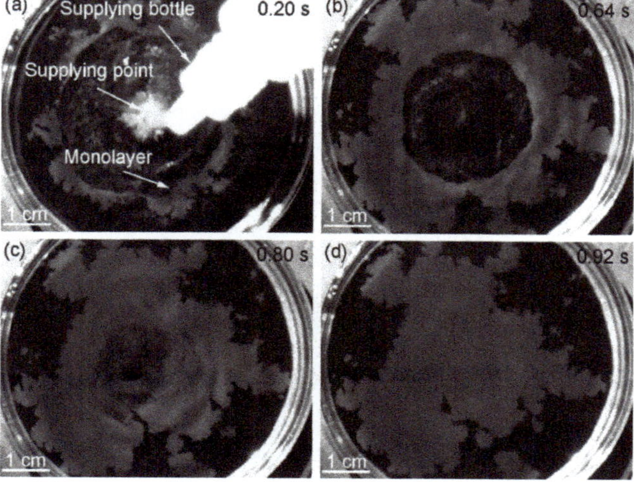

Figure 6. The appearance of the formation process of the monolayer at different time: (**a**) 0.20 s; (**b**) 0.64 s; (**c**) 0.80 s and (**d**) 0.92 s.

3.2. Influence of SDS Concentration in the Subphase on the Formation of the Monolayer

The research found that the quality of the monolayer was quiet dependent on the SDS concentration in the water subphase. The solvents of the silica suspensions were the mixtures of water and ethanol (1:1, *v/v*). The OTSU algorithm [25] was used to calculate the coverage rates of the monolayers on the silicon wafers (φ) in the insets of Figure 7a–c. In the treated pictures (Figure A1 in Appendix A), the monolayers and the substrates were divided into the white and black parts, respectively. By calculating the proportion of white pixels in the total pixels of the images, we got the coverage rates of the monolayers on the silicon wafers. For the hexagonally close-packed structure in an ideal state, φ is about 90.7% based on the geometrical calculation in Appendix B. Figure 7a shows a monolayer of silica particles formed at the deionized water surface. Most of the particles arranged in a disordered manner and φ was just 62.4%. The dominating attractive forces acting on the particles at the air-water interface are the van der Waals forces, the hydrophobic attractive forces, and the capillary forces [26]. Capillary forces are long range interactions with an effective range of up to several millimeters, while the ranges of hydrophobic forces and van der Waals forces are much shorter [26]. Particles brought into close vicinity by capillary forces immediately feel the hydrophobic forces and the van der Waals forces. The strong attractive forces cause the aggregation of particles and suppress the rearrangement of particles. When the SDS was added to the water prior to the spreading of the silica, the order and the coverage rate of the monolayer were greatly increased. As SDS molecules are amphiphilic, they accumulate at the air-water interface and act as a soft barrier. The particles rising to the interface push against the SDS molecules which in turn push the particles closer, therefore the coverage rate of the monolayer is increased. This phenomenon has been known as the piston oil effect [12]. Despite the equal sign of the charges, some SDS molecules are adsorbed at the surfaces of silica particles, which introduces more negative charges and enhances the electrostatic repulsion forces between particles [27]. The energy barrier for a close contact is increased. Moreover, SDS lowers the capillary forces by reducing the surface tension of water [28]. As a result, the repulsion forces could counteract the attractive forces, which enables particles to have sufficient mobility. The extrusion forces from the SDS arrange the particles into a crystalline monolayer. It is also found that the presence of surfactants makes the monolayer exhibit higher mechanical stability and less fracture at the edges. Thus, the addition of SDS has benefitted the monolayer transfer, too.

The monolayers formed with SDS concentrations between 0.1 and 0.7 mmol/L are shown in Figure 7b–d. When the concentration of SDS was 0.1 mmol/L, some defects, such as vacancies, dislocations and disordered regions still existed, and φ increased to 66.8% (Figure 7b). At the most appropriate concentration (0.4 mmol/L), a crystalline monolayer with the highest coverage rate (φ = 78.7%) was obtained (Figure 7c). As the concentration reached 0.7 mmol/L, particles stacked into multilayers (Figure 7d). Figure 8a shows that the maximum inner diameter of the ring-like structure was much smaller than that when no SDS was added in Figure 6a, which illustrated that the amount of SDS at the air-water interface was high enough to impede the spreading of the particles. The multilayers (the whiter regions in Figure 8b) were mainly found near the supplying point. The formation mechanism is shown in Figure 8c. At the end of the self-assembly process, there are a small number of silica suspensions under the air-water interface. The Marangoni convection is too weak to overcome the counterforce of SDS. The monolayer shrinks and covers the air-water interface near the supplying point. When those silica particles move up, they could only move to the bottom of the monolayer and are fixed under the van der Waals forces between particles.

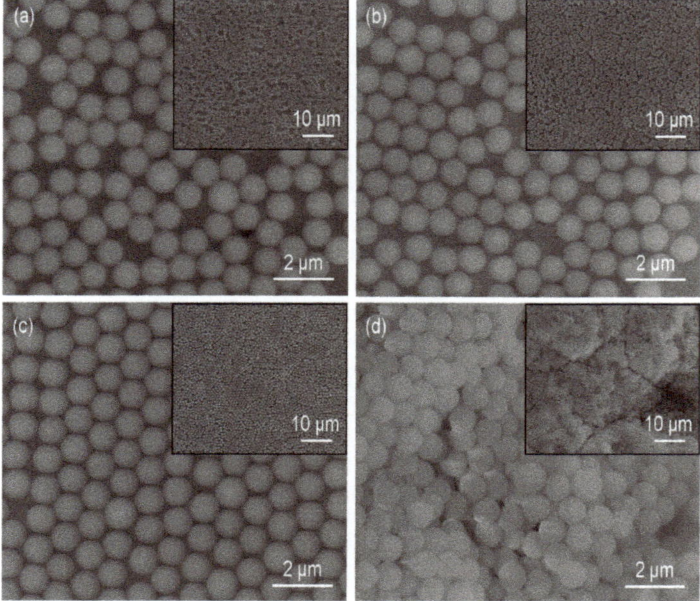

Figure 7. SEM images of monolayers formed at water surface with sodium dodecyl sulfonate (SDS) concentration of (**a**) 0; (**b**) 0.1; (**c**) 0.4 and (**d**) 0.7 mmol/L. The insets show lower magnifications of the same samples.

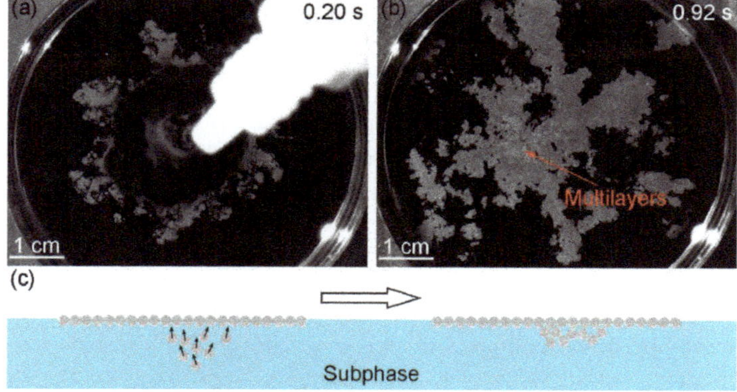

Figure 8. The appearance of the monolayer obtained with SDS concentration of 0.7 mmol/L at different time: (**a**) 0.20 s and (**b**) 0.92 s; (**c**) Schematic illustration of the formation mechanism of the multilayers. The solid arrows in (**c**) represent the motion direction of the particles.

3.3. Influence of Ethanol Concentration in the Subphase on the Formation of the Monolayer

Another parameter that could influence the formation of the monolayer was the ethanol concentration in the water subphase. The solvents of the silica suspensions were the mixtures of water and ethanol (1:1, *v/v*). With the increase in ethanol concentration in the subphase, the capillary forces decrease because the surface tension of the mixture becomes lower. The addition of ethanol could also decrease the hydrophobic attractive forces [29,30]. The regularity of the monolayer is improved as the attractive forces are counteracted more efficiently by the repulsive forces and the particles have more time to reach their minimum free energy position and crystalize into a hexagonally close-packed structure. Taking the case of ethanol concentration being 20 vol.%, the particles congregated when the Marangoni convection disappeared (Figure 9a). The monolayer was divided into pieces under the Marangoni convection induced by the ethanol evaporation near the container wall and then the pieces moved to the side where the convection was stronger (Figure 9b–d). This indicated the attractive forces between particles were weak, otherwise the monolayer would move as a whole. We also found that the slightly liquid sloshing on the surface led to the rearrangement of the particles, which was not observed without ethanol in the subphase. Thus, the addition of ethanol decreased the attractive forces between the particles. SEM images of monolayers formed at liquid surfaces with ethanol concentrations between 10 and 40 vol.% are shown in Figure 10. The treated insets of Figure 10a–c by the OTSU algorithm are shown in Figure A2 in Appendix A. When the ethanol concentration was 10 vol.%, the arrangement of particles was disordered (Figure 10a). The coverage rate φ was 65.7%, which was higher than that (φ = 62.4%) when no ethanol existed. As the ethanol concentration reached 20 vol.%, the order of the arrangement was improved and φ was 72.8% (Figure 10b). Figure 10c reveals that a monolayer with limited dislocations was obtained at the ethanol concentration of 30 vol.%. 74.6% of the substrate surface was covered by the silica monolayer. When the concentration of ethanol reached 40 vol.%, Figure 10d shows that there was a discretely distributed upper layer on the hexagonally ordered bottom layer, which formed a quasi-double layer. Each structural unit in the upper layer consisted of three particles arranged triangularly and every particle located in the interstice formed by three particles of the bottom layer. This kind of structure was also reported in previous papers [10,31,32]. The particles marked with black and blue characters made up the representative regions of the monolayer and the double layer, respectively. By analyzing the positional relationships between the particles of the double layer, the formation mechanism is similar with that in Reference [10]. At the concentration of 40 vol.%, some particles dive into the subphase after the addition of the suspension. In order to confirm this, we used PIV to observe the movements of particles from a side view after the suspension (1.5×10^{-5} *w/v*) was added to the subphase, as shown in Figure 11. The suspension was diluted in order to meet the requirement of PIV observation and the solvent remained unchanged. The initial time was the moment when the drop of the suspension contacted with the air-liquid interface (Figure 11a) and the particles moved downward with the suspension because of inertia (Figure 11b). Then, the particles moved up and approached the interface in Figure 11c. At 0.352 s, no particle was seen in Figure 11d, which illustrated the particles were at the interface. However, many particles were seen below the interface after 0.376 s in Figure 11e,f, which meant the particles could not be held at the interface. The reason is fully discussed in Section 3.4. During the transfer process, particles below the interface (6 and 6′ in Figure 10e) could be captured on the substrate and move to the bottoms of the interstices formed by every three neighboring particles in the monolayer. They push upper three particles 1, 2, 3 and 1′, 2′, 3′ (Figure 10d), respectively. Each of the particles 6 and 6′ comes into the monolayer and occupies the positions previously belonging to three particles. Particles 4, 5 and 4′, 5′ move towards particles 3 and 3′, respectively. The pentagonal structures are formed in the bottom layer.

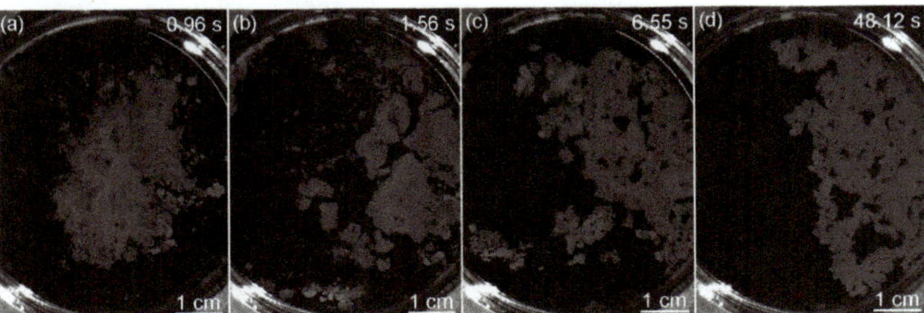

Figure 9. The appearance of the monolayer obtained with ethanol concentration of 20 vol.% at different time: (**a**) 0.96 s; (**b**) 1.56 s; (**c**) 6.55 s and (**d**) 48.12 s.

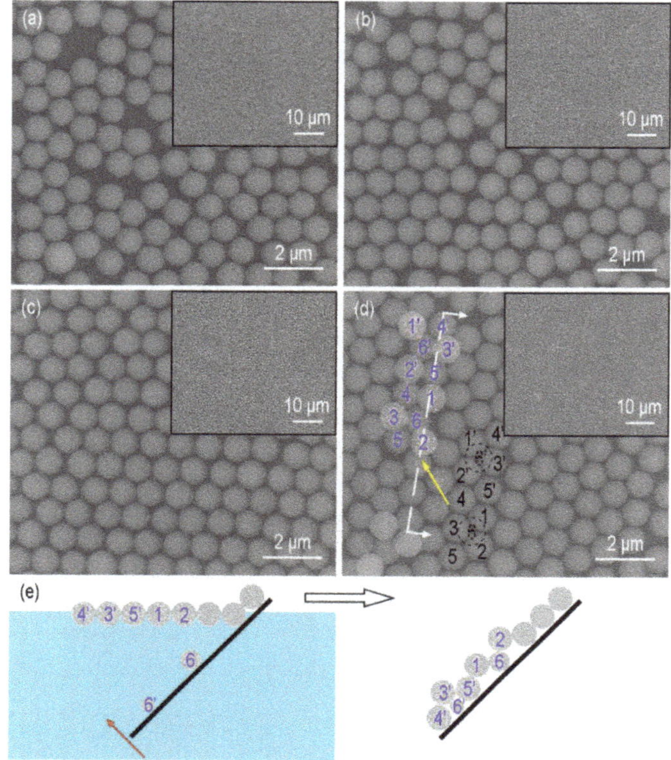

Figure 10. SEM images of monolayers formed at liquid surface with ethanol concentration of (**a**) 10; (**b**) 20; (**c**) 30 and (**d**) 40 vol.%. The insets show lower magnifications of the same samples; (**e**) Schematic illustration of the formation mechanism of the quasi-double layer, which is drew according to the cutting position (the white dashed line) and the projection direction (the white arrows) in (**d**).

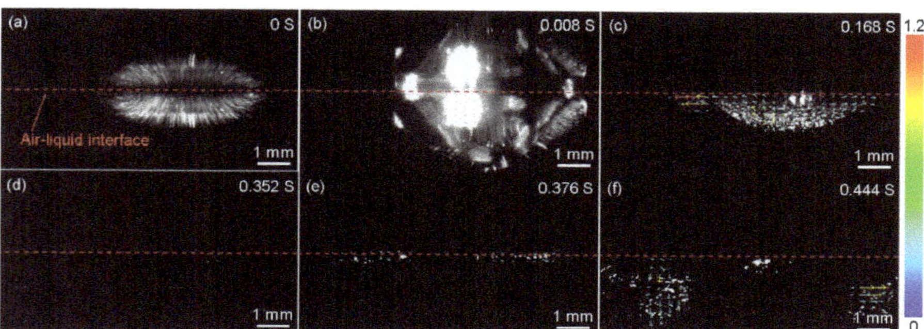

Figure 11. The movement process of particles after the suspension is added to the subphase containing 40 vol.% ethanol, which is captured by PIV. (**a–f**) correspond to the movements of particles at 0, 0.008, 0.168, 0.352, 0.376 and 0.444 s, respectively. The reflections above the interfaces are caused by the level fluctuations. The direction of the arrows indicates the motion direction of the particles. The color scale bar shows the magnitude of the velocity (cm/s).

3.4. Relationship between the Particle Wettability and the Formation of the Monolayer

The particle wettability can be properly described by the three-phase contact angle θ. Figure 12a–d show the side-view SEM images of monolayers obtained at the surfaces of subphases containing different concentrations of SDS. One could see horizontal boundaries representing the positions of subphase surfaces. PECA also formed around the parts of the particles below the boundaries, which reflected that the polymerization reaction could occur in the subphases. With no SDS in the subphase, the contact angle was 101°, which illustrated that the silica particles were hydrophobic (Figure 12a). The contact angle changed from 73° (Figure 12b) to 65° (Figure 12d) with the increase of the SDS concentration. Analogously, the contact angle went smaller with higher ethanol concentration as shown in Figure 12e–h. In other words, the particles were more immersed in the subphase. The surface tensions of different subphases are summarized in Table 1. It was worth mentioning that the surface tension of the subphase containing 0.4 mmol/L SDS was higher than that containing 10 vol.% ethanol, while the contact angles of particles at surfaces of these two subphases were similar. The tensions are related to the contact angle through Young's equation [33]

$$\gamma_{pa} = \gamma_{ps} + \gamma_{sa} \cos\theta \tag{1}$$

where γ is the appropriate interfacial tension and the subscripts a, p, s represent the air, particle and subphase, respectively; θ is the contact angle of a particle at a subphase surface. For different subphases, γ_{pa} was equal. Thus, the value of γ_{ps} of the subphase containing 0.4 mmol/L SDS was lower than that containing 10 vol.% ethanol. For the subphases containing 0.7 mmol/L SDS and 20 vol.% ethanol, the phenomenon was similar. One of the contribution factors is that the adsorption process of SDS changes the surface property of silica particles. The hydrophobic interactions between the alkyl chains of surfactants and the hydrophobic sites on the particles [34] lead to the conversion of particle surface state from the hydrophobicity to the hydrophilicity due to the ionic heads of the surfactants orienting towards the bulk solution. As mentioned above, some particles dove into the subphase when the ethanol concentration was up to 40 vol.%. The strength with which a particle is held at a subphase surface is related not only to θ but also to the surface tension γ_{sa}. Assuming the particle is small enough so that the effect of gravity is negligible and the subphase surface remains planar up to the contact line with the particle, the detachment energy E required to remove the particle from the surface into the subphase is given by [35]

$$E = \pi R^2 \gamma_{sa} (1 - \cos\theta)^2 \tag{2}$$

where R is the radius of the particle. The detachment energies of particles at different subphase surfaces are listed in Table 1. For the subphase containing 40 vol.% ethanol, the value of detachment energy (2.65×10^5 kT) is much smaller than that (160×10^5 kT) when no ethanol existed and the desorption of particles is the easiest. k is the Boltzmann constant and T is the temperature measured in Kelvin. However, it is considered that the particles are attached to the surface when the energy of adsorption is greater than the thermal energy kT [36]. Firstly, Equation (2) neglects the effect of the line tension, which is defined as the excess free energy per unit length of the line where the three phases meet. The detachment energies could be relatively low in the presence of the positive line tension [37]. The contribution of the line tension is important in the regime where the contact angle is larger than 120° or smaller than 60° [38]. Secondly, the impact of droplet on the surface when the suspension is added leads to the surface fluctuation, which is unfavorable to the adsorption of particles to the surface. Thirdly, the contact angle of interfacially absorbed particles has been found to possess a broad distribution [18,39]. Partial particles might have a contact angle lower than 40° and dive into the subphase. Although $\theta > 90°$ is not necessary, the particles could not float at the surface as long as the detachment energy is too small. These experiments also shows that the formation of the multilayers when the subphase containing 0.7 mmol/L SDS (Figure 7d) is not due to the desorption of particles, because the detachment energy is much higher than that when the subphase containing 40 vol.% ethanol (Figure 10d).

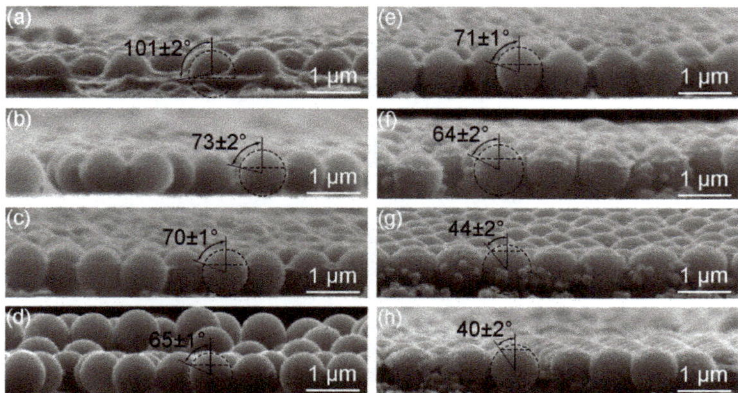

Figure 12. Side-view SEM images of the monolayers with the PECA. (**a**–**d**) correspond to monolayers obtained at SDS concentrations of 0, 0.1, 0.4, and 0.7 mmol/L; (**e**–**h**) correspond to monolayers obtained at ethanol concentrations of 10, 20, 30 and 40 vol.%. The data of θ are the average values of measurements made from five different particles.

Table 1. Contact angles and detachment energies of silica particles at surfaces of different subphases. The surface tensions of different subphases are listed.

Suphase	Surface Tension (mN/m)	Contact Angle (deg)	Detachment Energy ($\times 10^5$ kT)
deionized water	72.98 ± 0.01	101 ± 2	160
0.1 mmol/L SDS	67.77 ± 0.03	73 ± 2	52.4
0.4 mmol/L SDS	60.24 ± 0.07	70 ± 1	40.3
0.7 mmol/L SDS	48.05 ± 0.01	65 ± 1	24.8
10 vol.% ethanol	51.19 ± 0.04	71 ± 1	36.0
20 vol.% ethanol	40.80 ± 0.02	64 ± 2	19.9
30 vol.% ethanol	35.30 ± 0.07	44 ± 2	4.30
40 vol.% ethanol	31.30 ± 0.05	40 ± 2	2.65

3.5. Influence of Ethanol Concentration in the Suspension on the Formation of the Monolayer

We investigated the influence of ethanol concentration in the suspension on the formation of the monolayer by dispersing the particles in different compositions of ethanol/water while the weight percentage of particles in the mixture was fixed at 2.5%. The silica suspensions were added to the surfaces of deionized water. The treated insets of Figure 13b–d by the OTSU algorithm are shown in Figure A3 in Appendix A. When the volume percentage of ethanol was 25 vol.%, the particle film was a quasi-double layer structure consisting of a hexagonally close-packed bottom layer and a discretely distributed upper layer (Figure 13a). As mentioned above, the spreading of particles at the surface was outside-in due to the Marangoni convention. In Figure 14, the maximum velocity of particles induced by the low ethanol concentration was weak because the strength of the Marangoni convection depended on the surface tension gradient along the air-water interface [22]. Thus, the maximum inner diameter of the ring-like structure during the formation process of the monolayer decreased with the reduction of the ethanol concentration (Figure 14). At the ethanol concentration of 25 vol.%, the diameter was so small that there was not enough space for the subsequent particles to form into a monolayer. Figure 13b shows that there was enough ethanol for particles to spread into a monolayer at 50 vol.% concentration and φ was 62.4%. However, the arrangement of the monolayer was disordered because of the strong attractive forces acting at the particles. As the concentration reached 75 vol.%, φ decreased to 59.9% due to the further increase of the maximum inner diameter. Interestingly, the particles were arranged into more orderly structures (Figure 13c). Due to the low content of ethanol in one suspension drop, the influence of ethanol on the forces between particles is negligible. The important factor is the shearing stresses of the Marangoni convention exerted on the agglomerate. The increase of ethanol concentration improves the shearing forces, which could overcome the attractive forces. As a result, the particles in the agglomerate would change their positions to reduce the projected area against the interfacial flow and form into a hexagonally close-packed structure. This phenomenon is a kind of Kirkwood-Alder phase transition [40]. At the highest concentration, i.e., 100 vol.%, φ was only 57.1% (Figure 13d). Most importantly, the long range order was further improved which was even better than that in Figure 10c.

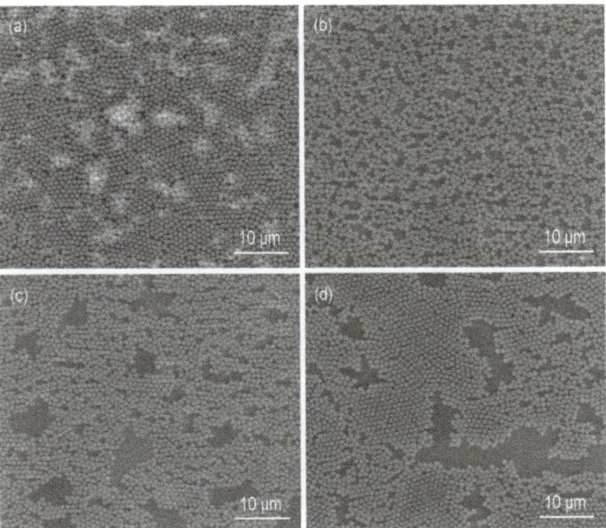

Figure 13. SEM images of monolayers formed by particles in different composition of ethanol/water: (**a**) 25%; (**b**) 50%; (**c**) 75% and (**d**) 100%.

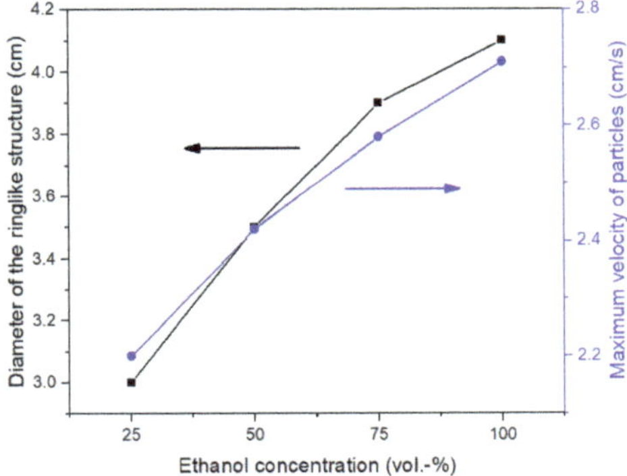

Figure 14. Influence of the ethanol concentration in the suspension on the maximum inner diameter of the ring-like structure (left-hand ordinate, ■) and on the maximum velocity of particles (right-hand ordinate, ●).

4. Conclusions

In summary, we have demonstrated a facile interfacial self-assembly procedure for the manufacture of large-area crystalline arrays of SiO_2 particles. The SiO_2 suspensions containing water and ethanol were directly added to water without other controls. The particles dove into water and subsequently moved up to the air-water interface attributed to the buoyancy and the tension gradient. On the one hand, the particles moved with the Marangoni convection of water resulting from the tension gradient. On the other hand, the upward motion of the particles relative to water appeared due to the tension gradient induced stresses. Once absorbed at the interface, particles formed a ring-like structure, which then transformed into a piece of monolayer. The optimum concentration of SDS (0.4 mmol/L) or ethanol (30 vol.%) in the water subphase helped SiO_2 particles self-assemble into a hexagonally close-packed array. By measuring the three-phase contact angles of particles through an interfacial polymerization of ethyl 2-cyanoacrylate, the wettability of the particles at different subphase surfaces was confirmed to be different. When the subphase was deionized water, the contact angle (101°) was the biggest and the monolayer was disordered. As for the subphase containing 40 vol.% ethanol, the contact angle (40°) was the smallest and the quasi-double layer formed. The reason was that the detachment energy was just 2.65×10^5 kT and the particles could not be steadily adsorbed at the air-liquid interface. Therefore, the appropriate wettability of particles was important to form crystalline monolayers. Moreover, the arrangements of the particles were affected by the ethanol concentrations in the suspensions, showing that the Marangoni convection played an important role in the self-assembly.

Supplementary Materials: The following video is available online at http://www.mdpi.com/2072-666X/9/6/297/s1, Video S1: The movement of particles after the addition of the suspension at a height of 1.46 m above the water surface.

Author Contributions: Conceptualization, D.F.; Project administration, D.W.; Supervision, J.W.; Writing—original draft, D.F.; Writing—review & editing, D.W. and J.W.

Funding: This research was funded by the National Natural Science Foundation of China Project (Grant Nos. 51775296, 51375253 and 51703116). The authors also acknowledge the support of this work from the National Laboratory for Information Science and Technology and State Key Laboratory of Tribology, Tsinghua University, China, under grant code SKLT2017C06.

Conflicts of Interest: The authors declare no conflict of interest.

Appendix A

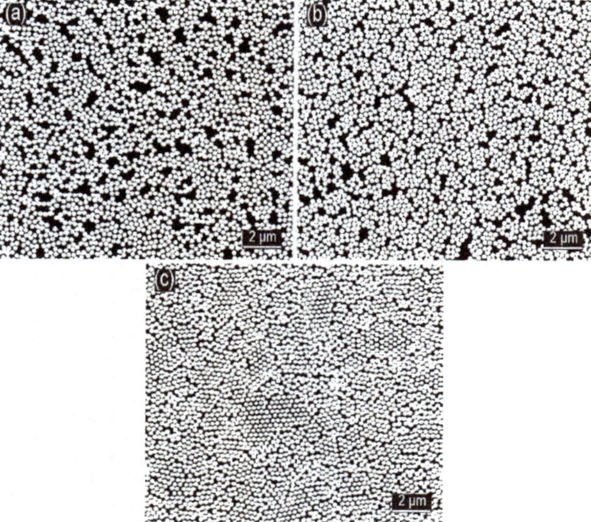

Figure A1. Treated SEM images of monolayers formed at water surface with SDS concentration of (**a**) 0; (**b**) 0.1 and (**c**) 0.4 mmol/L. The white parts are the monolayers and the black parts represent the substrates.

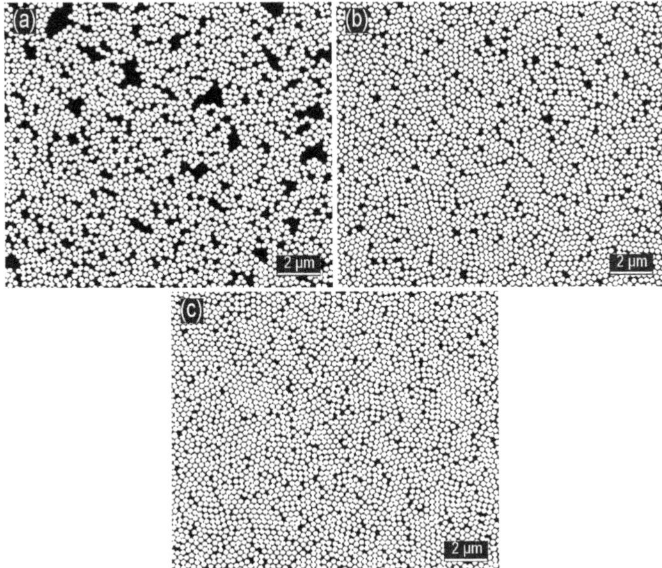

Figure A2. Treated SEM images of monolayers formed at liquid surface with ethanol concentration of (**a**) 10; (**b**) 20; (**c**) 30 vol.%. The white parts are the monolayers and the black parts represent the substrates.

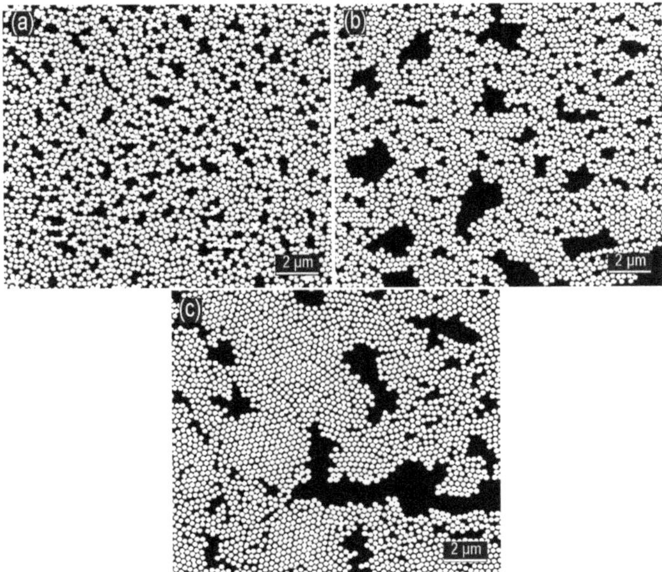

Figure A3. Treated SEM images of monolayers formed by particles in different composition of ethanol/water: (**a**) 50%; (**b**) 75% and (**c**) 100%. The white parts are the monolayers and the black parts represent the substrates.

Appendix B

Figure A4 shows the structural unit of the ideal hexagonally close-packed structure, which could be used to calculate the coverage rate of the ideal monolayer on the silicon wafer.

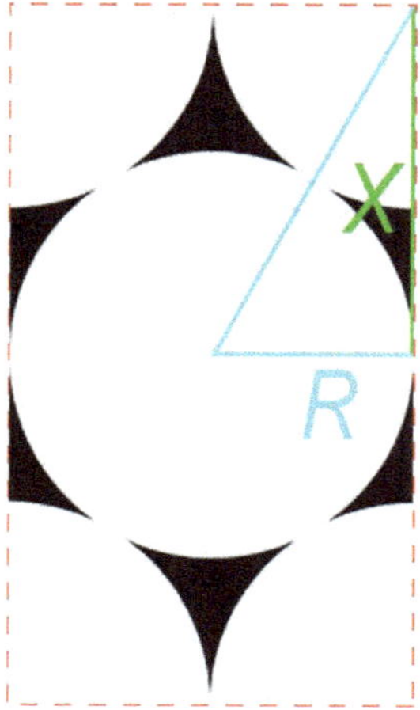

Figure A4. Structural unit of the ideal hexagonally close-packed structure.

Using Pythagoras, X can be calculated from the radius of the spheres as

$$X^2 + R^2 = (2R)^2 \tag{A1}$$

$$X = \left(4R^2 - R^2\right)^{1/2} = \sqrt{3}R \tag{A2}$$

The area S of the structural unit is given by

$$S = 4RX = 4\sqrt{3}R^2 \tag{A3}$$

The area S_0 of the particles in the structural unit is expressed as

$$S_0 = 2\pi R^2 \tag{A4}$$

Thus, the coverage rate φ of the ideal monolayer was

$$\varphi = \frac{S_0}{S} = \frac{\pi}{2\sqrt{3}} \approx 90.7\% \tag{A5}$$

References

1. Si, S.; Liang, W.; Sun, Y.; Huang, J.; Ma, W.; Liang, Z.; Bao, Q.; Jiang, L. Facile fabrication of high-density sub-1-nm gaps from Au nanoparticle monolayers as reproducible SERS substrates. *Adv. Funct. Mater.* **2016**, *26*, 8137–8145. [CrossRef]
2. Feng, D.; Weng, D.; Wang, B.; Wang, J. Laser pulse number dependent nanostructure evolution by illuminating self-assembled microsphere array. *J. Appl. Phys.* **2017**, *122*, 243102. [CrossRef]
3. Armstrong, E.; Khunsin, W.; Osiak, M.; Blömker, M.; Torres, C.M.S.; O'Dwyer, C. Ordered 2D colloidal photonic crystals on gold substrates by surfactant-assisted fast-rate dip coating. *Small* **2014**, *10*, 1895–1901. [CrossRef] [PubMed]

4. Hansson, P.M.; Skedung, L.; Claesson, P.M.; Swerin, A.; Schoelkopf, J.; Gane, P.A.; Rutland, M.W.; Thormann, E. Robust hydrophobic surfaces displaying different surface roughness scales while maintaining the same wettability. *Langmuir* **2011**, *27*, 8153–8159. [CrossRef] [PubMed]
5. Chen, J.; Dong, P.; Di, D.; Wang, C.; Wang, H.; Wang, J.; Wu, X. Controllable fabrication of 2D colloidal-crystal films with polystyrene nanospheres of various diameters by spin-coating. *Appl. Surf. Sci.* **2013**, *270*, 6–15. [CrossRef]
6. Ko, H.-Y.; Lee, H.-W.; Moon, J. Fabrication of colloidal self-assembled monolayer (SAM) using monodisperse silica and its use as a lithographic mask. *Thin Solid Films* **2004**, *447*, 638–644. [CrossRef]
7. Prevo, B.G.; Hon, E.W.; Velev, O.D. Assembly and characterization of colloid-based antireflective coatings on multicrystalline silicon solar cells. *J. Mater. Chem.* **2007**, *17*, 791–799. [CrossRef]
8. Lotito, V.; Zambelli, T. Approaches to self-assembly of colloidal monolayers: A guide for nanotechnologists. *Adv. Colloid Interface Sci.* **2017**, *246*, 217–274. [CrossRef] [PubMed]
9. Vogel, N.; Goerres, S.; Landfester, K.; Weiss, C.K. A convenient method to produce close-and non-close-packed monolayers using direct assembly at the air–water interface and subsequent plasma-induced size reduction. *Macromol. Chem. Phys.* **2011**, *212*, 1719–1734. [CrossRef]
10. Lu, Z.; Zhou, M. Fabrication of large scale two-dimensional colloidal crystal of polystyrene particles by an interfacial self-ordering process. *J. Colloid Interface Sci.* **2011**, *361*, 429–435. [CrossRef] [PubMed]
11. Wang, W.; Gu, B. Self-assembly of two-and three-dimensional particle arrays by manipulating the hydrophobicity of silica nanospheres. *J. Phys. Chem. B* **2005**, *109*, 22175–22180. [CrossRef] [PubMed]
12. Retsch, M.; Zhou, Z.; Rivera, S.; Kappl, M.; Zhao, X.S.; Jonas, U.; Li, Q. Fabrication of large-area, transferable colloidal monolayers utilizing self-assembly at the air/water interface. *Macromol. Chem. Phys.* **2009**, *210*, 230–241. [CrossRef]
13. Zhang, J.T.; Wang, L.; Lamont, D.N.; Velankar, S.S.; Asher, S.A. Fabrication of large-area two-dimensional colloidal crystals. *Angew. Chem. Int. Ed.* **2012**, *51*, 6117–6120. [CrossRef] [PubMed]
14. Li, C.; Hong, G.; Wang, P.; Yu, D.; Qi, L. Wet chemical approaches to patterned arrays of well-aligned ZnO nanopillars assisted by monolayer colloidal crystals. *Chem. Mater.* **2009**, *21*, 891–897. [CrossRef]
15. Gao, P.; He, J.; Zhou, S.; Yang, X.; Li, S.; Sheng, J.; Wang, D.; Yu, T.; Ye, J.; Cui, Y. Large-area nanosphere self-assembly by a micro-propulsive injection method for high throughput periodic surface nanotexturing. *Nano Lett.* **2015**, *15*, 4591–4598. [CrossRef] [PubMed]
16. Yu, J.; Yan, Q.; Shen, D. Co-self-assembly of binary colloidal crystals at the air−water interface. *ACS Appl. Mater. Interfaces* **2010**, *2*, 1922–1926. [CrossRef] [PubMed]
17. Vogel, N.; de Viguerie, L.; Jonas, U.; Weiss, C.K.; Landfester, K. Wafer-scale fabrication of ordered binary colloidal monolayers with adjustable stoichiometries. *Adv. Funct. Mater.* **2011**, *21*, 3064–3073. [CrossRef]
18. Vogel, N.; Ally, J.; Bley, K.; Kappl, M.; Landfester, K.; Weiss, C.K. Direct visualization of the interfacial position of colloidal particles and their assemblies. *Nanoscale* **2014**, *6*, 6879–6885. [CrossRef] [PubMed]
19. Comyn, J. Moisture cure of adhesives and sealants. *Int. J. Adhes. Adhes.* **1998**, *18*, 247–253. [CrossRef]
20. Lotito, V.; Zambelli, T. Self-assembly of single-sized and binary colloidal particles at air/water interface by surface confinement and water discharge. *Langmuir* **2016**, *32*, 9582–9590. [CrossRef] [PubMed]
21. Weekes, S.M.; Ogrin, F.Y.; Murray, W.A.; Keatley, P.S. Macroscopic arrays of magnetic nanostructures from self-assembled nanosphere templates. *Langmuir* **2007**, *23*, 1057–1060. [CrossRef] [PubMed]
22. Lu, H.-H.; Yang, Y.-M.; Maa, J.-R. Effect of artificially provoked marangoni convection at a gas/liquid interface on absorption. *Ind. Eng. Chem. Res.* **1996**, *35*, 1921–1928. [CrossRef]
23. Sha, Y.; Chen, H.; Yin, Y.; Tu, S.; Ye, L.; Zheng, Y. Characteristics of the marangoni convection induced in initial quiescent water. *Ind. Eng. Chem. Res.* **2010**, *49*, 8770–8777. [CrossRef]
24. Young, N.O.; Goldstein, J.S.; Block, M.J. The motion of bubbles in a vertical temperature gradient. *J. Fluid Mech.* **1959**, *6*, 350–356. [CrossRef]
25. Otsu, N. A threshold selection method from gray-level histograms. *IEEE Trans. Syst. Man. Cybern.* **1979**, *9*, 62–66. [CrossRef]
26. Wang, Y.; Zhou, W. A review on inorganic nanostructure self-assembly. *J. Nanosci. Nanotechnol.* **2010**, *10*, 1563–1583. [CrossRef] [PubMed]
27. Ahualli, S.; Iglesias, G.; Wachter, W.; Dulle, M.; Minami, D.; Glatter, O. Adsorption of anionic and cationic surfactants on anionic colloids: Supercharging and destabilization. *Langmuir* **2011**, *27*, 9182–9192. [CrossRef] [PubMed]

28. Kralchevsky, P.A.; Denkov, N.D. Capillary forces and structuring in layers of colloid particles. *Curr. Opin. Colloid Interface Sci.* **2001**, *6*, 383–401. [CrossRef]
29. Wang, J.; Li, Z.; Yoon, R.-H.; Eriksson, J.C. Surface forces in thin liquid films of n-alcohols and of water–ethanol mixtures confined between hydrophobic surfaces. *J. Colloid Interface Sci.* **2012**, *379*, 114–120. [CrossRef] [PubMed]
30. Nguyen, A.V.; Nalaskowski, J.; Miller, J.D.; Butt, H.-J. Attraction between hydrophobic surfaces studied by atomic force microscopy. *Int. J. Miner. Process.* **2003**, *72*, 215–225. [CrossRef]
31. Stavroulakis, P.I.; Christou, N.; Bagnall, D. Improved deposition of large scale ordered nanosphere monolayers via liquid surface self-assembly. *Mater. Sci. Eng. B* **2009**, *165*, 186–189. [CrossRef]
32. Lee, Y.-L.; Du, Z.-C.; Lin, W.-X.; Yang, Y.-M. Monolayer behavior of silica particles at air/water interface: A comparison between chemical and physical modifications of surface. *J. Colloid Interface Sci.* **2006**, *296*, 233–241. [CrossRef] [PubMed]
33. Neumann, A.W.; David, R.; Zuo, Y. *Applied Surface Thermodynamics*; CRC Press: Boca Raton, FL, USA, 1996.
34. Zhang, R.; Somasundaran, P. Advances in adsorption of surfactants and their mixtures at solid/solution interfaces. *Adv. Colloid Interface Sci.* **2006**, *123*, 213–229. [CrossRef] [PubMed]
35. Binks, B.; Lumsdon, S. Influence of particle wettability on the type and stability of surfactant-free emulsions. *Langmuir* **2000**, *16*, 8622–8631. [CrossRef]
36. Fernandez-Rodriguez, M.A.; Binks, B.P.; Rodriguez-Valverde, M.A.; Cabrerizo-Vilchez, M.A.; Hidalgo-Alvarez, R. Particles adsorbed at various non-aqueous liquid-liquid interfaces. *Adv. Colloid Interface Sci.* **2017**, *247*, 208–222. [CrossRef] [PubMed]
37. Bresme, F.; Oettel, M. Nanoparticles at fluid interfaces. *J. Phys. Condens. Matter* **2007**, *19*, 413101. [CrossRef] [PubMed]
38. Zeng, M.; Mi, J.; Zhong, C. Wetting behavior of spherical nanoparticles at a vapor–liquid interface: A density functional theory study. *Phys. Chem. Chem. Phys.* **2011**, *13*, 3932–3941. [CrossRef] [PubMed]
39. Snoeyink, C.; Barman, S.; Christopher, G.F. Contact angle distribution of particles at fluid interfaces. *Langmuir* **2015**, *31*, 891–897. [CrossRef] [PubMed]
40. Shishido, M.; Kitagawa, D. Preparation of ordered mono-particulate film from colloidal solutions on the surface of water and continuous transcription of film to substrate. *Colloids Surf. A* **2007**, *311*, 32–41. [CrossRef]

Article

High-Precision Solvent Vapor Annealing for Block Copolymer Thin Films

Gunnar Nelson, Chloe S. Drapes, Meagan A. Grant, Ryan Gnabasik, Jeffrey Wong and Andrew Baruth *

Department of Physics, College of Arts and Sciences, Creighton University, 2500 California Plaza, Omaha, NE 68178, USA; GunnarNelson@Creighton.edu (G.N.); ChloeDrapes@Creighton.edu (C.S.D.); MeaganGrant@Creighton.edu (M.A.G.); gnaba001@umn.edu (R.G.); JeffreyWong@Creighton.edu (J.W.)

* Correspondence: AndrewBaruth@Creighton.edu; Tel.: +1-402-280-2644

Received: 19 April 2018; Accepted: 25 May 2018; Published: 29 May 2018

Abstract: Despite its efficacy in producing well-ordered, periodic nanostructures, the intricate role multiple parameters play in solvent vapor annealing has not been fully established. In solvent vapor annealing a thin polymer film is exposed to a vapor of solvent(s) thus forming a swollen and mobile layer to direct the self-assembly process at the nanoscale. Recent developments in both theory and experiments have directly identified critical parameters that govern this process, but controlling them in any systematic way has proven non-trivial. These identified parameters include vapor pressure, solvent concentration in the film, and the solvent evaporation rate. To explore their role, a purpose-built solvent vapor annealing chamber was designed and constructed. The all-metal chamber is designed to be inert to solvent exposure. Computer-controlled, pneumatically actuated valves allow for precision timing in the introduction and withdrawal of solvent vapor from the film. The mass flow controller-regulated inlet, chamber pressure gauges, in situ spectral reflectance-based thickness monitoring, and low flow micrometer relief valve give real-time monitoring and control during the annealing and evaporation phases with unprecedented precision and accuracy. The reliable and repeatable alignment of polylactide cylinders formed from polystyrene-*b*-polylactide, where cylinders stand perpendicular to the substrate and span the thickness of the film, provides one illustrative example.

Keywords: block polymers; self-assembly; thin films; solvent vapor annealing; nanolithography

---

1. Introduction

Techniques to achieve periodic nanostructures via traditional "top down" methods, including photolithography, have become increasingly challenging within the semiconductor industry. Ultra-small feature production is approaching fundamental resolution limits (193 nm ultraviolet (UV) lithography, for example, recently reaching sub-30 nm features) [1–11]. One promising strategy is investigating "bottom up" approaches that rely on nanoscale self-assembly. In 2007, directed self-assembly was first considered as a potential scaling solution, according to the International Technology Roadmap for Semiconductors (ITRS) [11]. In their 2013 report, the directed self-assembly of complex structures with low anneal time, low defect density, and high reproducibility was identified as one of the "Grand Challenges" to extend Moore's law [11]. The directed self-assembly of block polymer (BP) thin films has become a particularly strong candidate to achieve sub-20 nm dimensions, where the size and morphology is controlled by varying the molecular weight of the constituent polymer blocks.

Due to thorough investigations over decades, the BP research community generically considers the bulk behavior of many BPs well known [12,13]. Bulk morphologies are characterized by the Flory–Huggins interaction parameter (χ), the degree of polymerization (N), the volume fractions of the constituent blocks (f), and the particular architecture [13]. Furthermore, much is known about bulk behavior under a variety of stimuli, whether thermal, solvent, or mechanical [12]. In particular, measured or known quantities can well predict the alignment of structures and morphologies within the bulk. On the contrary, the confinement of a thin film and the associated surface energy contributions introduces additional confounders [14,15]. In particular, the complexity of surface interactions (both with the substrate and free surface) impose new, often asymmetric, boundary conditions [16,17]. Regardless, techniques designed to promote ordering of BP thin films have progressed rapidly, driven by applications requiring long-range lateral ordering, uniformity in feature size and high placement precision. Most of the historical approaches to BP ordering have been largely unpredictable and are often slow or energy-intensive: including thermal annealing [18,19], electric field alignment [20,21], and incorporation of low surface energy midblocks [22,23]. More recently, the use of pre-patterned substrates (chemical or topographical) to act as a guide for BP assembly has been successfully incorporated but can be time-intensive and involves multiple lithographic approaches [24]. Therefore, it is not necessarily cost-effective for high-throughput applications. As a result, a maturing technique is solvent vapor annealing (SVA) [25].

SVA was originally introduced as an alternative to thermal annealing for BPs exhibiting thermal degradation, problematic thermally-driven transitions, or slow dynamics due to high molar mass [26–28]. The interest in SVA of BPs has grown well beyond this in recent years. It has been shown to optimize organization quickly due to the increased chain mobility, a possibly decreased χ (dependent on solvent polarity), and tunable surface energies [25]. In this process, a BP film is exposed to the vapors of one or more organic solvents, offering direct control over lyotropic transitions (cylinders to spheres, for example) while in the solvated state as well as during evaporation [29,30]. This technique also has the ability to reduce defect density dramatically [31], while improving lateral ordering, both at the free surface and into the bulk of a film. This ordering is achieved more quickly (by several orders of magnitude) and completely than previous methods [32]. Although much is known about the interactions of BPs with solvent in bulk [33], the effects in thin films exhibit different behaviors due to the presence of confining surfaces and the dynamic exchange with the solvent vapor atmosphere [17]. Thus, this technique continues to suffer from reliability problems and no standardized methods have become apparent. It is becoming increasingly clear that a continued understanding of the specific ordering mechanisms of a BP system is paramount, where the quest for generic understanding of any BP thin film system remains elusive.

Of critical importance to many technological applications is the ability to direct BP thin films to form cylinders that stand perpendicular to the surface, traverse through the thickness of the film, and laterally pack with hexagonal order. A recently demonstrated approach to nanolithography of a magnetic thin film, for example, utilized such BP-based lithography masks using a Damascene-like approach. This approach was able to synthesize hexagonally-packed magnetic nanodots with a diameter of 19 nm with high fidelity and retention of robust ferromagnetism [34]. Furthermore, this approach achieved diameter control, down to 14 nm, and the potential for high-temperature processing with an additional atomic layer deposition step of ZnO [35]. These techniques rely on the vertical alignment of cylinders. On the contrary, many current advances in BP alignment, including important work in solvent concentration gradients [36], have focused solely on the free surface of the film. This is insufficient to be a direct substitution for most traditional lithographic approaches [37]. Recent results reveal that optimized ordering of hexagonally-packed cylinders can potentially occur in seconds and extend through the film using SVA, but this approach can be somewhat unreliable [38], where near 100% success of forming the desired morphology has yet to be achieved. Advances in computer simulation as well as in situ X-ray and neutron scattering continue to both further understand this process and improve upon it [39–41]. While such an understanding of the self-assembly behavior is critical to the advancement of this field, utilizing a statistical approach to quantifying SVA with large sample sets, with a goal of overcoming reliability barriers, requires the ability to identify, measure and develop controls for all of the pertinent variables with reliable precision. This includes chamber pressure, solvent exposure time, solvent concentration in the film, solvent evaporation rate, solvent purity or combinations, ambient temperature, sample and solvent temperature, humidity [42], and film thickness to name a few. These are addressed in the present manuscript, with a primary focus on the role of controlling chamber pressure and solvent evaporation rates at fast time scales of ~10 ms. We note that this chamber does not actively control sample [41] or solvent temperature [43–45], or include multiple solvents [46], which have been shown important to controlling annealing kinetics. So, these results work to keep these parameters as consistent as possible. To that end, we present here our strict annealing protocols and our climate-controlled SVA chamber with computer-controlled solvent vapor flow and pressure management and in situ spectral reflectance-based solvent concentration (ϕ) measurements. This chamber allows us to fix potential variables while investigating only one. This leads to unprecedented control over the SVA process with a goal of systematic studies with high reliability and repeatability that may have advantages as BP SVA alignment moves from its current research phase (relying heavily on in situ X-ray and neutron scattering) into scaled-up processing. One critical and unprecedented advancement is the ability to stabilize solvent concentration within a BP thin film for an arbitrarily long time at a user-defined chamber pressure, a necessary prerequisite for any temporal studies of crystallization.

To date, there are three primary methods of controlled SVA in the literature with differing levels of complexity, as illustrated in Figure 1. These processes were recently reviewed by Posselt et al. and Gu et al. [32,47], where each method typically incorporates an optical interferometer to monitor thickness (i.e., swelling due to solvent uptake) in real time that is placed above the chamber and shines down on the sample surface. Briefly, in "jar annealing," depicted in Figure 1a, a solvent reservoir is placed in a sealed vessel with the BP film. The solvent vapor pressure, and thus the film swelling, is parameterized by the ratio of liquid solvent surface area to the chamber volume [48]. The solvent evaporation is either done via opening the lid, which is difficult to quantify, or introducing a leak until all solvent has evaporated [48,49]. A natural extension of this SVA method is the inclusion of inlet/outlet flow lines, as seen in Figure 1b, where the incorporation of an inert gas flow can serve to modify the vapor pressure and BP film swelling more directly [50]. However, there is still only limited control over the solvent evaporation rate. The next extension is displacing the solvent reservoir from inside the sealed chamber into a separate sealed reservoir, as shown in Figure 1c. In this SVA method, solvent is carried in the vapor phase by a carrier gas into the chamber. Both the absolute pressure and vapor pressure inside the chamber are controlled via the flowrates of the vapor line and a second inlet for an inert gas stream. This gives significant dynamic control of BP film swelling, although the highest achievable vapor pressure is reduced compared to Figure 1a,b due to a dilution of vapor from the carrier gas. The solvent evaporation rate is controlled in a similar fashion to the method depicted in Figure 1b, but without the presence of the solvent reservoir, thus offering more control. Similar methods have proliferated [30,51–53], but only few utilize real-time computer control [41,45] and there is currently an inability to maintain a fixed solvent concentration for an arbitrarily long time scale. The present manuscript describes a new evolution in the SVA method that builds on these existing models, adding additional control over solvent introduction into and evaporation out of a BP film with a high degree of reliability and reproducibility by controlling the vapor pressure through inlet/outlet flows.

Several recent in situ studies utilizing Grazing Incidence Small Angle X-ray and Neutron Scattering (GISAXS and GISANS) serve to motivate the ability to stabilize a BP film with a specific solvent concentration [41,54]. These studies have shown to be critical in identifying the role of the order-to-disorder transition in directed self-assembly, as well as the role of solvent polarity and temperature. Their results could be further extended with an increased level of control over solvent concentration during the anneal and during the solvent evaporation. For example, Gu et al. show a high degree of ordering in poly(styrene)-*b*-poly(2-vinyl pyridine) using GISAXS with an increasing solvent concentration [30]. Upon rapid evaporation, scanning electron micrographs (SEM) of the resultant films exhibit long correlation lengths of in-plane cylinders. Furthermore, recent results indicate that approaching a solvent concentration consistent with the order-to-disorder transition leads to the largest correlation lengths [32]. This solvent concentration is also consistent with enhanced defect removal [31]. Annealing at solvent concentrations above the order-disorder transition will, instead, result in a potentially disordered film, as previously shown [34]. In addition, Sinturel et al. revealed similar ordering in a solvated poly(styrene)-*b*-poly(lactide) film, where GISAXS was monitored with increasing solvent concentration [29]. The data shows that the correlation length of perpendicularly aligned cylinders in quickly dried films is largest for a high solvent concentration (below the order-to-disorder transition). Finally, Berezkin et al. recently showed how mixed solvents and the use of elevated temperature in a controlled SVA chamber (similar to Figure 1c) further enhance control over final morphologies. These GISAXS studies, while extremely insightful, rely on somewhat uncontrolled solvent exposure methods, as presented in Figure 1. In each case, no static solvent concentration (often described as swelling ratio) was achieved over any appreciable amount of time. Such control is a prerequisite for any temporal study of ordering kinetics and further optimization of an annealing protocol. Moving forward, especially when considering scaling up these techniques towards mass production, creating a highly controlled SVA chamber could assist in avoiding the use of continued in situ X-ray or neutron scattering techniques that are not easily extensible to scaling up. Instead, a highly controlled SVA chamber could establish reliability and repeatability that incorporates the critical parameters that have been established from these scattering techniques.

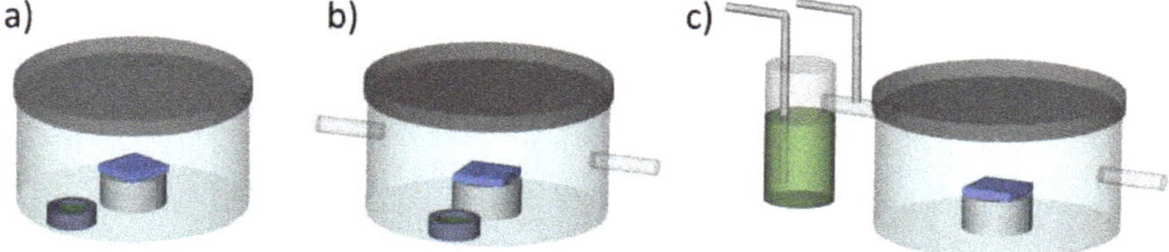

Figure 1. Primary methods of solvent vapor annealing; (**a**) "jar annealing," where a sealed chamber contains the sample and a solvent reservoir; (**b**) "jar annealing" with inlet/outlet lines for inert gas flow; (**c**) solvent vapor flow via a carrier gas through inlet/outlet lines. Block polymer films are depicted in blue, with liquid solvent depicted in green.

In addition to a controlled static solvent concentration, the evaporation rate serves as another complicating factor to be addressed. Solvent evaporation rate has been shown to strongly affect the surface morphology of BP films [52], where the evaporation has been shown to produce an ordering front that propagates through the film from the free surface to the substrate [55]. The timing of solvent removal dominates this effect. Enviable results were reported in 2004 from Kim et al. revealing defect-free ordering of cylinders at the free surface over large lateral length scales, where GISAXS data aimed to expose the role of this ordering front [55]. The ordering depends strongly on the exact trajectory through the BP phase diagram as solvent is removed [56], with changes in polymer dynamics and order–order transitions serving as potential complicating factors [57]. As such, conflicting results dictate that either slow [51,52] or fast [58] evaporation times may lead to ideal morphologies. Recently, studies of solvent concentration profiles [58] have suggested that the cylinder growth rate during evaporation is a product of the polymer chain mobility and the driving force to phase separate. Dynamical field theory simulations from Paradiso et al. further expose the role of rapid solvent evaporation and segregation strength (χ) on cylinder orientation [57]. It is increasingly clear that the ability to systematically remove solvent from the film, including at rapid time scales, is important to understanding final film morphologies. In particular, revealing how a given morphology propagates into the film during solvent evaporation, which is a necessary requirement for many nanolithographic applications. It is important to note that the role of solvent removal on rapid time scales (less than 100 ms) has not been successfully monitored using GISAXS, where integration times can be somewhat long (1–10 s) [32]; however, recent use of charge-coupled device (CCD) cameras have been shown to be effective for studying fast kinetics [47]. This opens the door for quantifying the effects of solvent removal in real time. Presently, in many of these studies, combining known structural properties from GISAXS in the solvated state with final morphologies imaged with atomic force microscopy (AFM) or SEM following rapid evaporation, as addressed above, have been used to reveal important relationships between evaporation trajectories and final morphologies. The present manuscript describes a systematically improved SVA method that offers a potential to elucidate multiple key variables in the SVA process and offers a level of control, reliability and reproducibility to enhance the understanding of ordering kinetics both during annealing and, critically, evaporation. Specifically, this improved SVA method monitors and controls chamber pressure, solvent exposure time, solvent concentration in the film, solvent purity and solvent evaporation rate, while concurrently monitoring ambient temperature and humidity.

2. Materials and Methods

2.1. Solvent Vapor Annealing

Optically polished substrates were single-crystalline silicon <111> wafers (n-type, 5 Ω·cm) with a native oxide layer (University Wafer, South Boston, MA, USA). No attempt was taken to remove the native oxide. Toluene (ACS Certified), tetrahydrofuran (THF) (Certified, contains 0.025% butylated hydroxytoluene as a preservative), acetone (99.5%), methanol (ACS Certified), and sodium hydroxide granules were all purchased from Fisher Chemical, Pittsburgh, PA, USA. 1,1,1,3,3,3-Hexamethyldisilazane (HMDS) (98%) and 3 Å, 4 to 8 mesh, 3333 were purchased from Acros Organics, Belgium, WI, USA.

Construction of the solvent vapor-annealing chamber (SAC) required chemically inert stainless-steel tubing and valves with Swagelok tube fittings. Pneumatically actuated stainless steel ball valves (1/4″ and 3/4″) utilizing Dow Corning M111 (heavy-consistency dimethyl silicone compound) lubricant and Modified PTFE packing (SS-T12-S-065-20 and SS-45S8-33C), stainless steel quarter-turn plug valves (1/4″) with Kalrez O-rings (SS-4P4T), a stainless steel low-flow metering valve (1/4″) utilizing a Kalrez O-ring (SS-SS4-KZ-VH), a stainless steel low-pressure (5 psig/34.5 kPa) unfilled pressure gauge (LP1-SS-254-5PSI), a stainless steel unfilled pressure gauge (60 psig/413.7 kPa, PGI-63C-PG60-LAOX), and 1/4″ and 3/4″ stainless steel tubing (SS-T4-S-035-20 and SS-T12-S-065-20, respectively) were purchased from the Swagelok Company, Saarland, OH, USA. The body and bubbler portions of the SAC utilized vacuum-grade Conflat flanges with oxygen-free copper gaskets. Stainless steel tees (275-150-CFT), 6-way cube (275-CUBE-OS), and all Conflat blanks (275-000-T) and Swagelok adaptors were purchased from LDS Vacuum Products, Inc., Longwood, FL, USA, or modified in-house. Conflat flanged zero length deep UV quartz (Corning HPFS 7980 Fused Silica) viewport was purchased from the Kurt J. Lesker Company, Jefferson Hills, PA, USA (VPZL-275Q). Dry N_2 inlet was achieved with a Drierite gas purifier, and was controlled via an Apex 500 SCCM mass flow controller with RS-232 digital control (Schoonover Inc., Canton, GA, USA). Dry N_2 pressure from the gas tank were monitored and controlled with a dual-stage gas regulator (0–344.7 kPa) and flowrates were monitored and controlled with a Panel-Mount Flowmeter (OMEGA Engineering, Stanford, CT, USA) for air, with a brass valve, with two different flow ranges (3–30 SCFH and 30–300 SCFH) from McMaster-Carr, Elmhurst, IL, USA. Pneumatic valve control was achieved with 3-way 1/4″ NPT, normally closed, 120 V, 100 psi solenoids, various One-touch fittings (1/4″) and 1/4″ nylon tubing purchased from the Swagelok Company, Saarland, OH, USA. Electric control of pneumatics was achieved with a National Instruments USB, 8 input, 12-Bit, 10 kS/s, Multifunction DAQ; 25 A, 250 V solid state relays; and LabVIEW 2016 software (National Instruments, Austin, TX, USA). Finally, film thickness was determined in situ with spectral reflectance via a Filmetrics F20-UV, San Diego, CA, USA, general-purpose film thickness measurement system with both halogen and deuterium sources.

2.2. Synthesis of Poly(Styrene)-Block-Poly(Lactide)

The synthesis of poly(styrene)-*block*-poly(lactide) (PS-*b*-PLA) is described fully elsewhere [34]. Succinctly, hydroxyl-terminated PS (Mn = 42.5 kDa) was synthesized via living anionic polymerization. The subsequent PLA was synthesized via ring opening transesterification polymerization (ROTEP) of D,L-lactide in dichloromethane at room temperature using 1,8-diazabicyclo[5.4.0]undec-7-ene (DBU) as a catalyst for 1 h. PS-*b*-PLA was obtained by precipitating in methanol after termination with benzoic acid. The final PS-*b*-PLA had a total Mn = 63 kDa, with a PLA volume fraction of 0.28 (by volume) yielding a cylindrical morphology with a polydispersity index of 1.08, as determined with ^{1}H Nuclear Magnetic Resonance (NMR) and Size-Exclusion Chromatography (SEC) [34].

2.3. Thin Film Preparation

Typical solutions of 1.5% (w/v) PS-*b*-PLA in toluene (non-selective solvent) were spin coated onto HMDS treated, natively-oxidized silicon wafers (20 mm × 20 mm). HMDS treatment of the Si wafers was carried out by ultrasonically cleaning substrates in organic solvents (acetone followed by methanol), treating them in a 1:5 (v/v) HMDS:toluene solution for 16 h, then rinsing in toluene to remove any excess HMDS that had not grafted fully to the substrate, and blowing dry with N_2 gas. Treated wafers were then placed in a 75 °C oven to remove any residual solvent. The films were spin coated at 1000–3000 rpm for 30 s, diced into 12–16 pieces (~5 mm × 5 mm), and immediately placed in a 75 °C (below the glass transition temperature of either block [34]) oven for drying. This process yielded ~60 nm thick films, dependent on exact spin speed and solution concentration. Samples were dried for a minimum of one day. Periodic atomic force micrographs indicate no apparent aging issues or annealing effects while storing films at this temperature.

For the present experiment, thin films were hot-loaded into a N_2 purged, over-pressurized solvent annealing chamber and immediately sealed to avoid water contamination. The samples were allowed to cool to room temperature and no further temperature control was implemented. This is described in full detail below. THF liquid was stored in a sealed Erlenmeyer flask over activated 3 Å molecular sieves for several days before introducing it to the SAC, resulting in water levels of 4.1 ppm or less [59]. Additional activated molecular sieves reside in the solvent reservoir of the SAC to continue active drying during subsequent solvent vapor anneals. Following solvent vapor exposure and complete solvent evaporation from the film, the samples were immediately transferred to a 0.05 M NaOH solution (H_2O:CH_3OH = 6:4 by volume) for PLA minority domain degradation and left to soak for 45 min, where the degradation rate is sensitive to molarity [60]. After removal, films were washed with deionized water/methanol for 5 min. By removing the minority component immediately, this serves to ensure the morphology of the film is immobilized for subsequent imaging. Furthermore, to remove any additional surface contamination, a 150 W, 50 KHz O_2 reactive ion etch (PE-50, PlasmaEtch, Inc., Carson, NV, USA) is employed for 10 s at 100 mTorr. This process removes ~2–3 nm of organic material. Samples at this stage were immediately imaged with atomic force microscopy (AFM), without any further modification.

2.4. Measuring Film Thickness

Film thickness was determined with a Filmetrics F20-UV (San Diego, CA, USA) general-purpose film thickness measurement system with both halogen and deuterium sources. Spectral reflectance data was taken at differing time intervals (as discussed below), between 10 ms–3 s with a 10–250 ms integration time. Experimental data was modeled over a spectral range of 270–900 nm with a three-layer model (Si + PS-*b*-PLA + air). We developed anticipated refractive index profiles based on known values (e.g., PS, n = 1.59; PLA, n = 1.482; THF, n = 1.407). Therefore, we expected an index of refraction of 1.55 for the neat film, dropping to 1.45 with increasing solvent (THF) concentration up to ϕ = 0.55 [34]. Samples not following this trend in refractive index with increasing solvent were aborted and disposed.

2.5. Atomic Force Microscopy (AFM)

Tapping mode AFM was performed on an Agilent 5420 microscope (Santa Clara, CA, USA) under ambient conditions using engagement setpoints between 0.9–0.95 of the free amplitude oscillation. The tapping mode cantilevers (BudgetSensors, Sofia, Bulgaria and Bruker, Billerica, MA, USA) had a resonant frequency of 300 kHz and a force constant of 40 N/m. For imaging the film/substrate interface, the PLA-removed thin films were placed upside down on double-sided transparent tape (ScotchBrand, St. Paul, MN, USA) and placed in liquid N_2 for 30 s. Following liquid N_2 exposure, the Si wafer was peeled away from the film providing access to the underside. These films were again exposed to a 10–20 s O_2 reactive ion etch (150 W 50 KHz in 100 mTorr) on the underside to remove any HMDS, a thin PS wetting layer or adhesive residue [34,61].

3. Results

3.1. Design of a Purpose-Built Solvent Vapor Annealing Chamber

The reproducibility of a final morphology and its propagation into solvent vapor-annealed BP films remains somewhat elusive, despite numerous advances in the field. To directly address this issue, we have designed, constructed and tested a purpose-built solvent vapor annealing chamber (SAC) that provides unparalleled control over introducing and maintaining precise solvent concentrations within the film during the annealing process as well as during the evaporation of solvent from the film. Concurrently, we maintain control over several additional necessary parameters (Table 1), including an exceptionally low dew point in the sample cell with active dry N_2 purging, solvent vapor flow rate, film thickness as a proxy for solvent concentration in the film [34], chamber pressure, solvent selectivity, and substrate surface preparation. The present investigation uses ~60 nm thick films of a prototypical BP, PS-*b*-PLA, that adopts a cylindrical morphology in the bulk. We use at room temperature a relatively neutral solvent, THF, for the anneal [29,34,62]; a PS-selective substrate, HMDS-functionalized Si; and maintain a low dew point (-100 °C) for the N_2 carrier gas, due to the hygroscopic nature of THF and PLA. The following SAC description is a significant evolution over initial work in developing these protocols, including the addition of pneumatically-actuated, computer controlled solvent flow rates and valve actuation [34,38]. In our method, a copper-gasket sealed, all-metal chamber controls the SVA climate (i.e., humidity, chamber pressure, and evaporation times). The all-metal construction, except for chemically inert modified polytetrafluoroethylene sealed ball valves and a perfluoroelastomer sealed low-flow metering valve, ensures negligible interaction with the solvent during the annealing process. In particular, THF can be particularly aggressive on traditional organic gaskets and lubricants. Following hundreds of hours of exposure to THF, no degradation to any valves in the chamber is evident.

The chamber, shown in Figure 2 (additional images are available in Appendix A), is connected to an actively dried N_2 line (dew point guaranteed to -100 °C) ①, which splits into two 1/4" stainless steel tubing paths. Path 1 (in red) purges the sample space before annealing and during sample loading; this ensures a low dew point during the SVA. Additionally, BP films are hot-loaded from a 75 °C oven into the purged, over-pressurized chamber via a Conflat-flanged door and immediately sealed to avoid water contamination and allowed to cool to room temperature. Path 1 is also vital to the controlled evacuation of solvent vapor from the SAC during the evaporation phase. N_2 flow is passed through an acrylic, block-style flowmeter (3–30 SCFH or 30–300 SCFH, dependent on flow rate chosen) to control and measure flow rates during solvent evaporation.

Table 1. A compilation of variables considered in the design of solvent vapor annealing protocols, along with their importance and the steps taken in the design to address them.

Variable	Importance	Protocols in Design
Humidity	Water is a polar solvent, which will modify the solubility. For example, water is PLA selective	Samples are stored in 75 °C oven Samples are hot loaded into chamber Vacuum-grade, copper gaskets and Swagelok seals throughout chamber Molecular sieve-dried solvent (tetrahydrofuran (THF)) Actively purged (dry N_2) sample cell Drierite Gas Purifier (−100 °C dewpoint)
Solvent Vapor Flow Rate	Flow rate is proportional to solvent uptake in film	Computer-controlled mass flow controller Low-flow metering outlet valve
Solvent Concentration	Solvent concentration in film during solvent vapor annealing (SVA) modifies mobility	In situ optical detection of solvent concentration, assuming proportional to film thickness, with 10–20 ms resolution
Solvent Evaporation Rate	Evaporation rate in linked to morphology alignment	Computer-controlled pneumatic valves Variable flow N_2 purge line
Initial Thickness	Role of commensurate thickness on morphology	Films are all spun cast at a constant spin speed from the same solution concentration
Vapor Pressure	Vapor pressure during SVA modifies solvent uptake and evaporation	High and low pressure gauges Low-flow metering outlet valve can finely adjust vapor pressure
Solvent Selectivity	Solvent selectivity modifies surface energy and polymer-polymer interactions	THF is a relatively neutral solvent for PS and PLA. It has slight PS selectivity
Surface Selectivity	Substrate preparation modifies surface energy	HMDS-functionalized Si substrate surface promotes PS (majority block) adhesion

Path 2 (in green) flow is governed by a mass flow controller ② connected to a LabVIEW-enabled computer (National Instruments, Austin, TX, USA). Flow rates range from 0–500 SCCM, dependent on intended anneal conditions. The flow of the metered, actively dried N_2 gas is computer controlled via solenoid actuated pneumatic valves ③ that control 80 psi (551 kPa) air flow to each of four pneumatic process valves ④. Following the flow controller, the N_2 gas is passed through a primary safety valve ④ (this protects the mass flow controller from liquid solvent exposure). The controlled gas flow continues on its way to a sealed, molecular-sieve-dried solvent reservoir ⑤, a stainless steel Conflat tee with view window (Figure A5). This reservoir contains an additional inlet tube with a normally closed plug valve that is used for liquid solvent loading without breaking any Conflat seals Figure A6). The solvent reservoir is backed by a safety reservoir ⑥, a stainless steel Conflat tee with viewing window. The safety reservoir will collect fluid solvent if backpressure is present, avoiding exposure to the mass flow controller. The safety reservoir contains an additional tube with a normally closed plug valve to remove any liquid solvent without opening any Conflat seals. Ultimately, the dry N_2 is bubbled through the solvent reservoir ⑤, which subsequently carries solvent in the vapor phase into the sample space ⑦, a 6-way Conflat cube (70 mm × 70 mm × 70 mm), via a computer-controlled, pneumatically-actuated chamber valve ④. All flow into the sample space exits through 3/4″ stainless steel tubing via a pneumatically-actuated ball valve ⑧ or a low-flow metering valve ⑨ to a fume hood. The computer controlled, pneumatically-actuated valves ④, with compressed air inlet ⑩ and 120 V solenoid valves ③, allow us to quickly initiate and terminate the SVA with a specified level of timing. Of critical importance, initial testing indicates a controlled SVA evaporation time down to 15 ms or any time longer with 10–20 ms temporal resolution. To our knowledge, this is the fastest recorded SVA evaporation time for a BP thin film.

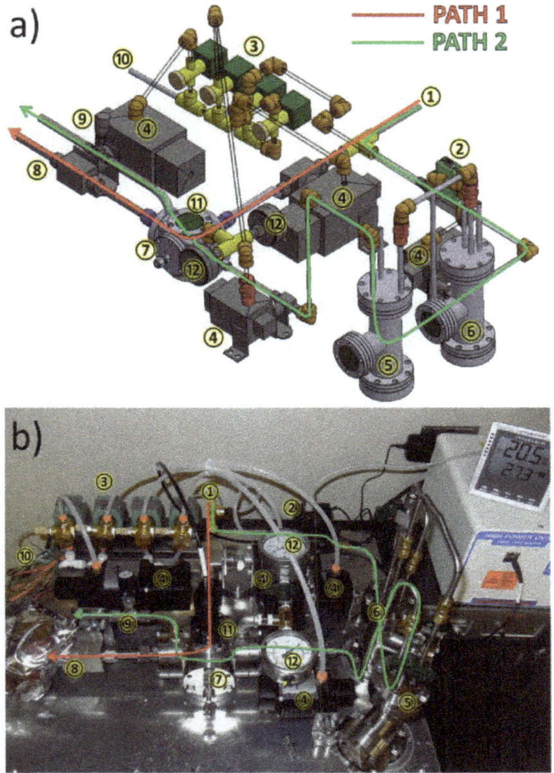

Figure 2. (**a**) Computer-aided design (CAD) and (**b**) real image of a computer-controlled, pneumatically actuated solvent vapor annealing chamber. More images can be found in Appendix A. Numbers and colored arrows are described in the text.

The BP sample resides in the sample space ⑦ on a custom-built mount (Figure A7). The sample mount features two recessed ports, which speeds sample loading and helps avoid direct flow from Paths 1 and 2 that can shift sample position. One recessed port holds the BP film, while the other holds a blank Si wafer for use as an optical standard. Recess ports are sized to hold up to an 8 mm × 8 mm BP film and a blank Si wafer. The chamber offers direct optical access to the sample space through a fused silica viewport ⑪, enabling continuous monitoring of the sample chamber with high precision pressure gauges ⑫ to directly measure chamber pressure and in situ spectral reflectance-based measurements of film thickness with 0.1 nm thickness and 10–20 ms temporal resolution. These thickness measurements are directly related to solvent volume concentration (ϕ) within the film by:

$$\phi = \frac{V_{solvent+film} - V_{film}}{V_{solvent+film}} = \frac{t_{solvent+film} - t_{film}}{t_{solvent+film}} = 1 - \frac{t_{film}}{t_{solvent+film}} \quad (1)$$

where ϕ is solvent concentration, V denotes volume and t is film thickness. As supported by direct observation, the areas for the two film states (i.e., swollen versus dry) are taken to be nominally identical and thus only a thickness measurement is required to obtain a real-time in situ probe of ϕ during SVA.

3.2. Theory of Operation

Through the active control of solvent vapor inlet and outlet flows, the solvent concentration within the BP film is controlled and monitored as a function of time. As shown in Figure 3, using an in situ measurement of ϕ, we divide the SVA process of a 60 nm PS-*b*-PLA film into four distinct time regimes. During the first three time regimes, thickness data was taken every 3 s with a 249 ms integration time. The initial regime (blue) includes the opening of the solvent reservoir to the sample chamber and an initial, constant N_2 inlet flow of 30–100 SCCM (dependent on desired chamber pressure, 50 SCCM in the present study), resulting in an exponential increase in ϕ. The thickness data is well modeled assuming a copolymer refractive index of 1.55, consistent with the neat PS-*b*-PLA film (Appendix B). This exponential region typically lasts ~60 s where the time dependence is well modeled with a single rate constant (0.03–0.1 s^{-1}) with 0.088 s^{-1} for Figure 3 (see Figure A8 for the exponential fit), which is dependent on exact inlet and outlet flow rates. Outlet flow rates are governed by a low flow micrometer relief valve, which is set to a flow coefficient, C_v, of 0.0002–0.002 (1–6 turns), with 0.0004 being used in Figure 3. This, along with inlet flow, dictates the chamber pressure, which was maintained below 3.5 kPa to obtain Figure 3. Following the displacement of residual N_2 in the chamber with solvent vapor, the solvent uptake enters a second regime (red). This regime involves the metered uptake of solvent into the film, indicated by a controlled, linear increase in thickness over a period of 2 min (highly tunable, based on relative flow rates), until the film reaches the targeted ϕ. The inlet and outlet flow rates remained the same for the second regime as given for the first regime in Figure 3, demonstrating this change in solvent uptake. The thickness data is well modeled assuming a copolymer refractive index that gradually approaches 1.45, consistent with a volumetric combination of neat PS-*b*-PLA and THF (Appendix B). During this phase, two possible methods drive solvent into the film and increase the thickness. If the inlet flow is higher than outlet flow, the increasing pressure in the chamber will increasingly force solvent into the film (a relatively fast mechanism). If the inlet and outlet flows are comparable, the increase in relative solvent concentration within the chamber leads to an increased uptake of solvent into the film (a relatively slow mechanism). Therefore, it is possible to swell a film with a chamber pressure that is nominally atmospheric. Dependent on desired chamber pressure and ultimate thickness, the inlet and outlet flow rates are tuned. Precise control is best achieved by adjusting the mass flow controller-regulated inlet. We have not seen any impact with the rate of swelling in this second regime on the final morphology of the film, other than a potential dependence on the pressure in the chamber (faster swelling is typically accomplished with an increase in chamber pressure). The thickness data in this regime is well modeled assuming a copolymer refractive index of 1.45 (Figure A10), consistent with a solvated PS-*b*-PLA film containing $\phi = 0.55$.

As the target ϕ is approached, decreasing the inlet flow or increasing the outlet flow causes ϕ to level off. The third regime (green) is characterized by an extremely constant (standard deviation is regularly less than $\Delta\phi = 0.002$) solvent concentration. This constant concentration can be maintained for nearly any specified anneal time (3 min is shown in Figure 3, but we have maintained similar consistency for more than an hour). It is maintained through slight manual variations in inlet/outlet flow rates, with the future potential of computer feedback control. Figure 3 was achieved with a fixed outlet flow, consistent with regimes 1 and 2 and by controlling the inlet flow between 10–30 SCCM to achieve a constant thickness (solvent concentration). In the final seconds of this period, the integration time and thickness acquisition interval are switched to 10 ms and 0 s, respectively. This is the fastest we can acquire spectral reflectance data and still get a high-fidelity model to extract thickness. This rapid data acquisition allows for close examination of the solvent evaporation period. In the fourth regime (magenta), evaporation of the solvent from the film occurs. Computer-controlled, pneumatically-actuated valves open Path 1 and close Path 2. Through the release of pressure in the chamber, in tandem with N_2 flow, the solvent vapor is released from the film and evacuated from the chamber. Details on timing control, which are dependent on Path 1 flow rates and chamber pressure, are given below. The initial evaporation of solvent from the film is not complete, where $\phi \approx 0.2$

typically remains in the film over the first 100 ms. This remaining solvent typically takes an additional 1.5–10 s to be fully removed from the film, dependent on the Path 1 inlet flow rate. At $\phi < 0.25–0.3$, the film is likely vitrified with exceptionally slow kinetics, consistent with recent GISAXS results [32]. Therefore, the final morphology is well locked-in during the initial, potentially fast, evaporation phase. In fact, atomic force micrographs of the free surface and the substrate surface (after removal from the Si substrate) of a PS-*b*-PLA film verify that a vertically-aligned cylindrical morphology can persist through an entire 60 nm film to the substrate surface (Figure 4). The PS-*b*-PLA film in Figure 4 was swollen with THF to $\phi = 0.55$ and the solvent was subsequently evaporated out of the film (down to $\phi = 0.2$, sufficient to vitrify the film) in 15 ms. No condensation of solvent was observed during any changes in pressure for the present investigation.

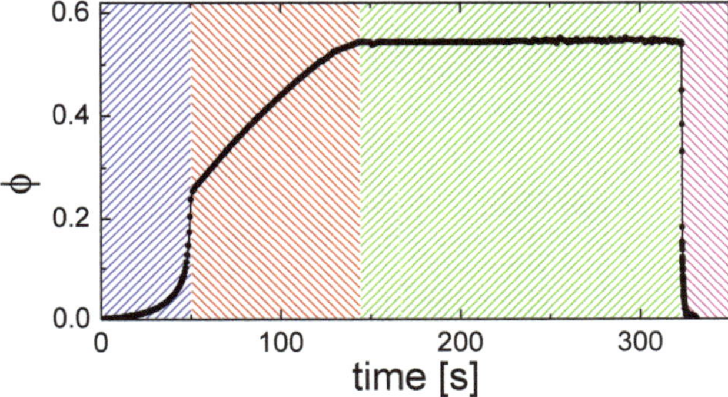

Figure 3. Solvent concentration in a BP film (ϕ) versus anneal time for a typical solvent vapor anneal of a 60 nm PS-*b*-PLA thin film. The blue indicates the initial solvent uptake regime. The red indicates the metered solvent uptake regime. The green indicates a fixed solvent concentration regime. The magenta indicates the solvent evaporation regime.

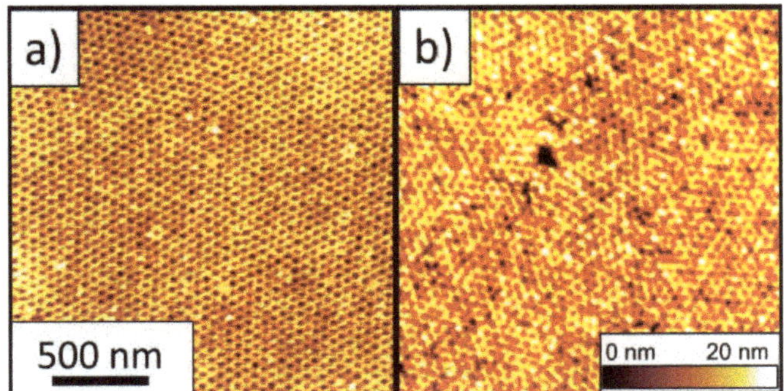

Figure 4. Atomic force topography micrographs of the free surface (**a**) and substrate surface (**b**) of a 60 nm PS-*b*-PLA film that was exposed to a THF concentration $\phi = 0.55$ and having a solvent evaporation occurring in 15 ms. Images were taken following hydrolytic removal of the PLA minority component and a brief (10 s) O_2 reactive ion etch. The false color height scale is 20 nm.

3.3. Solvent Evaporation

While a primary outcome of this purpose-built chamber is the ability to keep an arbitrary solvent concentration constant within a BP film for a specified time, another beneficial consequence is the ability to observe and control the solvent evaporation phase (fourth regime—solvent evaporation magenta regime—in Figure 3) with 10–20 ms temporal resolution. This high acquisition rate has revealed subtle differences in precise evaporation trajectories, dependent on exact chamber pressure and inlet/outlet flow rates. First, of importance to our specific chamber, is the timing between the inlet and outlet valves of Path 1. It is necessary to introduce a slight time delay between their actuation. This is detailed in Figure 5a. If the valves are open concurrently (i.e., 0 ms time delay), the finite impedance of the outlet tube causes a brief pressure spike in the chamber that drives residual solvent vapor from the chamber into the film. This increases the concentration within the film, possibly to disorder, ahead of the evaporation. This leads to inconsistent morphologies and is undesirable. If the outlet valve is opened far ahead of the inlet valve, a two-stage evaporation tends to take place. The opening of the outlet valve releases pressure from the chamber and leads to a drop in concentration within the film. Presumably, some solvent was retained in the film simply due to the finite pressure in the chamber during the anneal. Then, the opening of the inlet flow removes the remaining solvent in the chamber and the solvent is fully evaporated from the film. For example, a delay of 100 ms has two distinct evaporation trajectories, as seen in Figure 5a. For our specific system, a computer-controlled time delay of 25 ms leads to optimally fast evaporation without the associated pressure spike (Figure 5a).

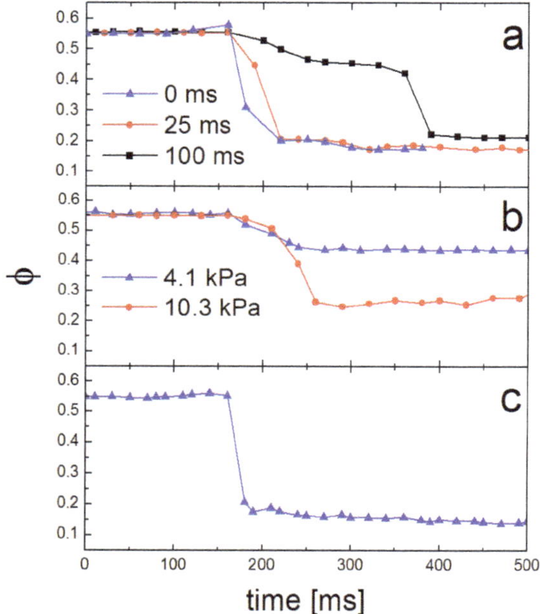

Figure 5. Solvent concentration (ϕ) versus anneal time during solvent evaporation; (**a**) solvent evaporation trajectories for three time delays between opening the chamber outlet to atmosphere and introducing an inlet flow of N_2; (**b**) solvent evaporation trajectories for an annealing chamber with a relatively high or low vapor pressure; (**c**) Solvent evaporation trajectory for a fast evaporation with high inlet flow and zero vapor pressure in the chamber.

Second, considering the role of chamber pressure during the evaporation phase; the release of pressure within the chamber as the outlet valve of Path 1 is opened leads to a release of solvent from the film. This is true even in the complete absence of any inlet N_2 flow, as shown in Figure 5b. As expected, the higher the pressure in the chamber, the more dramatic the decrease in solvent concentration in the film when the outlet valve is opened. In the case where pressure in the chamber approaches 10.3 kPa, the pressure drop to 0 kPa is sufficient to remove solvent down to $\phi = 0.25$. This concentration is sufficiently low to lock-in the morphology where only kinetically slow vitrification persists, discussed above. In the case where chamber pressure is somewhat lower, 4.1 kPa, the release of pressure to 0 kPa is insufficient to remove a sufficient amount of solvent to vitrify the film without additional inlet N_2 flow. Finally, considering inlet and outlet flows (Figure 5c), a higher Path 1 inlet flow rate leads to optimally fast solvent vapor removal from the chamber. This potentially leads to rapid solvent evaporation from the film (10–20 ms). This rapid evaporation causes a strong ordering front and tends to drive cylinder propagation perpendicular to the film (Figure 4). After reviewing 700 trials following these SVA protocols, including rapid evaporation (~15 ms) and low chamber pressure, 640 (91.4%) films showed full perpendicular alignment, 14 (2.0%) showed full in-plane alignment, and 46 (6.6%) exhibited a mixed, mostly perpendicular alignment with some in-plane cylinders evident. The latter two cases (in-plane cylinders or mixed) were regularly attributed to elevated ambient humidity in the lab. Clearly, the control over both chamber pressure and Path 1 flows provide extensive control over the solvent removal rate and its specific trajectory. The resultant morphologies and their propagation into the bulk of a film for these different trajectories are the subject of a forthcoming manuscript.

4. Conclusions

It is increasingly clear that annealing BP thin films is critical to achieving self-assembled nanostructures with a given morphology and lateral order. Solvent-based techniques have proven to be highly effective due to the dramatically increased polymer chain mobility while mitigating thermal degradation. In particular, chain mobility near the order disorder transition is optimally enhanced. In addition, the evaporation of solvent from a film is critical in determining the propagation of a given morphology into the bulk of the film. At present, there are three primary methods for incorporating solvent into a BP film; however, each is limited in its ability to directly control solvent concentration within the film and the solvent evaporation rate. These limitations have stifled investigation into the time-dependence of these effects. Furthermore, solvent-enhanced crystallization has evolved from BPs to other organic systems, with conjugated polymer/organic photovoltaics being one illustrative example [63], further indicating its potential efficacy. Therefore, we have presented a purpose-built solvent vapor annealing chamber that was designed and constructed to elucidate the role of key parameters involved in directed self-assembly in BP thin films with goals of enhanced reliability, repeatability and the eventual scaling up of the SVA process. Currently, the use of in situ scattering techniques, such as GISAXS or GISANS, are critical for understanding the ordering mechanisms of block polymer films. However, they may prove impractical for industrial applications. Therefore, transferring protocols developed with those techniques to purpose-built chambers, such as the one presented here, could be essential for proliferation. In particular, there is interest in observing and controlling the mechanisms for increased correlation lengths of self-assembled features and the growth propagation of those features into the bulk of the film in the final film state (i.e., following solvent evaporation). This level of control opens up possibilities for a variety of morphology controls. Moreover, in contrast to many current efforts, the technique does not involve extensive in situ monitoring with advanced scattering techniques, but could serve to enhance those techniques with higher temporal resolution and control. Rather, the present in situ monitoring relies on inexpensive, readily available optical techniques in conjunction with pneumatically-actuated, computer-controlled flow controllers and valves. Such low-cost and simple methods could prove useful for any future scaling-up of this process.

Author Contributions: Conceived and designed—G.N., R.G. and A.B. Performed the experiments—G.N., C.S.D., M.A.G., J.W. Analyzed the data—G.N., C.S.D., M.A.G., J.W. and A.B. Wrote the paper—G.N. and A.G.B. Supervised work—A.B. All authors discussed the results and implications and commented on the manuscript at all stages.

Acknowledgments: This work was supported by the National Science Foundation through the Nebraska EPSCoR FIRST Award under contract 1004094. Additional support came from the Omaha Public Power District and Creighton University's Center for Undergraduate Research and Scholarship. G.N. and R.G. acknowledge support from the NASA Nebraska Space Grant Consortium. C.S.D. acknowledges support from the Clare Boothe Luce foundation. We acknowledge mechanical expertise and insight from Brad Walters and Rustin Haase.

Conflicts of Interest: The authors declare no conflict of interest.

Appendix A

Figure A1. Front view of the solvent vapor annealing (SVA) chamber.

Figure A2. Right view of the SVA chamber.

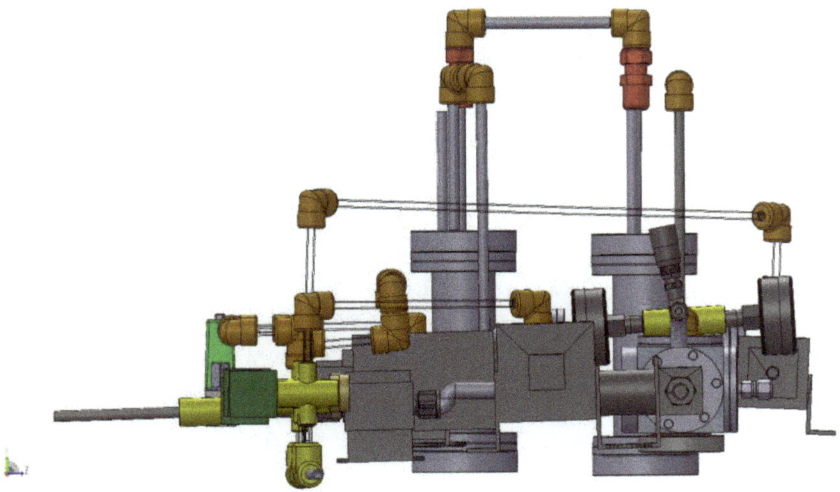

Figure A3. Left view of the SVA chamber.

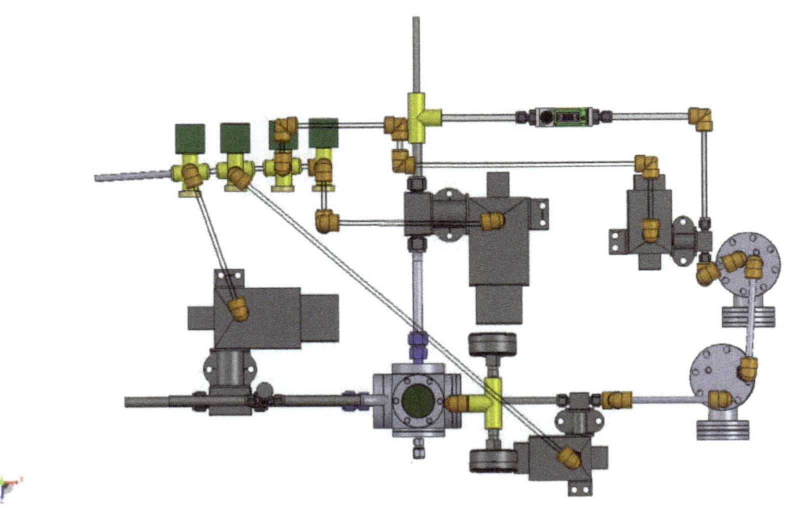

Figure A4. Top view of the SVA chamber.

Figure A5. Solvent reservoir with optical view port, liquid solvent, and molecular sieves.

Figure A6. Depiction of the solvent reservoir with three $\frac{1}{4}''$ tubing ports. One tube (**a**) is an N_2 inlet that extends to the bottom of the reservoir. One tube (**b**) is a solvent vapor outlet that only extends into the reservoir $\frac{1}{2}''$ to capture the vapor. One tube (**c**) is capped with a normally-closed plug valve. This is used to add liquid solvent without opening the Conflat flanges.

Figure A7. Custom designed sample holder. The block copolymer film sits in the center well, a silicon optical standard sits in the left well. A hole for background ultraviolet-visible (UV-V) light is spectroscopy subtraction is on the right. The sample holder rests on a two-pronged frame to slide the samples in and out of the main chamber and is sealed with a Conflat flange with copper gasket.

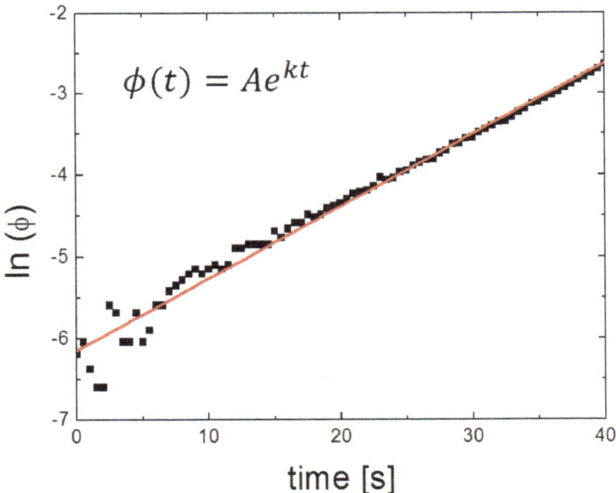

Figure A8. Best linear fit to the natural log of solvent concentration for the first 40 s (regime 1) of Figure 3, showing a single exponential rate constant of 0.088 ± 0.0013 s^{-1}. The R^2 value of the fit is 0.982.

Appendix B

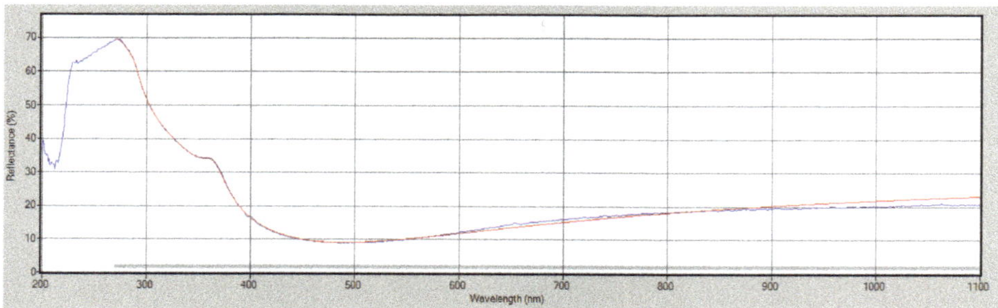

Figure A9. Spectral reflectance of a neat 60 nm polystyrene-*block*-polylactide film on silicon. The blue line is reflectance data; red line is the spectral reflectance model using an index of refraction of 1.55 for the copolymer layer.

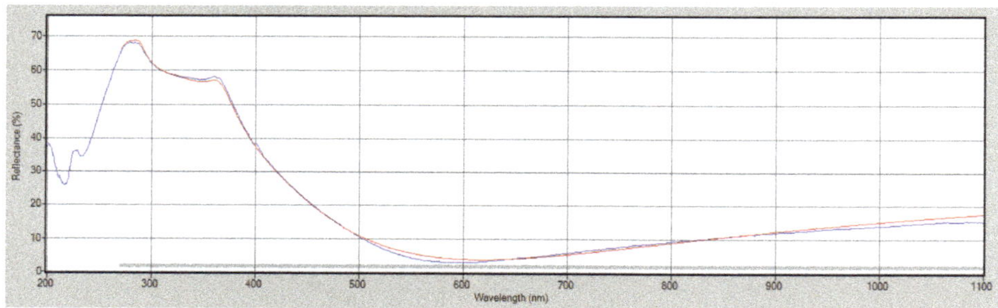

Figure A10. Spectral reflectance of a THF swollen polystyrene-*block*-polylactide film on Silicon. The film contains 55% THF solvent. The blue line is reflectance data; red line is the spectral reflectance model using an index of refraction of 1.45 for the copolymer layer.

References

1. Terris, B.D.; Thomson, T. Nanofabricated and self-assembled magnetic structures as data storage media. *J. Phys. D Appl. Phys.* **2005**, *38*, R199–R222. [CrossRef]
2. Borah, D.; Shaw, M.T.; Rasappa, S.; Farrell, R.A.; O'Mahony, C.; Faulkner, C.M.; Bosea, M.; Gleeson, P.; Holmes, J.D.; Morris, M.A. Plasma etch technologies for the development of ultra-small feature size transistor devices. *J. Phys. D Appl. Phys.* **2011**, *44*, 174012. [CrossRef]
3. Luo, Y.; Montarnal, D.; Kim, S.; Shi, W.; Barteau, K.P.; Pester, C.W.; Hustad, P.D.; Christianson, M.D.; Fredrickson, G.H.; Kramer, E.J.; et al. Poly(dimethylsiloxane-*b*-methyl methacrylate): A Promising Candidate for Sub-10 nm Patterning. *Macromolecules* **2015**, *48*, 3422–3430. [CrossRef]
4. Bang, J.; Jeong, U.; Ryu, D. Block copolymer nanolithography: Translation of molecular level control to nanoscale patterns. *Adv. Mater.* **2009**, *21*, 4769–4792. [CrossRef] [PubMed]
5. Bates, C.M.; Maher, M.J.; Janes, D.W.; Ellison, C.J.; Willson, C.G. Block copolymer lithography. *Macromolecules* **2014**, *47*, 2–12. [CrossRef]
6. Hamley, I.W. Ordering in thin films of block copolymers: Fundamentals to potential applications. *Prog. Polym. Sci.* **2009**, *34*, 1161–1210. [CrossRef]
7. Black, C.T.; Ruiz, R.; Breyta, G.; Cheng, J.Y.; Colburn, M.E.; Guarini, K.W.; Kim, H.-C.; Zhang, Y. Polymer self assembly in semiconductor microelectronics. *IBM J. Res. Dev.* **2007**, *51*, 605–633. [CrossRef]
8. Kim, H.-C.; Park, S.-M.; Hinsberg, W.D. Block copolymer based nanostructures: Materials, processes, and applications to electronics. *Chem. Rev.* **2010**, *110*, 146–177. [CrossRef] [PubMed]

9. Stoykovich, M.P.; Nealey, P.F. Block copolymers and conventional lithography. *Mater. Today* **2006**, *9*, 20–29. [CrossRef]
10. Guarini, K.W.; Black, C.T.; Yeung, S.H.I. Optimization of diblock copolymer thin film self assembly. *Adv. Mater.* **2002**, *14*, 1290–1294. [CrossRef]
11. International Technology Roadmap for Semiconductors (ITRS). 2013. Available online: https://www.semiconductors.org/main/2013_international_technology_roadmap_for_semiconductors_itrs/ (accessed on 18 April 2018).
12. Hamley, I.W. *The Physics of Block Copolymers*; Oxford University Press: Oxford, UK, 1998; ISBN 0198502184.
13. Bates, F.S. Polymer-polymer phase behavior. *Science* **1991**, *251*, 898–905. [CrossRef] [PubMed]
14. Anastasiadis, S.H.; Russell, T.P.; Satija, S.K.; Majkrzak, C.F. Neutron reflectivity studies of the surface-induced ordering of diblock copolymer films. *Phys. Rev. Lett.* **1989**, *62*, 1852–1855. [CrossRef] [PubMed]
15. Menelle, A.; Russell, T.P.; Anastasiadis, S.H.; Satija, S.K.; Majkrzak, C.F. Ordering of thin diblock copolymer films. *Phys. Rev. Lett.* **1992**, *68*, 67–70. [CrossRef] [PubMed]
16. Fasolka, M.J.M.; Mayes, A.M.A. Block copolymer thin films: Physics and applications. *Annu. Rev. Mater. Res.* **2003**, *31*, 323–355. [CrossRef]
17. Albert, J.N.L.; Epps, T.H., III. Self-assembly of block copolymer thin films. *Mater. Today* **2010**, *13*, 24–33. [CrossRef]
18. Olayo-Valles, R.; Lund, M.S.; Leighton, C.; Hillmyer, M.A. Large area nanolithographic templates by selective etching of chemically stained block copolymer thin films. *J. Mater. Chem.* **2004**, *14*, 2729–2731. [CrossRef]
19. Zhang, X.; Harris, K.D.; Wu, N.L.Y.; Murphy, J.N.; Buriak, J.M. Fast assembly of ordered block copolymer nanostructures through microwave annealing. *ACS Nano* **2010**, *4*, 7021–7029. [CrossRef] [PubMed]
20. Morkved, T.L.; Lu, M.; Urbas, A.M.; Ehrichs, E.E.; Jaeger, H.M.; Mansky, P.; Russell, T.P. Local control of microdomain orientation in diblock copolymer thin films with electric fields. *Science* **1996**, *273*, 931–933. [CrossRef] [PubMed]
21. Thurn-Albrecht, T. Ultrahigh-density nanowire arrays grown in self-assembled diblock copolymer templates. *Science* **2000**, *290*, 2126–2129. [CrossRef] [PubMed]
22. Kubo, T.; Wang, R.F.; Olson, D.A.; Rodwogin, M.; Hillmyer, M.A.; Leighton, C. Spontaneous alignment of self-assembled ABC triblock terpolymers for large-area nanolithography. *Appl. Phys. Lett.* **2008**, *93*, 133112. [CrossRef]
23. Kubo, T.; Parker, J.S.; Hillmyer, M.A.; Leighton, C. Characterization of pattern transfer in the fabrication of magnetic nanostructure arrays by block copolymer lithography. *Appl. Phys. Lett.* **2007**, *90*, 233113. [CrossRef]
24. Ruiz, R.; Kang, H.; Detcheverry, F.A.; Dobisz, E.; Kercher, D.S.; Albrecht, T.R.; de Pablo, J.J.; Nealey, P.F. Density multiplication and improved lithography by directed block copolymer assembly. *Science* **2008**, *321*, 936–939. [CrossRef] [PubMed]
25. Sinturel, C.; Vayer, M.; Morris, M.; Hillmyer, M.A. Solvent vapor annealing of block polymer thin films. *Macromolecules* **2013**, *46*, 5399–5415. [CrossRef]
26. Kelly, J.Y.; Albert, J.N.L.; Howarter, J.A.; Kang, S.; Stafford, C.M.; Epps, T.H.; Fasolka, M.J. Investigation of thermally responsive block copolymer thin film morphologies using gradients. *ACS Appl. Mater. Interfaces* **2010**, *2*, 3241–3248. [CrossRef] [PubMed]
27. Di, Z.; Posselt, D.; Smilgies, D.-M.; Papadakis, C.M. Structural rearrangements in a lamellar diblock copolymer thin film during treatment with saturated solvent vapor. *Macromolecules* **2010**, *43*, 418–427. [CrossRef] [PubMed]
28. Sun, Y.; Henderson, K.J.; Jiang, Z.; Strzalka, J.W.; Wang, J.; Shull, K.R. Effects of reactive annealing on the structure of poly(methacrylic acid)-poly(methyl methacrylate) diblock copolymer thin films. *Macromolecules* **2011**, *44*, 6525–6531. [CrossRef]
29. Sinturel, C.; Grosso, D.; Boudot, M.; Amenitsch, H.; Hillmyer, M.A.; Pineau, A.; Vayer, M. Structural transitions in asymmetric poly(styrene)-block-poly(lactide) thin films induced by solvent vapor exposure. *ACS Appl. Mater. Interfaces* **2014**, *6*, 12146–12152. [CrossRef] [PubMed]
30. Gu, X.; Gunkel, I.; Hexemer, A.; Gu, W.; Russell, T.P. An in situ grazing incidence X-ray scattering study of block copolymer thin films during solvent vapor annealing. *Adv. Mater.* **2014**, *26*, 273–281. [CrossRef] [PubMed]

31. Li, W.; Nealey, P.F.; de Pablo, J.J.; Müller, M. Defect removal in the course of directed self-assembly is facilitated in the vicinity of the order-disorder transition. *Phys. Rev. Lett.* **2014**, *113*, 168301. [CrossRef] [PubMed]
32. Gu, X.; Gunkel, I.; Hexemer, A.; Russell, T.P. Controlling domain spacing and grain size in cylindrical block copolymer thin films by means of thermal and solvent vapor annealing. *Macromolecules* **2016**, *49*, 3373–3381. [CrossRef]
33. Lodge, T.P.; Pudil, B.; Hanley, K.J. The full phase behavior for block copolymers in solvents of varying selectivity. *Macromolecules* **2002**, *35*, 4707–4717. [CrossRef]
34. Baruth, A.; Seo, M.; Lin, C.H.; Walster, K.; Shankar, A.; Hillmyer, M.A.; Leighton, C. Optimization of long-range order in solvent vapor annealed poly(styrene)-block-poly(lactide) thin films for nanolithography. *ACS Appl. Mater. Interfaces* **2014**, *16*, 13770–13781. [CrossRef] [PubMed]
35. Lin, C.-H.; Polisetty, S.; O'Brien, L.; Baruth, A.; Hillmyer, M.A.; Leighton, C.; Gladfelter, W.L.W.L. Size-tuned ZnO nanocrucible arrays for magnetic nanodot synthesis via atomic layer deposition-assisted block polymer lithography. *ACS Nano* **2015**, *9*, 1379–1387. [CrossRef] [PubMed]
36. Albert, J.N.L.; Bogart, T.D.; Lewis, R.L.; Beers, K.L.; Fasolka, M.J.; Hutchison, J.B.; Vogt, B.D.; Epps, T.H. Gradient solvent vapor annealing of block copolymer thin films using a microfluidic mixing device. *Nano Lett.* **2011**, *11*, 1351–1357. [CrossRef] [PubMed]
37. Luo, M.; Epps, T.H., III. Directed block copolymer thin film self-assembly: Emerging trends in nanopattern fabrication. *Macromolecules* **2013**, *46*, 7567–7579. [CrossRef]
38. Baruth, A.; Rodwogin, M.D.; Shankar, A.; Erickson, M.J.; Hillmyer, M.A.; Leighton, C. Non-lift-off block copolymer lithography of 25 nm magnetic nanodot arrays. *ACS Appl. Mater. Interfaces* **2011**, *3*, 3472–3481. [CrossRef] [PubMed]
39. Berezkin, A.V.; Papadakis, C.M.; Potemkin, I.I. Vertical domain orientation in cylinder-forming diblock copolymer films upon solvent vapor annealing. *Macromolecules* **2016**, *49*, 415–424. [CrossRef]
40. Berezkin, A.V.; Jung, F.; Posselt, D.; Smilgies, D.M.; Papadakis, C.M. Vertical vs. lateral macrophase separation in thin films of block copolymer mixtures: Computer simulations and GISAXS experiments. *ACS Appl. Mater. Interfaces* **2017**, *9*, 31291–31301. [CrossRef] [PubMed]
41. Berezkin, A.V.; Jung, F.; Posselt, D.; Smilgies, D.-M.; Papadakis, C.M. In situ tracking of composition and morphology of a diblock copolymer film with GISAXS during exchange of solvent vapors at elevated temperatures. *Adv. Funct. Mater.* **2018**, *28*, 1706226. [CrossRef]
42. Bang, J.; Kim, B.J.; Stein, G.E.; Russell, T.P.; Li, X.; Wang, J.; Kramer, E.J.; Hawker, C.J. Effect of humidity on the ordering of PEO-based copolymer thin films. *Macromolecules* **2007**, *40*, 7019–7025. [CrossRef]
43. Knoll, A.; Magerle, R.; Krausch, G. Phase behavior in thin films of cylinder-forming ABA block copolymers: Experiments. *J. Chem. Phys.* **2004**, *120*, 1105–1116. [CrossRef] [PubMed]
44. Zettl, U.; Knoll, A.; Tsarkova, L. Effect of confinement on the mesoscale and macroscopic swelling of thin block copolymer films. *Langmuir* **2010**, *26*, 6610–6617. [CrossRef] [PubMed]
45. Dehmel, R.; Dolan, J.A.; Gu, Y.; Wiesner, U.; Wilkinson, T.D.; Baumberg, J.J.; Steiner, U.; Wilts, B.D.; Gunkel, I. Optical imaging of large gyroid grains in block copolymer templates by confined crystallization. *Macromolecules* **2017**, *50*, 6255–6262. [CrossRef] [PubMed]
46. Gotrik, K.W.; Hannon, A.F.; Son, J.G.; Keller, B.; Alexander-Katz, A.; Ross, C.A. Morphology control in block copolymer films using mixed solvent vapors. *ACS Nano* **2012**, *6*, 8052–8059. [CrossRef] [PubMed]
47. Posselt, D.; Zhang, J.; Smilgies, D.-M.; Berezkin, A.V.; Potemkin, I.I.; Papadakis, C.M. Restructuring in block copolymer thin films: In situ GISAXS investigations during solvent vapor annealing. *Prog. Polym. Sci.* **2017**, *66*, 80–115. [CrossRef]
48. Jung, Y.S.; Ross, C.A. Solvent vapor induced tunability of self assembled block copolymer patterns. *Adv. Mater.* **2009**, *21*, 2540–2545. [CrossRef]
49. Bai, W.; Hannon, A.F.; Gotrik, K.W.; Choi, H.K.; Aissou, K.; Liontos, G.; Ntetsikas, K.; Alexander-katz, A.; Ross, C.A. Thin film morphologies of bulk-gyroid polystyrene-block- polydimethylsiloxane under solvent vapor annealing. *Macromolecules* **2014**, *47*, 6000–6008. [CrossRef]
50. Di, Z.; Posselt, D.; Smilgies, D.-M.; Li, R.; Rauscher, M.; Potemkin, I.I.; Papadakis, C.M. Stepwise swelling of a thin film of lamellae-forming poly(styrene-*b*-butadiene) in cyclohexane vapor. *Macromolecules* **2012**, *45*, 5185–5195. [CrossRef]

51. Son, J.G.; Gotrik, K.W.; Ross, C.A. High-aspect-ratio perpendicular orientation of PS-*b*-PDMS thin films under solvent annealing. *ACS Macro Lett.* **2012**, *1*, 1279–1284. [CrossRef]
52. Albert, J.N.L.; Young, W.-S.; Lewis, R.L.; Bogart, T.D.; Smith, J.R.; Epps, T.H. Systematic study on the effect of solvent removal rate on the morphology of solvent vapor annealed ABA triblock copolymer thin films. *ACS Nano* **2012**, *6*, 459–466. [CrossRef] [PubMed]
53. Bai, W.; Yager, K.G.; Ross, C.A. In Situ Characterization of the Self-Assembly of a Polystyrene–polydimethylsiloxane block copolymer during solvent vapor annealing. *Macromolecules* **2015**, *48*, 8574–8584. [CrossRef]
54. Müller-Buschbaum, P. GISAXS and GISANS as metrology technique for understanding the 3D morphology of block copolymer thin films. *Eur. Polym. J.* **2016**. [CrossRef]
55. Kim, S.H.; Misner, M.J.; Xu, T.; Kimura, M.; Russell, T.P. Highly oriented and ordered arrays from block copolymers via solvent evaporation. *Adv. Mater.* **2004**, *16*, 226–231. [CrossRef]
56. Nandan, B.; Vyas, M.K.; Böhme, M.; Stamm, M. Composition-dependent morphological transitions and pathways in switching of fine structure in thin films of block copolymer supramolecular assemblies. *Macromolecules* **2010**, *43*, 2463–2473. [CrossRef]
57. Paradiso, S.P.; Delaney, K.T.; García-Cervera, C.J.; Ceniceros, H.D.; Fredrickson, G.H. Block copolymer self assembly during rapid solvent evaporation: Insights into cylinder growth and stability. *ACS Macro Lett.* **2014**, *3*, 16–20. [CrossRef]
58. Phillip, W.; Hillmyer, M.; Cussler, E. Cylinder orientation mechanism in block copolymer thin films upon solvent evaporation. *Macromolecules* **2010**, *43*, 7763–7770. [CrossRef]
59. Williams, D.B.G.; Lawton, M. Drying of organic solvents: Quantitative evaluation of the efficiency of several desiccants. *J. Org. Chem.* **2010**, *75*, 8351–8354. [CrossRef] [PubMed]
60. Cummins, C.; Mokarian-Tabari, P.; Holmes, J.D.; Morris, M.A. Selective etching of polylactic acid in poly(styrene)-block-poly(D,L)lactide diblock copolymer for nanoscale patterning. *J. Appl. Polym. Sci.* **2014**, *131*, 40798. [CrossRef]
61. Zalusky, A.S.; Olayo-Valles, R.; Wolf, J.H.; Hillmyer, M.A. Ordered nanoporous polymers from polystyrene-polylactide block copolymers. *J. Am. Chem. Soc.* **2002**, *124*, 12761–12773. [CrossRef] [PubMed]
62. Vayer, M.; Hillmyer, M.A.; Dirany, M.; Thevenin, G.; Erre, R.; Sinturel, C. Perpendicular orientation of cylindrical domains upon solvent annealing thin films of polystyrene-*b*-polylactide. *Thin Solid Films* **2010**, *518*, 3710–3715. [CrossRef]
63. Miller, S.; Fanchini, G.; Lin, Y.-Y.; Li, C.; Chen, C.-W.; Su, W.-F.; Chhowalla, M. Investigation of nanoscale morphological changes in organic photovoltaics during solvent vapor annealing. *J. Mater. Chem.* **2008**, *18*, 306–312. [CrossRef]

Article

Self-Assembly Behavior and pH-Stimuli-Responsive Property of POSS-Based Amphiphilic Block Copolymers in Solution

Yiting Xu *, Kaiwei He, Hongchao Wang, Meng Li, Tong Shen, Xinyu Liu, Conghui Yuan and Lizong Dai *

Fujian Provincial Key Laboratory of Fire Retardant Materials, College of Materials, Xiamen University, Xiamen 361005, China; hekaiwei@stu.xmu.edu.cn (K.H.); 20720171150035@stu.xmu.edu.cn (H.W.); 20720171150098@stu.xmu.edu.cn (M.L.); 20720171150102@stu.xmu.edu.cn (T.S.); lxy212@xmu.edu.cn (X.L.); yuanch@xmu.edu.cn (C.Y.)

* Correspondence: xyting@xmu.edu.cn (Y.X.); lzdai@xmu.edu.cn (L.D.); Tel.: +86-592-2186178 (Y.X. & L.D.)

Received: 1 May 2018; Accepted: 17 May 2018; Published: 24 May 2018

Abstract: Stimuli-responsive polymeric systems containing special responsive moieties can undergo alteration of chemical structures and physical properties in response to external stimulus. We synthesized a hybrid amphiphilic block copolymer containing methoxy polyethylene glycol (MePEG), methacrylate isobutyl polyhedral oligomeric silsesquioxane (MAPOSS) and 2-(diisopropylamino)ethyl methacrylate (DPA) named MePEG-*b*-P(MAPOSS-*co*-DPA) via atom transfer radical polymerization (ATRP). Spherical micelles with a core-shell structure were obtained by a self-assembly process based on MePEG-*b*-P(MAPOSS-*co*-DPA), which showed a pH-responsive property. The influence of hydrophobic chain length on the self-assembly behavior was also studied. The pyrene release properties of micelles and their ability of antifouling were further studied.

Keywords: stimuli-responsive polymer; polyhedral oligomeric silsesquioxane; self-assembly; protein adsorption resistance; controlled release

---

1. Introduction

Recently, stimuli-responsive polymers have attracted great attention because of their unique properties. The characteristics of these smart polymers can be adjusted by temperature, light, pH values, ionic strength, enzyme proteins, antigens, and so on [1–6]. Among stimuli-sensitive copolymers, temperature- or pH-sensitive copolymers, mainly consisting of 2-diisopropylaminoethyl methacrylate (DPA) [7], 2-(dimethylamino)ethyl methacrylate (DMAEMA) [8–10] and *N*-isopropylacryl amide (NIPAM) [11] blocks, have been extensively investigated for their potential application in targeted drug delivery.

Block copolymers are composed of two or more chemical components connected by chemical bonds. The structure of block copolymer micelles is easily regulated by changing the environmental conditions, so the self-assembly of block copolymers has attracted wide interest in the field of self-assembly. Amphiphilic block copolymers spontaneously self-assemble into well-organized nanostructures in selective solvents. Depending on many factors such as water content, copolymer concentration and architecture of the amphiphilic macromolecules, the self-assembled structures could be of various morphologies, ranging from star-like spherical micelles to multicompartment nanostructures [12–18].

Polyhedral oligomeric silsesquioxane (POSS), classified as a unique inorganic silica nanoparticle with a uniform cubic structure, has been widely incorporated into polymer matrices to produce novel organic/inorganic hybrid materials. The POSS-based polymer acts as a hybrid material with excellent mechanical and inorganic properties. POSS with inert vertex groups on the silicon-oxygen cage renders it quite hydrophobic, therefore promoting the formation of a well-patterned micelles structure to control drug release [19–21]. The aggregation of POSS moieties, which are nanoscale at about 1.5 nm, can form a regular and stable hydrophobic core in micelles. The characteristic structure makes the encapsulating of hydrophobic drugs in micelles more advantageous. There is increasing interest in the synthesis, self-assembly and applications of stimuli-responsive POSS-based amphiphilic macromolecules. Recent advances in the field of synthetic polymer chemistry, especially in controlled radical polymerizations, such as ATRP and various click reactions, have paved the way for the introduction of POSS into well-defined copolymer architectures. Matyjaszewski et al. were the first to report POSS containing well-defined P(MA-POSS)-*b*-poly(*n*-butyl acrylate)-*b*-P(MA-POSS) triblock copolymers via ATRP, and their self-assembly behaviors in films [22]. Deng et al. prepared block copolymers of MA-POSS and PMMA by reversible addition-fragmentation chain transfer (RAFT) polymerization and studied their self-assembly to gain nanostructured hybrid polymer networks [23]. Cheng et al. prepared an amphiphilic copolymer, polystyrene-(carboxylic acid-functionalized POSS) (PS-APOSS), and the assembled structure including vesicles, worm-like cylinders and spheres was obtained by changing the ionization degree of the carboxylic acid [24]. He et al. synthesized amphiphilic POSS-containing copolymers poly(acrylic acid)-*co*-poly(acrylate-POSS) via ATRP, and the self-assembled structure could be tuned in aqueous solution by regulating the molar ratio of poly(acrylic acid) to poly(acrylate-POSS) in the copolymer [25]. This exciting discovery provides a totally wide and novel window for designing amphiphilic polymers due to ATRP being a simple and powerful method to synthesize diverse polymers.

In this work, a novel hybrid amphiphilic block copolymer was prepared with MePEG as the backbone connecting another block with MAPOSS and DPA units via ATRP. The synthetic route is described in Scheme 1. Owning to the unique amphiphilic architecture, MePEG-*b*-P(MAPOSS-*co*-DPA) could form micelles. The self-assembly and pH-sensitive behaviors of copolymers in aqueous solution were studied by dynamic light scattering (DLS) and transmission electron microscopy (TEM), including the cladding and release of pyrene. Meanwhile, the antifouling ability of the assembly was systematically studied.

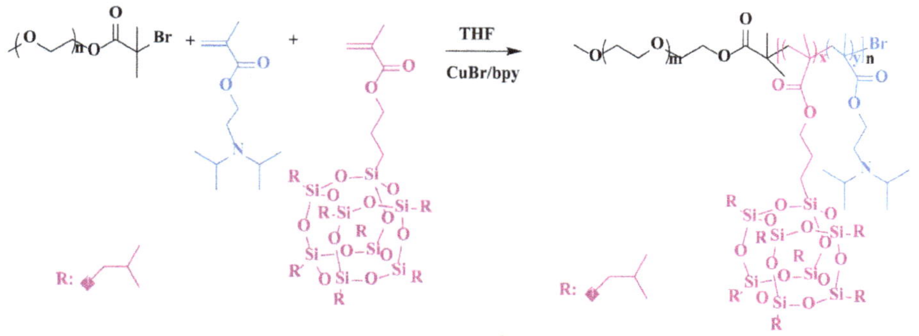

Scheme 1. Synthetic way of MePEG-*b*-P(MAPOSS-*co*-DPA).

2. Experimental Section

2.1. Materials

2-(Diisopropylamino)ethyl methacrylate (DPA) (97%, Aladdin Co., Shanghai, China) was purified by basic alumina columns prior to use. Methoxy polyethylene glycol (MePEG 5000) was freeze-dried for 24 h to remove water. Methacrylate isobutyl POSS (MAPOSS) purchased from Hybrid Plastic Co.

Styrene (Sinoreagent Co., Shanghai, China) was purified by passing through a column filled with basic Al_2O_3 prior to use. Other chemicals were purchased from Sinoreagent Co. and used as received.

2.2. Synthesis of MePEG-Br Macroinitiator

The MePEG-Br was synthesized using the methods described in the literature [26]. Typically, the methoxy polyethylene glycol (MePEG) (5.0 g, 1.0 mmol) and triethylamine (TEA) (2.2 g, 4.0 mmol) were dissolved in dichloromethane (DCM) (20.0 mL) and cooled with an ice-salt bath below 0 °C. Then, bromoisobutyryl bromide (BiBB) (0.368 g, 3.2 mmol) dissolved in DCM (2.0 mL) was slowly added dropwise to the reaction solution. The solution was slowly warmed to room temperature and reacted for 24 h. The product was dissolved in distilled water and DCM removed by rotary evaporation. Finally, the macroinitiator MePEG-Br was obtained as a white powder. The product was characterized by ^{1}H-NMR spectroscopy (Figure S1).

^{1}H-NMR ($CDCl_3$, ppm, TMS) d: 3.39 ppm [3H, $\underline{CH_3}$-O-], 3.66 ppm [2H, -(O$\underline{CH_2}$)n-], 4.34 ppm [2H, -(OCH_2)n-$\underline{CH_2}$O-], 1.95 ppm [6H, -OCOC($\underline{CH_3}$)$_2$Br].

2.3. Synthesis of MePEG-b-P(MAPOSS-co-DPA)

A series of amphiphilic copolymers MePEG-*b*-P(MAPOSS-*co*-DPA) via ATRP was synthesized. MePEG-Br (2.96 g, 0.5 mmol), DPA (0.15 g, 0.03 mol), MAPOSS (5.11 g, 5.5 mmol), 2-dipyridyl (bpy) (0.15 g, 1 mmol) and anhydrous 1,4-dioxane (2.0 mL) were added into a 10-mL Schlenk tube, followed by three freeze-pump-thaw cycles. Then, CuBr (0.07 g, 0.5 mmol) was added into the Schlenk tube under the protection of argon, followed by a freeze-pump-thaw cycle. Next, this Schlenk tube was heated to 70 °C, and polymerization was performed for 24 h under continuous stirring. Then, the reaction solution was passed through the Al_2O_3 column for the removal of the catalyst. Finally, the solvent was removed by rotary evaporation, and the product was dried until constant weight in a vacuum oven at 40 °C. The preparation of MePEG-*b*-P(MAPOSS-*co*-DPA) with different contents of MAPOSS was then performed in a similar way.

2.4. Self-Assembly Procedure of MePEG-b-P (MAPOSS-co-DPA) Amphiphilic Copolymer

A series of micellar solutions was prepared by solution volatilization-induced self-assembly [27]. The POSS-based amphiphilic copolymer MePEG-*b*-P(MAPOSS-*co*-DPA) was dissolved in tetrahydrofuran (THF) to get the polymer mother liquor. Then, the liquid was added dropwise to a certain volume of ultrapure water, and the mixture was stirred at ambient temperature until the THF was completely volatilized to get the amphiphilic copolymer micelles. Typical self-assembly solutions were prepared as follows:

MePEG-*b*-P(MAPOSS-*co*-DPA) (5 mg) was dissolved in 1 mL of THF. Then, the solution was gradually diluted by 5 mL of deionized water. The solution was stirred overnight to completely remove THF at room temperature. The final solution of self-assembled micelles was used for the following characterizations. A series of different concentrations of hydrochloric acid and sodium hydroxide was selected to adjust the pH of the micellar solution.

2.5. Preparation of Pyrene Encapsulated Micelles

Zero-point-one milliliters of pyrene in acetone (10 mg/mL) were added to the serum vial, and acetone was evaporated under continuous flow of argon. Then, 25 mL of polymer micelle solution (1 mg/mL) were added and sonicated for 1 h to ensure encapsulation of the pyrene. Finally, after the loaded pyrene was removed by filtration, a pyrene-loaded micelle solution was successfully prepared.

2.6. Preparation of Micelles and Fetal Bovine Serum Mixed Solution for Anti-Protein Adsorption Performance Test

The amphiphilic copolymer MePEG-*b*-P(MAPOSS-*co*-DPA) was dissolved in THF, and the resulting polymer mother liquor was slowly added dropwise to the fetal bovine serum (FBS) solution until the THF was completely volatilized. Then, the mixed solution was obtained.

2.7. Characterization

The ^{1}H-NMR measurements were carried out on a Bruker AV400 MHz NMR spectrometer (Bruker, Geneva, Switzerland) at room temperature with tetramethylsilane (TMS) as the internal standard and $CDCl_3$ as a solvent. Fourier-transform infrared spectrometry (FTIR) spectra were recorded on Nicolet Avatar 360 FTIR (Thermo Fisher Scientific, Shanghai, China). The molecular weight was measured by gel permeation chromatography (GPC) on the APC Installation Kit. Data were obtained using the PSS WinGPC system. THF was used as the eluent at a flow rate of 0.5 mL/min. A series of low polydispersity polystyrene standards were used for the GPC calibration.

The optical transmittance of the MePEG-*b*-P(MAPOSS-*co*-DPA) solution was obtained on a UV-2550 spectrometer (SHIMADZU, Kyoto, Japan) at 500 nm. Transmission electron microscopy (TEM) images were conducted using a JEM2100 transmission electron microscope (JEOL, Tokyo, Japan) with an accelerating voltage of 200 kV. One drop of micelles solution was placed on a copper-mesh coated with carbon and then air-dried before measurement. Zeta potential measurements and dynamic light scattering (DLS) were recorded on a Zetasizer NanoZS Instrument (Malvern Instruments, Malvern, UK) at a wavelength of 500 nm. Fluorescence spectroscopy was conducted on the F7000 fluorescence spectrophotometer (Hitachi, Tokyo, Japan). Both the excitation and emission slit widths were 2.5 nm, and the scan rate was 240 nm/min.

3. Results and Discussion

3.1. Synthesis of MePEG-b-P(MAPOSS-co-DPA)

The macroinitiator MePEG-Br was used to initiate the ATRP polymerization of MAPOSS and DPA monomers. In this paper, three kinds of amphiphilic copolymers with different contents of MAPOSS segments were designed and synthesized, named as BCP1, BCP2, BCP3, respectively. Experimental conditions and results of amphiphilic polymer with MAPOSS and DPA are shown in Table 1.

Table 1. Experiments and results of ATRP polymerization of MePEG-Br, MAPOSS and DPA.

Sample Name	[MePEG-Br]:[DPA]:[MAPOSS] [a]	[MePEG-Br]:[DPA]:[MAPOSS] [b]	Time (h)	$\overline{Mn}$ (^{1}H-NMR)	$\overline{Mn}$ (GPC)	PDI
BCP1	1:60:11	1:50.06:10.60	24	25,839	26,314	1.09
BCP2	1:60:13	1:53.84:12.02	24	28,010	30,681	1.27
BCP3	1:60:15	1:57.66:13.86	24	30,566	35,908	1.23

[a] The feed molar ratio of monomers; [b] the composition ratio in copolymer determined by ^{1}H-NMR spectra.

^{1}H-NMR spectra of MePEG-*b*-P(MAPOSS-*co*-DPA) with various molecular weight are shown in Figure S2. The characterization analysis of BCP2 was used as an example. The MAPOSS structural unit has one characteristic signal at 0.61 ppm (Peak b and Peak c). The new signals at 2.65 ppm (Peak d) and 3.02 ppm (Peak e) correspond to hydrogen in methylene ($-NCH_2-$) and methine (-NCH-), respectively, which are attached to tertiary amine groups in the DPA structure. The integral ratio of Peak d and Peak e is 1:1, which is consistent with the theoretical ratio. The methylene peak attached to the MePEG backbone ($-CH_2-COO-$) appears at 3.66 ppm (Peak a). What is more, there is no characteristic double bond peak appearing at 5.5–6.5 ppm, indicating that there was no residual monomer. This indicates that the two monomers were successfully introduced into the polymer molecular chain.

FTIR (Figure S3) was used to further certify the successful synthesis of the hybrid POSS-based copolymers. Peaks at 1096 cm^{-1} are attributed to asymmetric vibrational stretching of the cage-shaped skeleton, indicating that MAPOSS has been successfully introduced into the copolymer. The characteristic peaks at 2865 cm^{-1} and 972 cm^{-1} correspond to C-H stretching vibration on -NCH- and C-N vibration, respectively, implying the successful synthesis of the copolymer. The C=O stretching vibration of MAPOSS and DPA occurs at 1725 cm^{-1}. GPC analysis of the samples also exhibited that the molecular weight is relatively uniform with well-defined and relatively narrow polydispersity (Figure S4). The GPC curve of each copolymer has a certain tail, mainly due to the high content of MePEG structural units in MePEG-*b*-P(MAPOSS-*co*-DPA). The -C-O-C- structure in the main chain makes the copolymer easily adsorbed by the stationary phase of the ultra-high-efficiency polymer column, resulting in the increased leaching time and trailing of the GPC curve.

3.2. Self-Assembly Behavior of MePEG-b-P(MAPOSS-co-DPA) in Aqueous Solution

To explore the influence of the hydrophobic ratio on the self-assembly system, we chose three amphiphilic copolymers (BCP1, BCP2 and BCP3) with the same MePEG units and different hydrophobic content. The segment lengths of MePEG and DPA structural units are the same, while the proportion of the MAPOSS segment is different. The TEM photos of the copolymer micelles and the results of the DLS test are shown in Figure 1. From the DLS results, it can be seen that the micelle size of BCP1 is 223 nm. As the proportion of the hydrophobic segment gradually increases, the micelle size of BCP2 and BCP3 decreased to 188 nm and 158 nm, respectively. The hydrodynamic diameter of the micelle measured by DLS is usually larger than the micelle observed by TEM, because DLS measures swollen micelles in aqueous solutions, and the sizes observed by TEM are derived from the dried micelles.

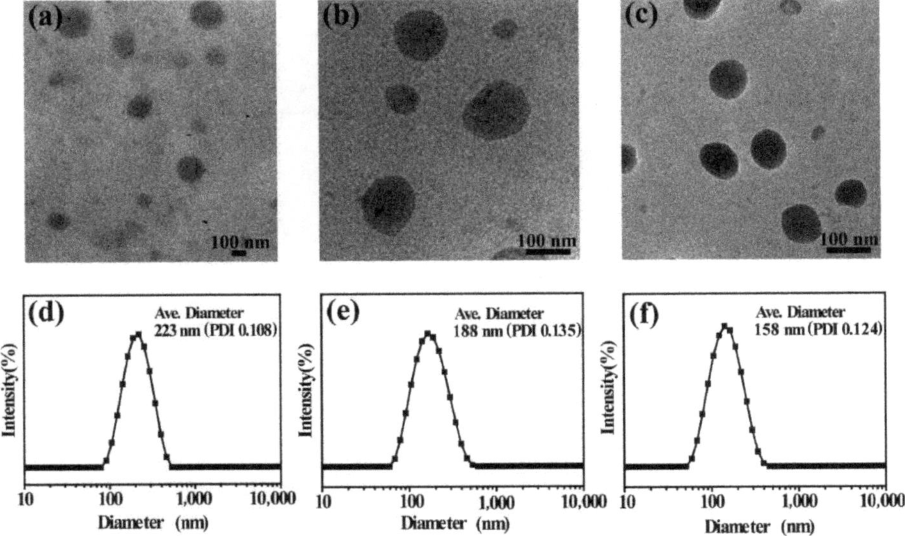

Figure 1. TEM images and DLS curves of BCP1 (**a**,**d**), BCP2 (**b**,**e**) and BCP3 (**c**,**f**) micelles.

The self-assembly behavior of MePEG-*b*-P(MAPOSS-*co*-DPA) in aqueous solution was investigated. With the increase of the MAPOSS unit number, the nanometer size of micelles decreases in sequence. This is because the MAPOSS has a strong hydrophobicity, and its increased content can significantly enhance the hydrophobicity of the copolymer chain in solution. Therefore, the copolymer is driven by a relatively larger hydrophobic effect in the process of forming a gel, resulting in a smaller micelle size. The result is consistent with the report of Zhang et al. [28].

The critical micelle concentrations (CMC) of BCP1, BCP2 and BCP3 are also measured by a fluorescence method with hydrophobic pyrene as a fluorescent probe. Fluorescence spectroscopy is utilized to monitor the self-assembly process (Figure S5). The intensity ratio (I_1/I_3) of the peaks located at 373 nm and 383 nm from the pyrene emission spectra can be taken into account to know the local environment of pyrene. A lower value of I_1/I_3 indicates that pyrene is located in the hydrophobic environment. From the concentration variable experiment, the I_1/I_3 values reflect the CMC value [29,30]. The CMC of BCP1, BCP2 and BCP3 is found to been 0.016 mg/mL, 0.015 mg/mL and 0.012 mg/mL, respectively. Despite the three units ratio being pretty similar in BCP1, BCP2 and BCP3, the small increase of POSS content improved the hydrophobicity of the whole polymer chain, resulting in the lowest CMC of BCP3. Furthermore, the longer polymer chain length possesses a stronger intermolecular interaction. These factors can increase the driving force of the self-assembly process, which means that this amphiphilic copolymer can form micelles at a lower concentration.

In order to explore the effect of initial concentration on the self-assembly of MePEG-*b*-P(MAPOSS-*co*-DPA) solution, the micellar size of amphiphilic block copolymers with initial concentrations was investigated, and the results are shown in Figure 2. It is found that the size of micelles increase with the increasing initial concentrations, indicating that the size of micelles was significantly affected by the initial concentration of the amphiphilic copolymer solution. Ultraviolet transmittance is a common method to assist in the characterization of micelle size in solution. When the micelle size of the same sample increases with the increasing concentration, the transmittance to ultraviolet light of specific wavelength decreases. The ultraviolet transmittance of BCP1, BCP2 and BCP3 micelles at different initial concentrations is shown in Figure 2d. As the initial concentration of the amphiphilic copolymer solution increases, the corresponding UV transmittance becomes smaller. The change of UV transmittance with different initial concentrations is consistent with the DLS results. From this phenomenon, we can conclude that the initial concentration of the amphiphilic copolymer solution affects the micellar size. Copolymer micelles with higher initial concentrations favor larger micellar size.

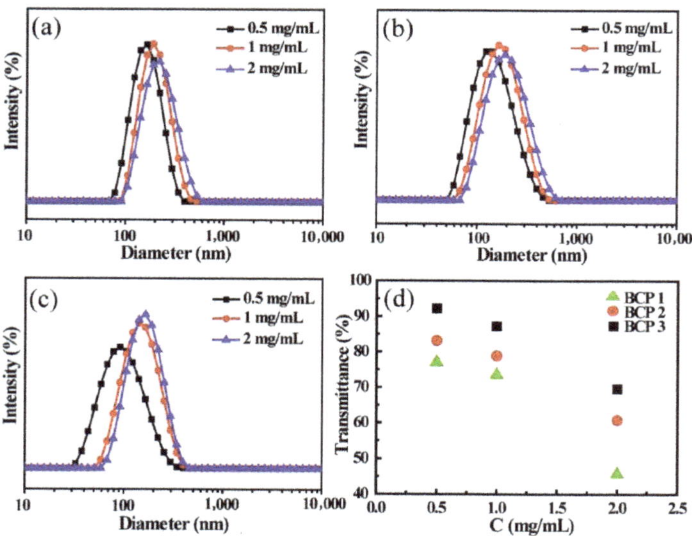

Figure 2. DLS curves of BCP1 (**a**), BCP2 (**b**) and BCP3 (**c**) micelles at various initial concentration of 0.5 mg/mL, 1 mg/mL and 2 mg/mL; (**d**) optical transmittance at 500 nm with different concentration for BCP1, BCP2 and BCP3.

3.3. pH-Responsive Behavior of MePEG-b-P (MAPOSS-co-DPA) Copolymers

DPA is typically used as a monomer for the preparation of pH-responsive polymers for changing the hydrophilicity under different pH conditions. The DLS curves and micelle size distribution of the amphiphilic copolymer micelles under different pH values are shown in Figure 3. It shows that the micelle size changes with the pH value of the aqueous solution, and the samples with a different proportion of MAPOSS segments have a similar tendency of diameter variations. That is, with the decreasing of pH values in an acid solution system, the size of the micelles of the copolymers increases gradually. Once the pH decreases from 3–2, the micelles slightly decrease. On the other hand, when the pH of the solution increases from 3–12, the micelle size decreases gradually.

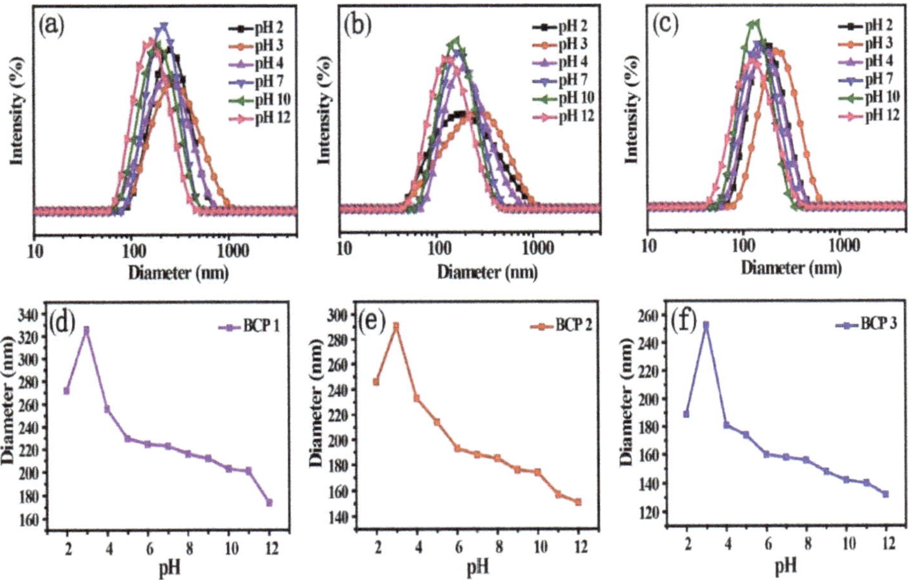

Figure 3. DLS plots of amphiphilic copolymer aggregates and schematic of the aggregate size variation of BCP1 (**a**,**d**), BCP2 (**b**,**e**) and BCP3 (**c**,**f**) under different pH values of aqueous solution.

In an acid solution system, the protonation of the tertiary amine group in DPA units is high, which make DPA become hydrophilic and diffuse to the shell. There is a strong electrostatic repulsive force between DPA units, leading to DPA units' greater attraction to water molecules. Therefore the hydrodynamic size of the micelles increases in the acidic system. By further decreasing the pH, the effect can become stronger. However, DPA segments become hydrophobic due to deprotonation in alkaline conditions, which induces the micelles to shrink. The hydrodynamic size of micelles decreases with increasing pH value.

Once the pH dropped to two, a strong acidic system causes some of the POSS cages to fall off and the micelles to disassemble, resulting in a smaller average micelle size. This is because of the hydrolysis of ester groups of POSS segments on the polymer chain [31]. POSS has a deep contrast under the observation of TEM, so black spots can be observed in Figure 4.

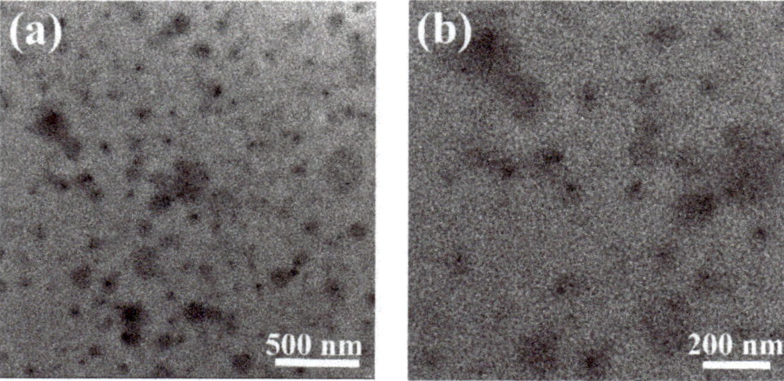

Figure 4. TEM image of micelles in aqueous solution with pH = 2: (**a**) lower magnification and (**b**) higher magnification.

The surface zeta potential of amphiphilic copolymers MePEG-*b*-P(MAPOSS-*co*-DPA) micelles at different pH values was determinated by Zetasizer NanoZS Instrument. As shown in Figure 5, the zeta potential on the micelle surface is positive in acidic systems, as well as negative in alkaline systems. The isoelectric point (pI) values of BCP1, BCP2 and BCP3 micelles are 7.85, 7.74 and 7.71. It can be seen that with the increase of hydrophobic MAPOSS structure units, the relative proportion of DPA to MAPOSS units is 4.72, 4.48 and 4.16, thus resulting in a lower pI value. Under acidic conditions, the zeta potential on the micelle surface is positive due to the protonation of DPA structure units. However, protonated DPA segments become deprotonated under alkaline condition, resulting in the negative zeta potential on the micelle surface.

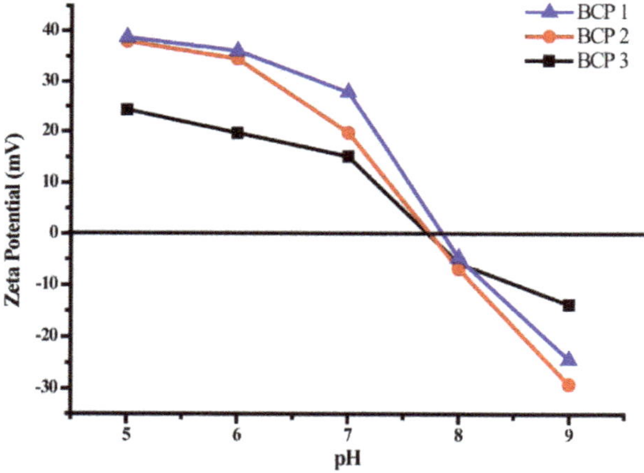

Figure 5. Zeta potential of amphiphilic block copolymers aggregates BCP1, BCP2 and BCP3.

3.4. Controlled Release of Pyrene from Amphiphilic Block Copolymers Micelles

Drug release is an important application for responsive polymer, and the preparation of fluorescent molecules is the primary means to evaluate drug release efficiency [32–35]. Herein, the release of pyrene as a drug model from the micelles by adjusting the pH value is investigated. During the self-assembly process of copolymer micelles, pyrene is added in the aqueous solution. As can be seen from Figure 6, the size of the micelles encapsulating pyrene does not change obviously, and the PDI increased slightly. The self-assembled copolymer BCP2 formed in aqueous solution has a micelle size of 188 nm and a PDI of 0.135. When the micelles are loaded with fluorene molecules, the particle size is 191.8 nm, and the PDI is 0.153. Similarly, the sizes of BCP1 and BCP3 micelles do not change much before and after encapsulation. It is indicated that the process of loading the hydrophobic molecules does not affect the self-assembly process of the polymer.

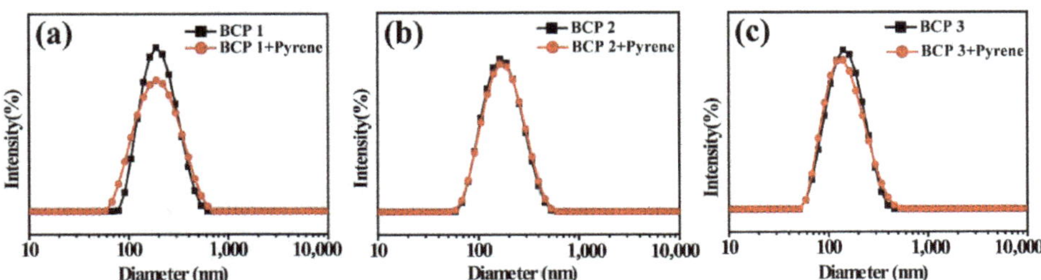

Figure 6. DLS curves of (**a**) BCP1, (**b**) BCP2 and (**c**) BCP3 micelles before and after encapsulating pyrene.

The conditions of pH = 3 and pH = 7 were selected to study the encapsulation behavior of micelles on pyrene. It can be seen from Figure 7 that when the pH is seven, the fluorescence intensity of BCP1 micelles encapsulated with pyrene is the highest, and the intensity of BCP2 and BCP3 decreases sequentially. The result is reasonable because the micelle size of BCP3 is the smallest, resulting in less encapsulated pyrene molecules; therefore, the corresponding fluorescence intensity decreases. When the pH is at seven, MePEG-*b*-P(MAPOSS-*co*-DPA) micelles contain more fluorescent molecules and thus have higher fluorescence intensity. Once the pH decreased from 7–3, the micelles loaded less fluorescent molecules and had lower fluorescence intensity.

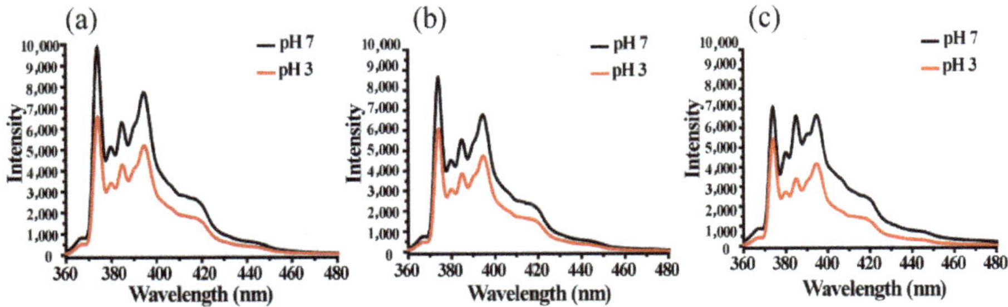

Figure 7. The fluorescent spectra of the micellar solution under different pH values: (**a**) BCP1; (**b**) BCP2; (**c**) BCP3.

From the above results, we can conclude that the micelle had a relatively compact core because of the low protonation of DPA when the pH was seven near the isoelectric point of the micelles. This is favorable for micelle core loading of guest molecules stably. However, under acidic conditions (pH = 3), the tertiary amine groups of the DPA segment are protonated. The micelles are in a swelling state in the aqueous solution, which facilitates the release of the pyrene.

The fluorescence intensity of MePEG-*b*-P(MAPOSS-*co*-DPA) micelles loaded with pyrene in different pH varies with time as shown in Figure 8. In order to reflect the effect of acid conditions on the release rate and degree of pyrene micelles, the release of pyrene from BCP1, BCP2 and BCP3 micelles with different contents of MAPOSS is determined by recording the fluorescence spectra with respect to time, as shown in Figure S6. In the fluorescence spectra of pyrene, the peak intensity of I_1 and I_3 represents the hydrophilic environment and the hydrophobic environment, respectively [36]. Therefore, the cumulative fraction of released pyrene can be estimated from the ratio of $(I_{30} - I_3)/I_{30}$.

I_{30} and I_3 represent the initial fluorescence intensity and fluorescence intensity at a certain time, respectively. It is easy to find that when the pH is seven, the fluorescence intensity of BCP1, BCP2 and BCP3 decreases slightly within 8 h, indicating that the micelles can load pyrene stably. When the pH drops to three, the fluorescence intensity decreases with time. Notably, the fluorescence intensity decreases faster within 1 h from the beginning. In other words, polymer micelles BCP1 released the most pyrene molecules within 8 h, followed by BCP2 micelles and BCP3 micelles in an acidic environment, and their cumulative release percentages were decreased, which were 62.33%, 61.07% and 52.79% respectively.

In conclusion, the proportion of MAPOSS structural units increases sequentially in BCP1, BCP2 and BCP3. The hydrophobicity of the micelles also increases; the aggregation in the micelle core is more compact; and the fluorescent molecules are more difficult to release. However, tertiary amine groups of DPA units are positively charged under acidic conditions and become hydrophilic, which induce DPA to diffuse to the shell. The electrostatic repulsion of DPA units results in a large degree of looseness in the micelle. Pyrene molecules are easily released through the loose shell. Based on the above results, we present the mechanism diagram of the release of pyrene from MePEG-*b*-P(MAPOSS-*co*-DPA) micelles. As shown in Figure 9, the encapsulation and release of the fluorescent molecules are controlled by the loosening or compaction of the micelles. When the pH is seven (near the equipotential point of micelle), there is a compact micelle because of the low protonation of DPA. This is conducive to the stabilization of pyrene loaded in the micellar cores. However, DPA segments in hydrophobic blocks are protonated under acidic conditions and diffuse to the shell. The same type of charge repulsion occurs between the positively-charged DPA segments, and the molecular chain is more stretched. The micelles are in a swollen state in the aqueous solution, which facilitates the release of the pyrene in the micelles.

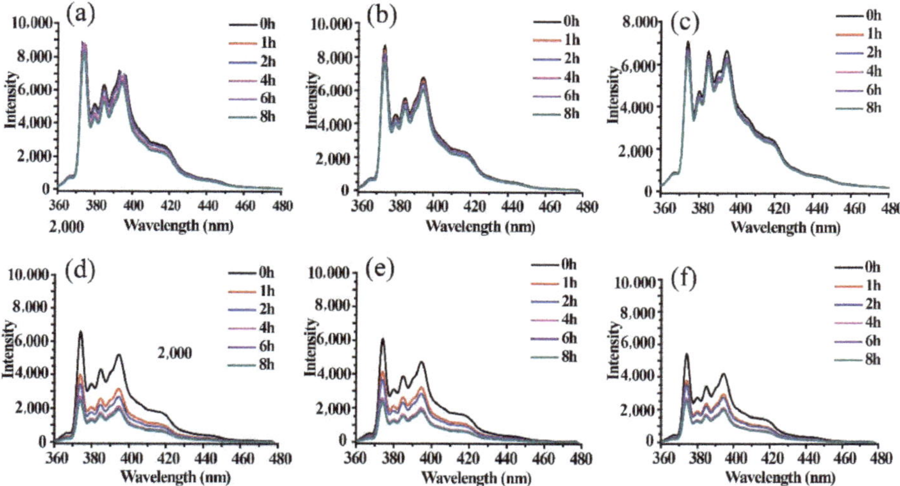

Figure 8. The fluorescent emission spectra of BCP1, BCP2 and BCP3 micelles encapsulated with pyrene in aqueous solution when pH = 7 (**a–c**) and pH = 3 (**d–f**) under an excitation wavelength of 336 nm along time.

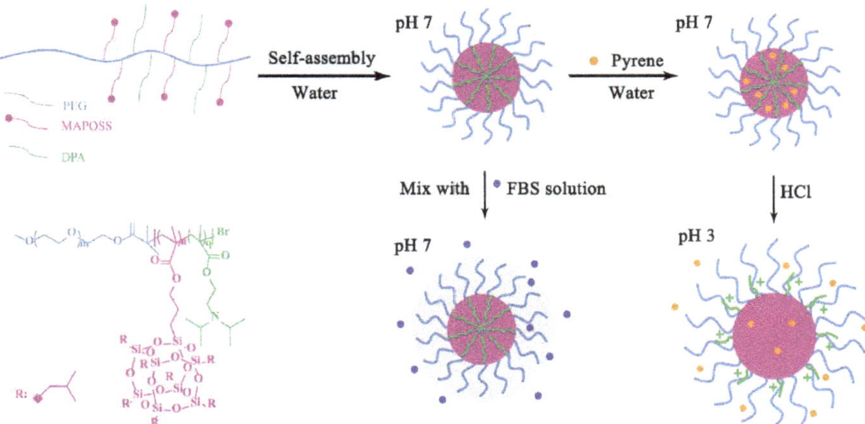

Figure 9. The schematic illustration of the self-assembly of MePEG-*b*-P(MAPOSS-*co*-DPA) micelles.

3.5. Anti-Protein Adsorption Properties of MePEG-b-P(MAPOSS-co-DPA) Micelles

The amphiphilic copolymers MePEG-*b*-P(MAPOSS-*co*-DPA) contained MePEG segments, which have a certain effect on the anti-protein adsorption ability of the system [37–43]. The anti-protein adsorption properties of the copolymer are investigated by mixing FBS solution with copolymers' micellar solution. The DLS results of the copolymers with different content of MAPOSS segments with FBS at pH = 7 are shown in Figure 10. The particle sizes of BCP1, BCP2 and BCP3 micelles are all increased, but to a very small extent, and the PDI is also slightly increased. The results showed that the micelles could effectively inhibit the further adsorption of protein, and the micelles had a certain ability to resist protein adsorption.

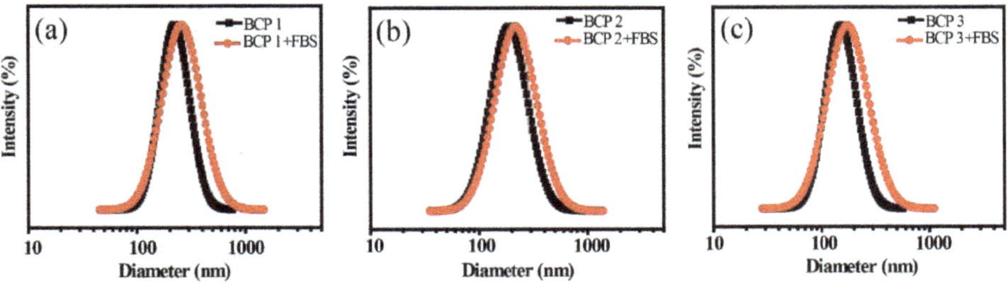

Figure 10. DLS results of micelles before and after mixing with FBS: (**a**) BCP1; (**b**) BCP2; (**c**) BCP3.

The ability of the amphiphilic copolymer micelles to resist proteins was mainly provided by the MePEG; however, DPA is often used as a protein adsorbent because of its ability to adsorb proteins. Therefore, there is a competitive relationship between the adsorption capacity of MePEG against protein and the protein adsorption capacity of DPA in the amphiphilic copolymer. The hydrodynamic size variation of micelles versus time in FBS solution is shown in Figure 11. As we can see, a slight increase of the hydrodynamic sizes of BCP1, BCP2 and BCP3 after mixing with FBS occurs within 32 h, and then, the micelle sizes keep stable. That is, when the adsorption equilibrium of the protein is reached, the three micellar sizes increased only by 20~30 nm compared to that before the adsorption of protein, indicating a certain ability of the micelle to resist protein. From the above results, we can conclude that MePEG segments in the micelles act as a shell layer; MAPOSS and DPA units are hydrophobic as the core during the formation of micelles in near-neutral media. At the same time, the molecular weight of the PEG used in the system is 5000, which is much larger than that of DPA content. Although the hydrophobic DPA proportion of the core is decreased gradually in the BCP1, BCP2 and BCP3 micelles, the ability of the anti-protein is mainly dominated by the MePEG content of the shell. It is calculated that the three copolymer micelles are coated by a long hydrophilic MePEG shell, suggesting the ability of the micelles to resist protein change a little.

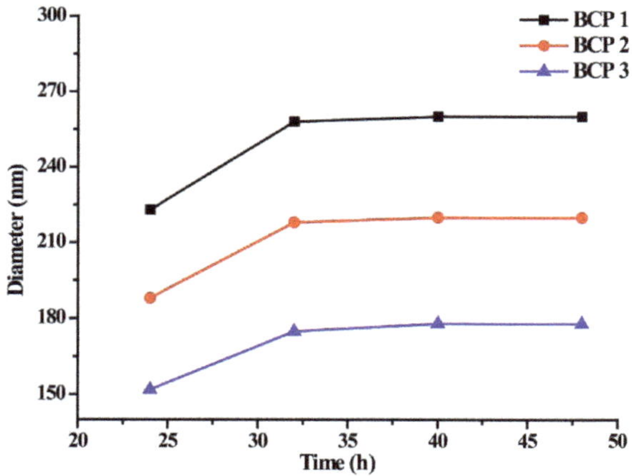

Figure 11. Hydrodynamic size of micelles versus time in FBS solution.

Therefore, we believe that the interaction between the MePEG segment of amphiphilic copolymers and water molecule effectively inhibits the further adsorption of protein by micelles, which results in a slow increase of the micelle size, and micelles have anti-protein adsorption properties. The schematic of the mixing of micelles and FBS solution is also shown in Figure 9. Amphiphilic copolymers containing MePEG segments act as an anti-protein adsorption material, and its anti-protein adsorption mechanism is related to hydration. The combined water on the surface of the material prevents the material from adsorbing protein. The amphiphilic copolymers first combine a large number of water molecules around it by hydrogen bonding, forming a "hydration layer" of physical and energy barriers [44], preventing the micelles from adsorbing protein.

4. Conclusions

A hybrid amphiphilic block copolymer containing MAPOSS, MePEG and DPA via ATRP polymerization was synthesized. The results of ^{1}H-NMR, FTIR and gel permeation chromatography indicated the well-defined structures of the copolymers. The copolymer has the ability to self-assemble into a spherical micelle in response to pH. With the increase of the pH value from 2–12, the size of micelle increases first, then decreases, indicating an outstanding pH-sensitive response. At the same pH, the more MAPOSS units, the tighter the micelle and the smaller the particle size of the micelle. With the fluorescent molecule pyrene as a hydrophobic drug model, MePEG-*b*-P(MAPOSS-*co*-DPA) can realize controlled release of fluorescent molecular pyrene in an acidic environment. The MAPOSS unit content affects the release percentage and release rate of pyrene. With the content of the MAPOSS units increasing, the degree of release of the fluorescent molecule pyrene from the micelle becomes smaller and the release rate becomes slower. The BCP3 micelles in an acidic environment show the smallest cumulative release percentages (52.79%). This is important for polymers with anti-protein adsorption ability to prevent biological pollution, which eventually averts the host immune response. The MePEG-*b*-P(MAPOSS-*co*-DPA) micelles containing MePEG segments effectively inhibit the micelles from adsorbing proteins and have good anti-protein adsorption ability.

Supplementary Materials: The following are available online at http://www.mdpi.com/2072-666X/9/6/258/s1, Figure S1: ^{1}H-NMR spectrum of the MePEG-Br macroinitiator, Figure S2: ^{1}H-NMR spectra of MePEG-*b*-P(MAPOSS-*co*-DPA) with various molecular weights, Figure S3: FTIR spectrum of MePEG-*b*-P(MAPOSS-*co*-DPA), Figure S4: GPC traces of MePEG-*b*-P(MAPOSS-*co*-DPA), Figure S5: Fluorescence-emission spectrogram of pyrene in an aqueous solution of amphiphilic block copolymer: (a) BCP1; (c) BCP2 and (e) BCP3 with different concentrations (mg/mL); relationship between I_1/I_3 of pyrene and amphiphilic block copolymer concentration: (b) BCP1, (d) BCP2 and (f) BCP3, Figure S6: Cumulative pyrene release percentage of pyrene encapsulated by micelles of BCP1, BCP2 and BCP3 when pH = 7 and pH = 3.

Author Contributions: Y.X. and L.D. conceived of and designed the experiments. K.H. performed the experiments. H.W., M.L. and T.S. analyzed the data. X.L. and C.Y. contributed reagents/materials/analysis tools. K.H. wrote and revised the paper.

Acknowledgments: The authors acknowledge the National Natural Science Foundation of China (51273164, 51541307) for funding. Financial support from the Scientific, Xiamen Science and Technology Major Project (3502Z20171002) and the Science and Technology Major Project of Fujian Province (2018HZ0001-1) is also gratefully acknowledged.

Conflicts of Interest: The authors declare no conflict of interest.

References

1. Rodriguezhernandez, J.; Checot, F.; Gnanou, Y.; Lecommandoux, S. Toward 'smart' nano-objects by self-assembly of block copolymers in solution. *Prog. Polym. Sci.* **2005**, *30*, 691–724. [CrossRef]
2. Ganta, S.; Devalapally, H.; Shahiwala, A.; Amiji, M. A review of stimuli-responsive nanocarriers for drug and gene delivery. *J.Control. Release* **2008**, *126*, 187–204. [CrossRef] [PubMed]
3. Wu, S.; Zhang, Q.J.; Bubeck, C. Solvent effects on structure, morphology, and photophysical properties of an azo chromophore-functionalized polydiacetylene. *Macromolecules* **2010**, *43*, 6142–6151. [CrossRef]
4. Roy, D.; Cambre, J.N.; Sumerlin, B.S. Future perspectives and recent advances in stimuli-responsive materials. *Prog. Polym. Sci.* **2010**, *35*, 278–301. [CrossRef]
5. Lee, J.; Yang, H.; Park, C.H.; Cho, H.H.; Yun, H.; Kim, B.J. Colorimetric thermometer from graphene oxide platform integrated with red, green, and blue emitting, responsive block copolymers. *Chem. Mater.* **2016**, *28*, 3446–3453. [CrossRef]
6. Xu, Y.; Huang, J.; Li, Y.; Wang, M.; Cao, Y.; Yuan, C.; Zeng, B.; Dai, L. A novel hybrid polyhedral oligomeric silsesquioxane-based copolymer with zwitterion: Synthesis, characterization, self-assembly behavior and PH responsive property. *Macromol. Res.* **2017**, *25*, 817–825. [CrossRef]
7. Wu, W.; Wang, W.G.; Li, S.; Wang, J.T.; Zhang, Q.J.; Li, X.H.; Luo, X.L.; Li, J.S. Physiological ph-triggered morphological transition of amphiphilic block copolymer self-assembly. *J. Polym. Res.* **2014**, *21*, 494. [CrossRef]
8. Jiang, X.; Zhang, J.; Zhou, Y.; Xu, J.; Liu, S. Facile preparation of core-crosslinked micelles from azide-containing thermoresponsive double hydrophilic diblock copolymer via click chemistry. *J. Polym. Sci. Part A* **2008**, *46*, 860–871. [CrossRef]
9. Liu, Y.; Cao, X.; Luo, M.; Le, Z.; Xu, W. Self-assembled micellar nanoparticles of a novel star copolymer for thermo and pH dual-responsive drug release. *J. Colloid Interface Sci.* **2009**, *329*, 244–252. [CrossRef] [PubMed]
10. Vamvakaki, M.; Palioura, D.; Spyros, A.; Armes, S.P.; Anastasiadis, S.H. Dynamic light scattering vs 1h nmr investigation of ph-responsive diblock copolymers in water. *Macromolecules* **2006**, *39*, 5106–5112. [CrossRef]
11. Kim, K.H.; Kim, J.; Jo, W.H. Preparation of hydrogel nanoparticles by atom transfer radical polymerization of n-isopropylacrylamide in aqueous media using peg macro-initiator. *Polymer* **2005**, *46*, 2836–2840. [CrossRef]
12. Mai, Y.; Eisenberg, A. Self-assembly of block copolymers. *Chem. Soc. Rev.* **2012**, *41*, 5969–5985. [CrossRef] [PubMed]
13. Tada, Y.; Yoshida, H.; Ishida, Y.; Hirai, T.; Bosworth, J.K.; Dobisz, E.; Ruiz, R.; Takenaka, M.; Hayakawa, T.; Hasegawa, H. Directed self-assembly of poss containing block copolymer on lithographically defined chemical template with morphology control by solvent vapor. *Macromolecules* **2011**, *45*, 292–304. [CrossRef]
14. Lejeune, E.; Drechsler, M.; Jestin, J.; Müller, A.H.E.; Chassenieux, C.; Colombani, O. Amphiphilic diblock copolymers with a moderately hydrophobic block: Toward dynamic micelles. *Macromolecules* **2010**, *43*, 2667–2671. [CrossRef]
15. Cho, H.K.; Cheong, I.W.; Lee, J.M.; Kim, J.H. Polymeric nanoparticles, micelles and polymersomes from amphiphilic block copolymer. *Korean J. Chem. Eng.* **2010**, *27*, 731–740. [CrossRef]

16. Blanazs, A.; Armes, S.P.; Ryan, A.J. Self-assembled block copolymer aggregates: From micelles to vesicles and their biological applications. *Macromol. Rapid Commun.* **2009**, *30*, 267–277. [CrossRef] [PubMed]
17. Zhu, J.; Hayward, R.C. Spontaneous generation of amphiphilic block copolymer micelles with multiple morphologies through interfacial instabilities. *J. Am. Chem. Soc.* **2008**, *130*, 7496–7502. [CrossRef] [PubMed]
18. Letchford, K.; Burt, H. A review of the formation and classification of amphiphilic block copolymer nanoparticulate structures: Micelles, nanospheres, nanocapsules and polymersomes. *Eur. J. Pharm. Biopharm.* **2007**, *65*, 259–269. [CrossRef] [PubMed]
19. Liu, H.; Zheng, S. Polyurethane networks nanoreinforced by polyhedral oligomeric silsesquioxane. *Macromol. Rapid Commun.* **2005**, *26*, 196–200. [CrossRef]
20. Liu, L.; Tian, M.; Zhang, W.; Zhang, L.; Mark, J.E. Crystallization and morphology study of polyhedral oligomeric silsesquioxane (poss)/polysiloxane elastomer composites prepared by melt blending. *Polymer* **2007**, *48*, 3201–3212. [CrossRef]
21. Lewicki, J.P.; Pielichowski, K.; Jancia, M.; Hebda, E.; Albo, R.L.F.; Maxwell, R.S. Degradative and morphological characterization of poss modified nanohybrid polyurethane elastomers. *Polym. Degrad. Stab.* **2014**, *104*, 50–56. [CrossRef]
22. Pyun, J.; Matyjaszewski, K.; Wu, J.; Kim, G.-M.; Chun, S.B.; Mather, P.T. Aba triblock copolymers containing polyhedral oligomeric silsesquioxane pendant groups: Synthesis and unique properties. *Polymer* **2003**, *44*, 2739–2750. [CrossRef]
23. Deng, Y.; Bernard, J.; Alcouffe, P.; Galy, J.; Dai, L.; Gérard, J.-F. Nanostructured hybrid polymer networks from in situ self-assembly of raft-synthesized poss-based block copolymers. *J. Polym. Sci. Part A* **2011**, *49*, 4343–4352. [CrossRef]
24. Yu, X.; Zhong, S.; Li, X.; Tu, Y.; Yang, S.; Van Horn, R.M.; Ni, C.; Pochan, D.J.; Quirk, R.P.; Wesdemiotis, C.; et al. A giant surfactant of polystyrene-(carboxylic acid-functionalized polyhedral oligomeric silsesquioxane) amphiphile with highly stretched polystyrene tails in micellar assemblies. *J. Am. Chem. Soc.* **2010**, *132*, 16741–16744. [CrossRef] [PubMed]
25. Wang, Z.; Tan, B.; Hussain, H.; He, C. pH-responsive amphiphilic hybrid random-type copolymers of poly (acrylic acid) and poly(acrylate-POSS): synthesis by ATRP and self-assembly in aqueous solution. *Colloid Polym. Sci.* **2013**, *291*, 1803–1815. [CrossRef]
26. Nguyen, V.H.; Shim, J.-J. Ionic liquid-mediated synthesis and self-assembly of poly(ethylene glycol)-block-polystyrene copolymer by atrp method. *Colloid Polym. Sci.* **2014**, *293*, 617–623. [CrossRef]
27. Liu, Y.; Liu, B.; Nie, Z. Concurrent self-assembly of amphiphiles into nanoarchitectures with increasing complexity. *Nano Today* **2015**, *10*, 278–300. [CrossRef]
28. Weian Zhang, B.F.; Walther, A. Synthesis via raft polymerization of tadpole-shaped organic/inorganic hybrid poly(acrylic acid) containing polyhedral oligomeric silsesquioxane (poss) and their self-assembly in water. *Macromolecules* **2009**, *42*, 2563–2569. [CrossRef]
29. Aguiar, J.; Carpena, P.; Molina-Bolıvar, J.A.; Carnero Ruiz, C. On the determination of the critical micelle concentration by the pyrene 1:3 ratio method. *J. Colloid Interface Sci.* **2003**, *258*, 116–122. [CrossRef]
30. Kalyanasundaram, K.; Thomas, J.K. Environmental effects on vibronic band intensities in pyrene monomer fluorescence and their application in studies of micellar systems. *J. Am. Chem. Soc.* **1977**, *99*, 2039. [CrossRef]
31. Xu, Y.; Chen, M.; Xie, J.; Li, C.; Yang, C.; Deng, Y.; Yuan, C.; Chang, F.-C.; Dai, L. Synthesis, characterization and self-assembly of hybrid ph-sensitive block copolymer containing polyhedral oligomeric silsesquioxane (poss). *React. Funct. Polym.* **2013**, *73*, 1646–1655. [CrossRef]
32. Zhu, Y.; Shi, J.; Shen, W.; Dong, X.; Feng, J.; Ruan, M.; Li, Y. Stimuli-responsive controlled drug release from a hollow mesoporous silica sphere/polyelectrolyte multilayer core-shell structure. *Angew. Chem.* **2005**, *117*, 5213–5217. [CrossRef]
33. Mertoglu, M.; Garnier, S.; Laschewsky, A.; Skrabania, K.; Storsberg, J. Stimuli responsive amphiphilic block copolymers for aqueous media synthesised via reversible addition fragmentation chain transfer polymerisation (raft). *Polymer* **2005**, *46*, 7726–7740. [CrossRef]
34. Li, Y.; Lokitz, B.S.; Armes, S.P.; Mccormic, C.L. Synthesis of reversible shell cross-linked micelles for controlled release of bioactive agents†. *Macromolecules* **2006**, *39*, 2726–2728. [CrossRef]
35. Stefani, S.; Kurniasih, I.; Sharma, S.; Böttcher, C.; Servin, P.; Haag, R. Triglycerol-based hyperbranched polyesters with an amphiphilic branched shell as novel biodegradable drug delivery systems. *Polym. Chem.* **2016**, *7*, 887–898. [CrossRef]

36. Luo, S.; Han, M.; Cao, Y.; Ling, C.; Zhang, Y. Temperature- and ph-responsive unimolecular micelles with a hydrophobic hyperbranched core. *Colloid Polym. Sci.* **2011**, *289*, 1243–1251. [CrossRef]
37. Dalsin, J.L.; Hu, B.H.; Lee, B.P.; Messersmith, P.B. Mussel adhesive protein mimetic polymers for the preparation of nonfouling surfaces. *J. Am. Chem. Soc.* **2003**, *125*, 4253–4258. [CrossRef] [PubMed]
38. Huber, D.L.; Manginell, R.P.; Samara, M.A.; Kim, B.I.; Bunker, B.C. Programmed adsorption and release of proteins in a microfluidic device. *Science* **2003**, *301*, 352–354. [CrossRef] [PubMed]
39. Krishnan, S.; Weinman, C.J.; Ober, C.K. Advances in polymers for anti-biofouling surfaces. *J. Mater. Chem.* **2008**, *18*, 3405–3413. [CrossRef]
40. Chen, S.; Li, L.; Zhao, C.; Zheng, J. Surface hydration: Principles and applications toward low-fouling/nonfouling biomaterials. *Polymer* **2010**, *51*, 5283–5293. [CrossRef]
41. Banerjee, I.; Pangule, R.C.; Kane, R.S. Antifouling coatings: Recent developments in the design of surfaces that prevent fouling by proteins, bacteria, and marine organisms. *Adv. Mater.* **2011**, *23*, 690–718. [CrossRef] [PubMed]
42. Callow, J.A.; Callow, M.E. Trends in the development of environmentally friendly fouling-resistant marine coatings. *Nat. Commun.* **2011**, *2*. [CrossRef] [PubMed]
43. Campoccia, D.; Montanaro, L.; Arciola, C.R. A review of the biomaterials technologies for infection-resistant surfaces. *Biomaterials* **2013**, *34*, 8533–8554. [CrossRef] [PubMed]
44. Chen, H.; Yuan, L.; Song, W.; Wu, Z.; Li, D. Biocompatible polymer materials: Role of protein–surface interactions. *Prog. Polym. Sci.* **2008**, *33*, 1059–1087. [CrossRef]

Article

Microstructure Formation of Functional Polymers by Evaporative Self-Assembly under Flexible Geometric Confinement

Xiangmeng Li [1,2,*], **Xijing Zhu** [1,2] **and Huifen Wei** [3]

[1] Shanxi Province Key Laboratory of Advanced Manufacturing Technology, North University of China, Taiyuan 030051, Shanxi, China; zxj161501@nuc.edu.cn

[2] Institute of Precision & Special Manufacturing, School of Mechanical Engineering, North University of China, Taiyuan 030051, Shanxi, China

[3] Academy of Science and Technology, North University of China, Taiyuan 030051, Shanxi, China; whf_nuc@163.com

* Correspondence: xmli123@nuc.edu.cn; Tel.: +86-0351-3921355

Received: 30 January 2018; Accepted: 8 March 2018; Published: 12 March 2018

Abstract: Polymer microstructures are widely used in optics, flexible electronics, and so forth. We demonstrate a cost-effective bottom-up manner for patterning polymer microstructures by evaporative self-assembly under a flexible geometric confinement at a high temperature. Two-parallel-plates confinement would become curve-to-flat shape geometric confinement as the polydimethylsiloxane (PDMS) cover plate deformed during solvent swelling. We found that a flexible cover plate would be favorable for the formation of gradient microstructures, with various periodicities and widths obtained at varied heights of clearance. After thermal annealing, the edge of the PMMA (Poly-methylmethacrylate) microstructures would become smooth, while the RR-P3HT (regioregular-poly(3-hexylthiophene)) might generate nanocrystals. The morphologies of RR-P3HT structures included thick films, straight lines, hierarchical stripes, incomplete stripes, and regular dots. Finally, a simple field-effect transistor (FET) device was demonstrated with the RR-P3HT micropattern as an active layer.

Keywords: polymer; microstructure; nanocrystalline; flexible geometric confinement; evaporative self-assembly; field-effect transistor

---

1. Introduction

Evaporative self-assembly under geometric confinement is a simple bottom-up facial manner for fabricating microstructures of various nanomaterials which are widely used in applications of optics, optoelectronics, flexible electronics, bioengineering, and so forth [1–9]. In recent years, functional polymers have been patterned into highly ordered microstructures using many kinds of geometric confinement. As reported by Lin et al., stable and fix shapes of geometric confinement including two parallel plates, curve-to-flat and wedge-to-flat confinement, were able to form regular stripes in large area [1–4,10–12]. To the best of our knowledge, most of the abovementioned studies have reported on the self-assembly under a rigid geometric confinement rather than a flexible one. It is easy to figure out that the opening of the geometric confinement would influence the evaporative rate. Hence, it is desirable for the opening of the geometric confinement to be varied in an easy manner. Moreover, it would take long time to produce a polymer microstructure at low temperature due to the low evaporation rate, while a high temperature could enhance the evaporative induced self-assembly process of patterning polymeric microstructures. In addition, we have proposed in previous work that low temperatures were not favorable for the regular stripe formation of nanoparticles as well as for other nanomaterials [13]. Therefore, the effect of high temperatures in evaporative self-assembly should be further studied, especially for some solvents with high boiling points, in order that the formation of microstructure can be promoted.

Poly-methylmethacrylate (PMMA) as a kind of structural polymer, and regioregular-poly(3-hexylthiophene) (RR-P3HT) as a functional polymer with semiconducting property, were commonly used [9,14–18]. For instance, PMMA gratings could be used as diffractive components for optical devices or as a template for replicating other functional patterns [11,19]. The latter one, RR-P3HT, is widely used in organic solar cells and often used as an electron donor in the blend of P3HT: PCBM with PCBM as the acceptor [18,20,21]. Han et al. have reported highly oriented nanofibrils of such semiconducting materials by blading deposition in a microfluidic dragging manner [22,23]. Recently, Ding et al. reported a regular P3HT grating microstructure by using nanoimprinting lithography for studying the growing behavior of muscle cells [24]. Lin et al. proposed stripe-like structures by evaporative self-assembly of P3HT in a capillary tube [25].

In this study, we demonstrate a cost-effective method for fabricating polymeric micro patterns under flexible geometric confinement at a high temperature. Pure materials of PMMA and RR-P3HT are used. A flexible polydimethylsiloxane (PDMS) cover plate and a silicon wafer are used to establish a flexible geometric confinement. Then, the polymer solution is allowed to evaporate on a hotplate. With the time-elapsed evaporation, the solvent could be absorbed by the PDMS materials and the flat cover would become curved and bumped downwards, thus leading to a variable clearance for solvent evaporation. It is easier to remove the solvent than that of the parallel-plate confinement, because the entrapped solution for the latter can form denser vapor of the solvent and slow down the evaporation. In addition, the effect of gap height is also considered as an important influential factor for the formation of varied polymer microstructures. Finally, we demonstrate a simple field-effect transistor (FET) device using the RR-P3HT pattern as an active layer [26].

2. Materials and Methods

2.1. Materials

PMMA (950 k) powder and RR-P3HT (87 k) powder were both purchased from Sigma-Aldrich (Shanghai, China). PMMA solutions with concentrations of 5–10% were obtained by resolving the PMMA powder in toluene, and the mixture was stirred for at least 10 h in fume hood. The RR-P3HT solutions with concentrations of 1–3 wt % were obtained by mixing 0.05 g powder into chlorobenzene after vigorously stirring under ambient condition for 1 h in fume hood.

2.2. Experimental Process

A thoroughly cleaned silicon wafer and glass-slides as substrate was placed on a hotplate. Two smaller pieces of 250 to 1000 μm-thick metal foils were used as spacers. As for the flexible geometric confinement, a PDMS plate 2 mm in thickness was placed over on the spacers to form a bridge and clearance (Figure 1). As for the rigid geometric confinement in comparison, a glass slide as a cover plate, was put onto the substrate to form a wedge-shape clearance (Figure S1c). After that, a drop of polymer solution with certain volume was supplied to the clearance, and allowed to evaporate. The temperature was adjusted to 110 °C for PMMA solution, while it was 120 °C for RR-P3HT solution. Once the polymer solutions were totally evaporated, microstructures could be observed on the substrate. In order to make it smoother, we further thermally annealed the as-prepared PMMA patterns on a 200 °C hotplate for half an hour. For comparison, the RR-P3HT was evaporated in both ambient and nitrogen atmosphere, and then the two samples were thermally annealed at 140 °C under nitrogen atmosphere for one hour.

2.3. FET Device Fabrication Based on RR-P3HT

Interdigital electrodes were fabricated on a thermally oxidized high-doped silicon wafer using ultraviolet (UV)-lithography, magnetic sputtering (of titanium and platinum in 100 nm thickness) and lift-off process (to remove the photoresist and the extra metal). After that, RR-P3HT patterns were fabricated over the interdigital electrodes using the proposed flexible geometric confined evaporative self-assembly approach. Finally, the simple FET device based on the RR-P3HT pattern was demonstrated.

2.4. Characterization

The morphologies of the polymer microstructures were characterized by field emission scanning electron microscope (FESEM), laser scanning confocal microscope (LSCM), and optical microscope. The nanostructure of the RR-P3HT were characterized by atomic force microscope (AFM, BRUKER, Billerica, MA, USA) using tapping mode, the scanning range was 2 × 2 μm and 5 × 5 μm, with a frequency of 0.7 Hz and 512 scanning lines per square. The absorbance spectra of the RR-P3HT films on glass slide were characterized using UV–vis Spectrometer (UV-3600, Shimadzu, Kyoto, Japan), with wavelength range of 310–800 nm. The electric test of the FET device was performed using a semiconductor analyzer (Keysight Technologies, Santa Rosa, CA, USA).

3. Results and Discussion

3.1. Formation of the Flexible Geometric Confinement

Figure 1 illustrates the flexible cover plate deformation and thereafter the formation of curve-to-flat shape geometric confinement gradually as soon as the polymer solution was absorbed by PDMS. In the initial state of the cover-substrate set-up, the polymer solution did not evaporate fast because of the limited clearance. PDMS is a good absorber for the organic solvent, thus leading to an increment of the concentration of the initial polymer solution. Therefore, the PDMS cover would become curved gradually with time elapsed evaporation, and the clearance became larger at the outside region and smaller in the center region. By this means, the polymer solution could have a large evaporation rate at the outmost side, and retreat slowly with solvent evaporation. It would take a short time to form a curved shape of PDMS cover plate (Figure S2). We could consider the process of solvent absorption and PDMS swelling as a simple manner to make a curve-to-flat shape geometric confinement as reported Lin et al. [10,12]. Meanwhile, a high temperature near the boiling point of solvent would enhance the evaporative rate, thus leading to faster polymer deposition.

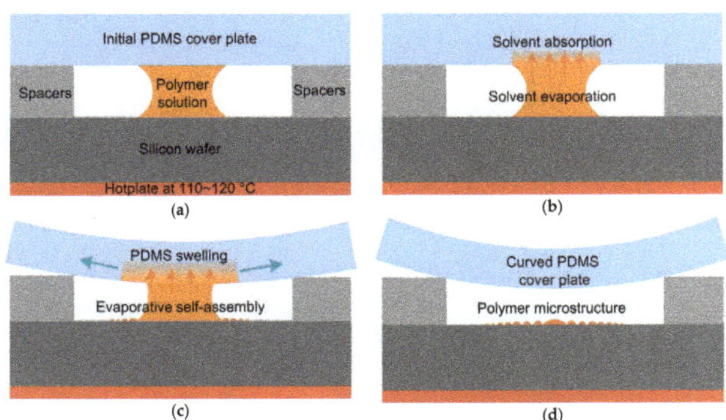

Figure 1. Schematic illustration of evaporating polymer solution in a flexible geometric confinement on a heated silicon wafer in the (**a**) initial state, (**b**) solvent absorption, (**c**) PDMS swelling, and (**d**) eventually curved state of the PDMS cover plate.

3.2. Effect of Evaporative Openning on the Patterning of PMMA Microstructure

Figure 2a shows the circular rings of PMMA resulted from evaporating a single droplet of toluene solution in the center of the clearance, where the confinement was narrowed down (see also Figure S1a). In comparison, Figure 2b shows comparable regular stripes generated by evaporative-assembly from a meniscus on the vertical wedge-shape confinement which had a very larger open confinement (see also Figure S1b).

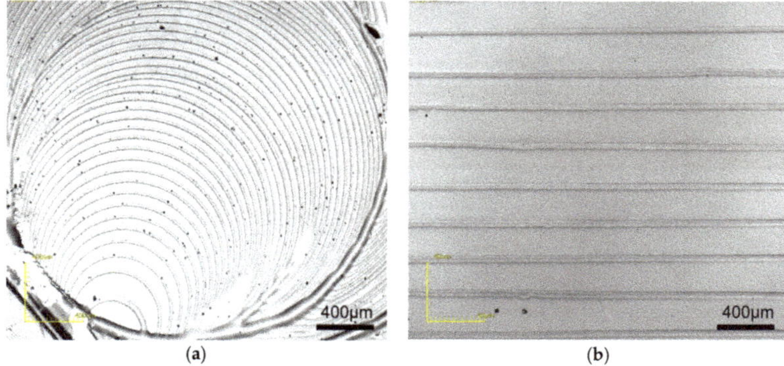

Figure 2. Patterns of PMMA obtained by evaporating (**a**) a single droplet in the center of the clearance and (**b**) wedge meniscus of PMMA solution on silicon wafer at 110 °C hotplate.

When the PMMA solution was trapped in a two parallel plates, the formation would be different. By adjusting the height of the evaporative opening, the widths and periodicities of the PMMA patterns could be both changed (Figure 3 and Figure S2). The evaporation rate would be increased with increasing the gap height of the clearance of evaporative opening. The relationship between the gap height and structure dimension is shown in Table 1. The width of the PMMA microstructure would increase with larger gap heights of clearance. Meanwhile, the density of the PMMA microstructure would be reduced with a higher gap of clearance for evaporation.

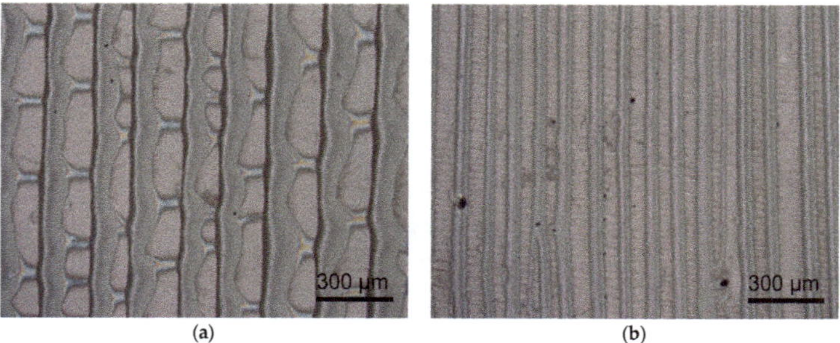

Figure 3. PMMA patterns obtained by evaporative self-assembly at varied gap heights of clearance: (**a**) 1000 μm and (**b**) 250 μm.

Table 1. Dimensional change of patterned PMMA microstructure with the gap height of clearance.

Gap Height/μm	Vertical Periodicity/μm	Vertical Width/μm	Lateral Periodicity/μm	Lateral Width/μm
250	80	35	25	3
500	100	45	120	20
1000	250	125	300	60

Figure 4 demonstrates the obtained stripe patterns with high ordered crater-like structure by gelation under high temperature assisted evaporation. We can see that the microstripes have ordered distribution at both the lateral direction and the vertical direction. After being annealed on 200 °C hotplate, the PMMA patterns turn round shape at the edge due to the reflow under surface tension. Therefore, the PMMA patterns become shorter in height and larger in width (Figure 4b,d). At the same time, the wave-shaped patterns in the vertical direction become not as apparent as the patterns that without annealing.

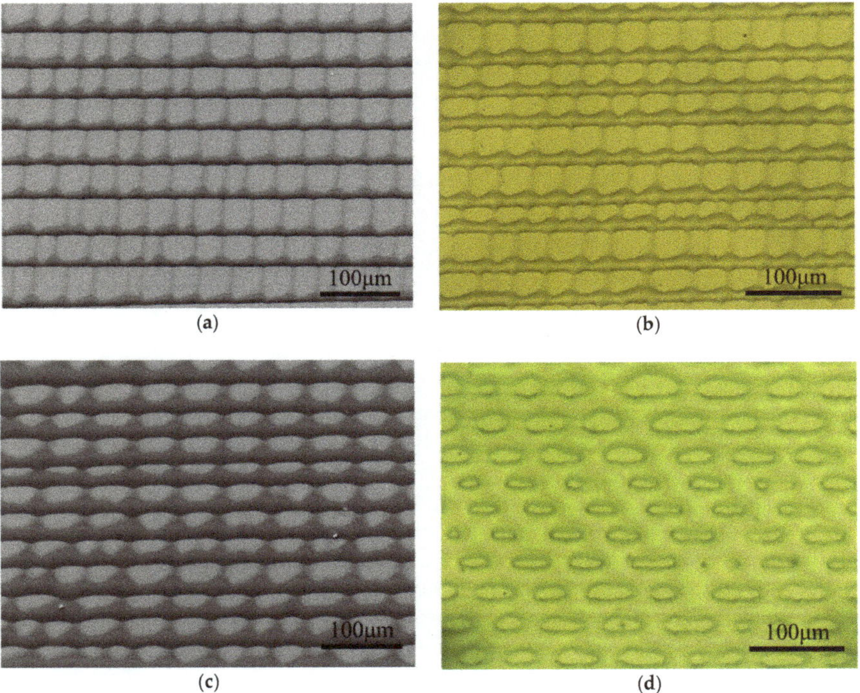

Figure 4. Morphologies of the PMMA microstructures. (**a**,**c**) SEM images of the as-prepared PMMA patterns obtained at varied gap heights of clearance; (**b**,**d**) optical microscope images of PMMA patterns after 200 °C annealing corresponding to (**a**,**c**), respectively.

The mechanism for the facial formation of the polymer microstructure by the evaporative manner has been reported by many researchers [1,2,27–29]. The formation of PMMA at the vertical direction should be attributed to the coffee-ring effect. Meanwhile, the formation at the lateral direction should be due to the Marangoni effect under confined evaporation [30,31]. High temperature would lead to difference of surface tension at the liquid–air interface, and therefore drive the reflow of PMMA solution and the reshape of patterned stripes during evaporation. The stick-slip motion of the evaporating front could be helpful for the generation of the polymer patterns. There are several advantages for using high temperature to promote evaporative self-assembly formation. One of the greatest advantages is the enhancement of generation of the coffee-ring patterns under high evaporation temperature. On the other hand, a complex structure would be also obtained due to the Marangoni effect at high temperature-driven interfacial tension. The flexible geometric confinement would favor patterning polymer microstructures with various periodicities and widths.

3.3. Patterning of RR-P3HT Microstructures

Figure 5 shows the morphologies of the RR-P3HT microstructures. The widths of RR-P3HT stripes are thicker in the central region than that at the outside region due to the varied clearance of the geometric confinement. There are many casually distributed dense microdots surrounding the straight stripes with larger width obtained at the higher concentration region (Figure 5a). In contrast, the region between the finer stripes formed at low-concentration appears much cleaner (Figure 5b).

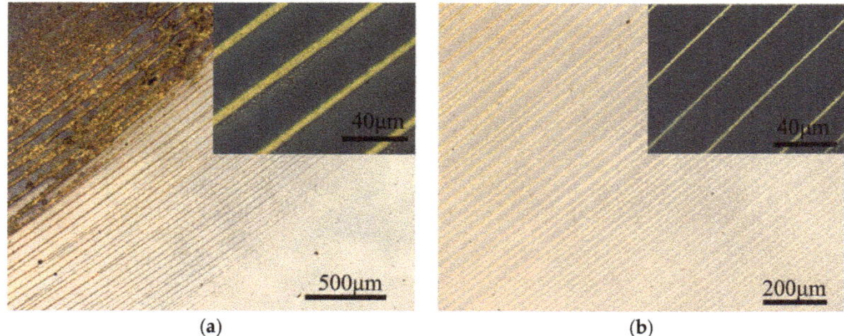

(a) (b)

Figure 5. Optical microscope images of RR-P3HT patterns formed by evaporation under flexible geometric confinement, in the central region (**a**) and outside region (**b**).

The concentration is the highest in the innermost region, and thus leading to a stacked thick film of RR-P3HT. Figure 6 shows the obtained RR-P3HT microstructures with various morphologies at different area (Figure S4). Figure 6a depicts the whole view of the obtained patterns of RR-P3HT microstructures with circular loops of various morphologies. The periodicity and width both becomes smaller ranging from the central region to the outmost region (Figure 6b). There are some small regions with regular stripes, while a dust would change the morphology of the stripe formation (Figure S4b). In the region not far away from the center, the widths become finer with few dots around the stripes (Figure S4c). At the outmost region where the evaporation occurred first, we can find some dotted patterns with regular distribution in both the vertical and lateral directions. In a certain region near the middle, we also find some incomplete stripes with casually distributed dilute microdots between the stripes (Figure 6c,d). Table 2 also lists the varied periodicities and widths at different distances from the interior region.

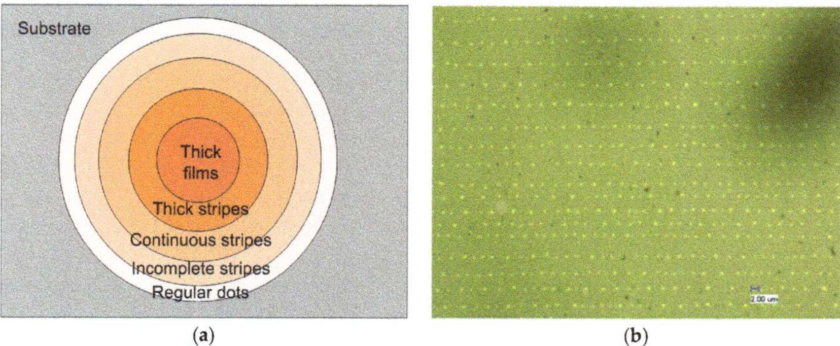

(a) (b)

Figure 6. *Cont.*

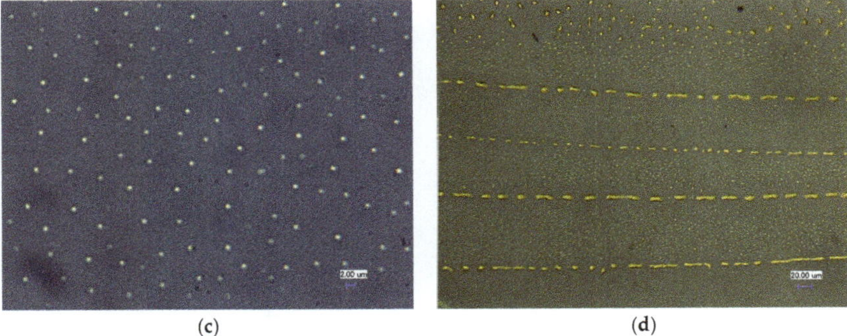

(c) (d)

Figure 6. (**a**) Diagram depicting the various morphologies of RR-P3HT patterns formed by evaporative self-assembly under flexible geometric confinement. Optical microscope images of RR-P3HT patterns formed by evaporation under flexible geometric confinement: (**b**) regular dots, (**c**) irregularly distributed dots in between the stripes as seen in (**d**) incomplete stripes.

Table 2. Widths and periodicities of RR-P3HT microstructure distant from the central region using solution of low concentration of 1 wt %.

Distance/mm	Width/μm	Periodicity/μm
1	6	60
2	2.5	42
3	1	18
4	0.6	9

Different from the results reported by Lin et al. [1,2,12], the microstructure has various morphologies with increasing periodicities and widths simultaneously from the outmost to the center. The generation of dot-patterns should be due to the low concentration at the beginning. With solvent evaporating, the concentration at the three-phase contact line would become higher, so that the accumulation of the polymeric molecules would be increased, thus leading to a larger width. The main mechanism for formation of different morphologies in different regions can be depicted as below. The regularly distributed microdots were formed by gelation as part of the pinning-triple contact line in a rather low concentration. In contrast, the irregularly distributed microdots should be attributed to locally sputtering during receding of the evaporating front under the thermal effect. The surface tension driven Marangoni effect may also lead to the formation of the microdots by thermal re-flow [30,31].

We also investigate the self-assembly of RR-P3HT with higher concentration solutions. Figure 7 compares the RR-P3HT patterns that resulted from the proposed flexible geometric confinement with that from rigid wedge-shape geometric confinement. For the former, straight dense gradient stripes were obtained (Figure 7a,b). For the latter, thick films would be found in the middle region, while finer dense gradient stripes surround the thick film (Figure 7c). The variation of periodicities and widths for RR-P3HT patterns was also demonstrated in Figure 7d. In comparison with the rigid glass-slide cover plate, the flexible PDMS cover plate would form a curve shape via solvent swelling which would attract most of the solution to the middle part, leaving a small amount for the surrounding area, thus leading to a shorter range of stripe pattern.

Nanostructure can be also characterized with atom force microscope (AFM), depicting that the RR-P3HT microstructure would generate nanocrystals in forms of nanofibrils (Figure 8a) or nanochains with alignment orientation to the microstripe (Figure 8b) after thermal annealing. Meanwhile, the absorbance spectra also show a blue shift of the peak value (Figure 8d), indicating the formation of larger amount of nanocrystals. In contrast, the nanostructure in the unannealed RR-P3HT patterns are random distributed and less crystallized due to the fast evaporation (Figure 8c).

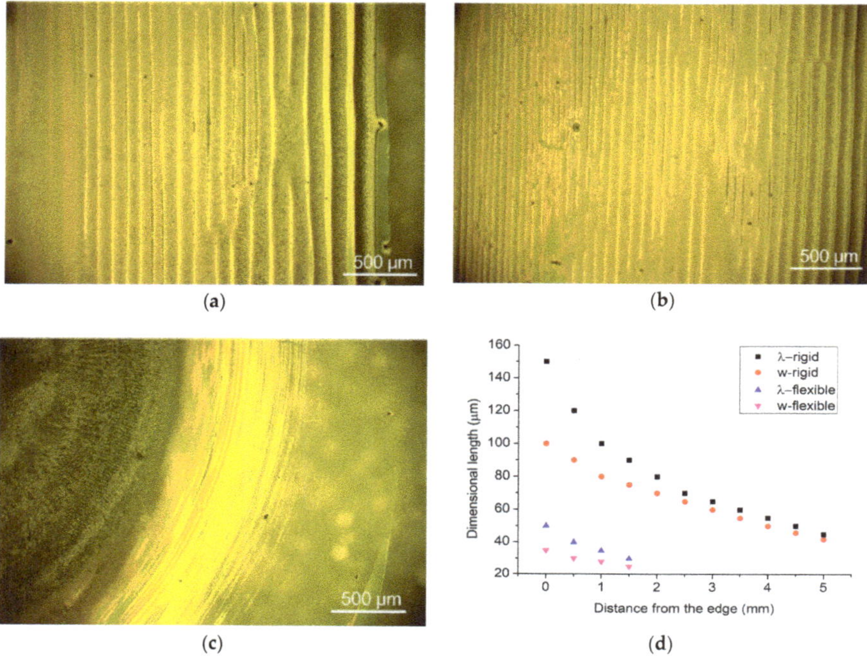

Figure 7. Optical microscope images of RR-P3HT obtained by evaporative self-assembly under rigid confinement (**a**,**b**) and flexible confinement (**c**) with a high solution concentration of 3 wt %. A thicker stripe pattern was obtained (**a**) near the edge and (**b**) distant from the edge; (**c**) very thick film surrounded by finer stripe patterns. (**d**) Comparison of periodicities (λ) and widths (w) of RR-P3HT stripe patterns resulted from the rigid and flexible cover plate.

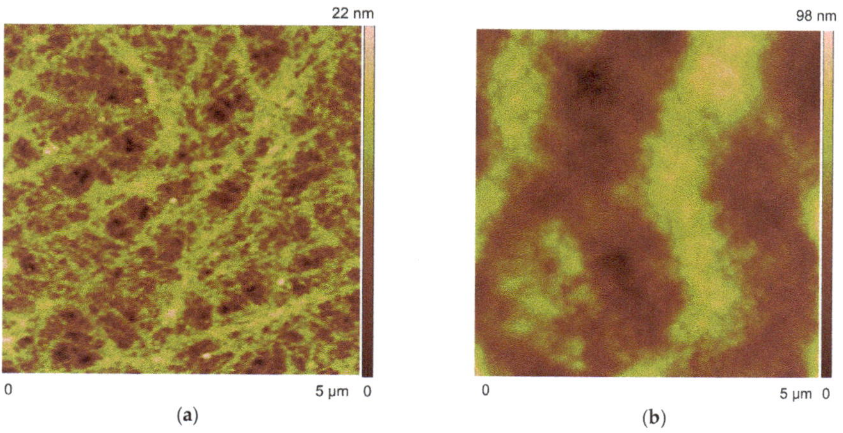

Figure 8. *Cont.*

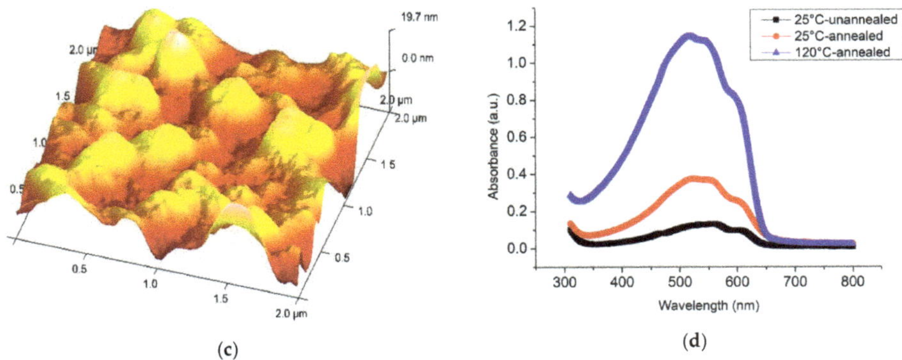

Figure 8. Characterization of nanostructure of RR-P3HT patterns. Atomic force microscope (AFM) images of the obtained RR-P3HT patterns: (**a**) nanofibrils observed on the stripes obtained by evaporative self-assembly at 120 °C and thermal annealing at 140 °C under nitrogen atmosphere; (**b**) nanochains observed in between two stripes of (**a**); (**c**) random distributed microstructure obtained at ambient condition at 25 °C without thermal annealing. (**d**) Absorbance spectra of the RR-P3HT patterns before and after thermal annealing, depicting the blue shift with thermal annealing.

The resultant RR-P3HT patterns can be used as an organic thin film transistor structure [5]. Interdigital electrodes were fabricated on a thermal-oxidized highly-doped p-type silicon wafer by using photolithography and lift-off methods. The silicon wafer was considered as a gate as well as a substrate. Meanwhile, the 500 nm-thick SiO_2 layer was used as dielectric, with the 100 nm-thick platinum interdigital electrodes as both source and drain. Figure S5a demonstrates the patterned RR-P3HT on digital electrodes. Figure 9 shows the performance of the FET device fabricated with the RR-P3HT stripe patterns as an active layer, using the bottom electrode mode (Figure S5b).

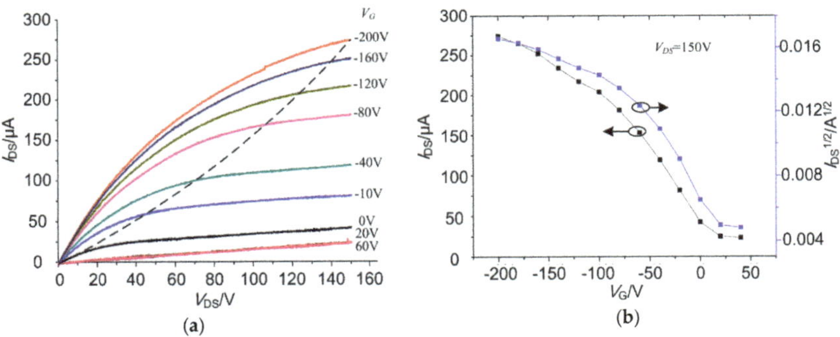

Figure 9. Performance of the RR-P3HT-based FET device: (**a**) output characteristic curve and (**b**) transfer characteristic curve.

By simple estimation from both the output and transfer characteristic curves, we could obtain a carrier mobility of about 0.01 cm^2/Vs, with threshold voltage of about 25 V, and the on/off current ratio of no larger than 10^2 (Table 3). However, the above device performance is far from being applicable. The reason for the deterioration performance of RR-P3HT microstructures should be attributed to the high temperature which would render oxidation of the P3HT molecules. This should be the very reason for limited application of such vulnerable functional polymers like RR-P3HT in the form of microstructure patterning. Moreover, the large dielectric thickness of the FET device would lead to a rather high threshold voltage and working voltages. Better results are believed to be obtained if more RR-P3HT molecules were generated into finer crystalline by slower evaporation and thermal annealing in protective atmosphere [22,32,33]. For example, the whole experimental set-up could be carried out in a vacuum chamber while being performed at a high temperature. Moreover, better polymer crystalline would be generated by using a directional growth method like blade-deposition at a low temperature [20,34]. Nevertheless, we believe that many other polymers can be also patterned by employing the evaporative self-assembly manner using such a flexible geometric confinement.

Table 3. Performance of the FET device with the RR-P3HT micropattern as an active layer.

μ (cm^2/Vs)	V_{th} (V)	I_{on}/I_{off}
~0.01	~25	$<10^2$

4. Conclusions

In summary, we propose a flexible geometric confinement for evaporative self-assembly with variable clearance to form stripe patterns of structural and functional polymeric materials. PMMA can be formed into microstructures—such as circular rings, straight stripes, and hierarchical stripes—with various periodicities and widths under different height of clearance for evaporative assembly. Thermally enhanced evaporation would result in formation of complex structures under flexible geometric confinement. Moreover, a semiconducting type polymer RR-P3HT can be patterned into gradient microstructures, with various morphologies of thick film, stripes, discontinuous stripes, regularly-distributed dots, etc. RR-P3HT nanocrystals and polymer alignment could be found upon thermal annealing at 140 °C under nitrogen atmosphere. Finally, a simple FET device was demonstrated with the RR-P3HT patterns as functional layer on interdigital thin-film electrodes. Although it would be a simple method for generating a curve-to-flat confinement as well as improving the formation of gradient polymer microstructures, there are still limitations to the proposed flexible geometric confinement. For instance, the swelling of the PDMS cover plate would be a non-linear process during the absorption and deformation. The thermal control over the evaporative self-assembly process could not be so accurately. Moreover, the reproducibility seems not so good for such a flexible confinement by means of thermally enhanced evaporation and solvent swelling. Nonetheless, the evaporative self-assembly under flexible geometric confinement could pave the way for patterning many other functional materials for practical applications.

Supplementary Materials: The following are available online at www.mdpi.com/2072-666X/9/3/124/s1.

Acknowledgments: This work is supported by NSFC fund (grant No. 51705479), Science Foundation of Shanxi province for Youths (grant Nos. 201701D221128, 201701D221167), and the Project of Science Foundation of North University of China (grant No. 2017005).

Author Contributions: Xiangmeng Li and Xijing Zhu conceived and designed the experiments; Huifen Wei performed the experiments; Xiangmeng Li and Xijing Zhu analyzed the data; Xiangmeng Li and Huifen Wei contributed reagents/materials/analysis tools; Xiangmeng Li wrote the paper.

Conflicts of Interest: The authors declare no conflict of interest.

References

1. Xu, J.; Xia, J.; Hong, S.W.; Lin, Z.; Qiu, F.; Yang, Y. Self-assembly of gradient concentric rings via solvent evaporation from a capillary bridge. *Phys. Rev. Lett.* **2006**, *96*, 066104. [CrossRef] [PubMed]
2. Han, W.; Lin, Z.Q. Learning from "Coffee Rings": Ordered Structures Enabled by Controlled Evaporative Self-Assembly. *Angew. Chem. Int. Ed.* **2012**, *51*, 1534–1546. [CrossRef] [PubMed]
3. Byun, M.; Bowden, N.B.; Lin, Z. Hierarchically organized structures engineered from controlled evaporative self-assembly. *Nano Lett.* **2010**, *10*, 3111–3117. [CrossRef] [PubMed]
4. Han, W.; Byun, M.; Lin, Z. Assembling and positioning latex nanoparticles via controlled evaporative self-assembly. *J. Mater. Chem.* **2011**, *21*, 16968–16972. [CrossRef]

5. Mei, J.; Diao, Y.; Appleton, A.L.; Fang, L.; Bao, Z. Integrated materials design of organic semiconductors for field-effect transistors. *J. Am. Chem. Soc.* **2013**, *135*, 6724–6746. [CrossRef] [PubMed]
6. Shimoni, A.; Azoubel, S.; Magdassi, S. Inkjet printing of flexible high-performance carbon nanotube transparent conductive films by "coffee ring effect". *Nanoscale* **2014**, *6*, 11084–11089. [CrossRef] [PubMed]
7. Tortorich, R.P.; Choi, J.W. Inkjet Printing of Carbon Nanotubes. *Nanomaterials* **2013**, *3*, 453–468. [CrossRef] [PubMed]
8. Ishikawa, F.N.; Chang, H.K.; Ryu, K.; Chen, P.C.; Badmaev, A.; Gomez De Arco, L.; Shen, G.; Zhou, C. Transparent electronics based on transfer printed aligned carbon nanotubes on rigid and flexible substrates. *ACS Nano* **2009**, *3*, 73–79. [CrossRef] [PubMed]
9. Oh, J.Y.; Shin, M.; Lee, T.I.; Jang, W.S.; Lee, Y.-J.; Kim, C.S.; Kang, J.-W.; Myoung, J.-M.; Baik, H.K.; Jeong, U. Highly Bendable Large-Area Printed Bulk Heterojunction Film Prepared by the Self-Seeded Growth of Poly(3-hexylthiophene) Nanofibrils. *Macromolecules* **2013**, *46*, 3534–3543. [CrossRef]
10. Xu, J.; Wang, J.; Mitchell, M.; Mukherjee, P.; Jeffries-El, M.; Petrich, J.W.; Lin, Z. Organic-inorganic nanocomposites via directly grafting conjugated polymers onto quantum dots. *J. Am. Chem. Soc.* **2007**, *129*, 12828–12833. [CrossRef] [PubMed]
11. Park, W.K.; Kim, T.; Kim, H.; Kim, Y.; Tung, T.T.; Lin, Z.; Jang, A.R.; Shin, H.S.; Han, J.H.; Yoon, D.H.; Yang, W.S. Large-scale patterning by the roll-based evaporation-induced self-assembly. *J. Mater. Chem.* **2012**, *22*, 22844–22847. [CrossRef]
12. Lin, Z.; Granick, S. Patterns formed by droplet evaporation from a restricted geometry. *J. Am. Chem. Soc.* **2005**, *127*, 2816–2817. [CrossRef] [PubMed]
13. Li, X.; Wang, C.; Shao, J.; Ding, Y.; Tian, H.; Li, X.; Wang, L. Periodic parallel array of nanopillars and nanoholes resulting from colloidal stripes patterned by geometrically confined evaporative self-assembly for unique anisotropic wetting. *ACS Appl. Mater. Interfaces* **2014**, *6*, 20300–20308. [CrossRef] [PubMed]
14. Powroźnik, P.; Stolarczyk, A.; Wrotniak, J.; Jakubik, W. Study of Poly(3-hexyltiophene) Polymer Sensing Properties in Nerve Agent Simulant (DMMP) Detection. *Proceedings* **2017**, *1*, 448. [CrossRef]
15. Wang, Y.; Lieberman, M.; Hang, Q.; Bernstein, G. Selective binding, self-assembly and nanopatterning of the Creutz-Taube ion on surfaces. *Int. J. Mol. Sci.* **2009**, *10*, 533–558. [CrossRef] [PubMed]
16. Smith, M.K.; Singh, V.; Kalaitzidou, K.; Cola, B.A. Poly(3-hexylthiophene) nanotube array surfaces with tunable wetting and contact thermal energy transport. *ACS Nano* **2015**, *9*, 1080–1088. [CrossRef] [PubMed]
17. Ting, Y.-H.; Liu, C.-C.; Park, S.-M.; Jiang, H.; Nealey, P.F.; Wendt, A.E. Surface Roughening of Polystyrene and Poly(methyl methacrylate) in Ar/O2 Plasma Etching. *Polymers* **2010**, *2*, 649–663. [CrossRef]
18. Singh, S.; Vardeny, Z.V. Ultrafast Transient Spectroscopy of Polymer/Fullerene Blends for Organic Photovoltaic Applications. *Materials* **2013**, *6*, 897–910. [CrossRef] [PubMed]
19. Sun, W.; Yang, F.Q. Self-Organization of Unconventional Gradient Concentric Rings on Precast PMMA Films. *J. Phys. Chem. C* **2014**, *118*, 10177–10182. [CrossRef]
20. Wang, H.; Xu, Y.; Yu, X.; Xing, R.; Liu, J.; Han, Y. Structure and Morphology Control in Thin Films of Conjugated Polymers for an Improved Charge Transport. *Polymers* **2013**, *5*, 1272–1324. [CrossRef]
21. Kakogianni, S.; Andreopoulou, A.; Kallitsis, J. Synthesis of Polythiophene–Fullerene Hybrid Additives as Potential Compatibilizers of BHJ Active Layers. *Polymers* **2016**, *8*, 440. [CrossRef]
22. Gao, X.; Han, Y.-C. P3HT stripe structure with oriented nanofibrils enabled by controlled inclining evaporation. *Chin. J. Polym. Sci.* **2013**, *31*, 610–619. [CrossRef]
23. Sun, Y.; Han, Y.; Liu, J. Controlling PCBM aggregation in P3HT/PCBM film by a selective solvent vapor annealing. *Chin. Sci. Bull.* **2013**, *58*, 2767–2774. [CrossRef]
24. Wang, K.; Li, X.; Wang, C.; Qian, M.; Ding, G.; Liu, J. Vapor-assisted room temperature nanoimprinting-induced molecular alignment in patterned poly(3-hexylthiophene) nanogratings and its stability during thermal annealing. *RSC Adv.* **2017**, *7*, 40208–40217. [CrossRef]
25. Sun, Y.; Lin, Y.; Su, Z.; Wang, Q. One-step assembly of multi-layered structures with orthogonally oriented stripe-like patterns on the surface of a capillary tube. *Phys. Chem. Chem. Phys.* **2017**, *19*, 23719–23722. [CrossRef] [PubMed]
26. Leite, G.V.; Van Etten, E.A.; Forte, M.M.C.; Boudinov, H. Degradation of current due to charge transport in top gated P3HT—PVA organic field effect transistors. *Synth. Met.* **2017**, *229*, 33–38. [CrossRef]
27. Gao, X.; Xing, R.-B.; Liu, J.-G.; Han, Y.-C. Uniaxial alignment of poly(3-hexylthiophene) nanofibers by zone-casting approach. *Chin. J. Polym. Sci.* **2013**, *31*, 748–759. [CrossRef]

28. Li, W.; Lan, D.; Wang, Y. Dewetting-mediated pattern formation inside the coffee ring. *Phys. Rev. E* **2017**, *95*, 042607. [CrossRef] [PubMed]
29. Deegan, R.D.; Bakajin, O.; Dupont, T.F.; Huber, G.; Nagel, S.R.; Witten, T.A. Capillary flow as the cause of ring stains from dried liquid drops. *Nature* **1997**, *389*, 827–829. [CrossRef]
30. Ristenpart, W.D.; Kim, P.G.; Domingues, C.; Wan, J.; Stone, H.A. Influence of substrate conductivity on circulation reversal in evaporating drops. *Phys. Rev. Lett.* **2007**, *99*, 234502. [CrossRef] [PubMed]
31. Hu, H.; Larson, R.G. Marangoni effect reverses coffee-ring depositions. *J. Phys. Chem. B* **2006**, *110*, 7090–7094. [CrossRef] [PubMed]
32. Zhao, K.; Xue, L.; Liu, J.; Gao, X.; Wu, S.; Han, Y.; Geng, Y. A new method to improve poly(3-hexyl thiophene) (P3HT) crystalline behavior: decreasing chains entanglement to promote order-disorder transformation in solution. *Langmuir* **2010**, *26*, 471–477. [CrossRef] [PubMed]
33. Oh, J.Y.; Shin, M.; Lee, T.I.; Jang, W.S.; Min, Y.; Myoung, J.-M.; Baik, H.K.; Jeong, U. Self-Seeded Growth of Poly(3-hexylthiophene) (P3HT) Nanofibrils by a Cycle of Cooling and Heating in Solutions. *Macromolecules* **2012**, *45*, 7504–7513. [CrossRef]
34. He, M.; Li, B.; Cui, X.; Jiang, B.; He, Y.; Chen, Y.; O'Neil, D.; Szymanski, P.; Ei-Sayed, M.A.; Huang, J.; Lin, Z. Meniscus-assisted solution printing of large-grained perovskite films for high-efficiency solar cells. *Nat. Commun.* **2017**, *8*, 16045. [CrossRef] [PubMed]

MDPI
St. Alban-Anlage 66
4052 Basel
Switzerland
Tel. +41 61 683 77 34
Fax +41 61 302 89 18
www.mdpi.com

Micromachines Editorial Office
E-mail: micromachines@mdpi.com
www.mdpi.com/journal/micromachines

www.ingramcontent.com/pod-product-compliance
Lightning Source LLC
LaVergne TN
LVHW070921170726
843515LV00005B/836

* 9 7 8 3 0 3 9 2 8 5 0 6 8 *